Photocatalysts - New Perspectives

Edited by Nasser S. Awwad, Saleh Saeed Alarfaji and Ahmed Alomary

Published in London, United Kingdom

Photocatalysts - New Perspectives
http://dx.doi.org/10.5772/intechopen.100789
Edited by Nasser S. Awwad, Saleh Saeed Alarfaji and Ahmed Alomary

Contributors
Yamuna Ezhumalai, Prabakaran Kumaresan, Thirupathy Jayapalan, Mohd Muslim, Musheer Ahmad, Tanu Mittal, Satoshi Yoshimoto, Masamitsu Ichihashi, Moemi Yoshida, Ana Cristina Estrada, Tito Trindade, Joana Lúcia Lopes, Thi Huong Nguyen, Minh Thanh Vu, Ngoc Son Nguyen, Lagnamayee Mohapatra, Jun Ha Park, Manuel Alberto Flores-Hidalgo, Diana Barraza-Jiménez, Hugo Iván Flores-Hidalgo, Sandra Iliana Torres-Herrera, Raúl Armando Olvera-Corral, Mohammed Muhanna Mohammed, Rozyanty Rahman, Marinela Barci, Wu Hao, Hazlini Dzinun, Ahmad Danish Haikal Bin Sha'ari, Muhammad Amjad Noor Azlan, Nurul Amila Diyanah Kamarul Bharin, Norsyazwani Binti Md Nizam, Hui Ean Lim, Sakkarapalayam Murugesan Senthil Kumar, Selvaraj Venkateshwaran, Karuppiah Selvakumar, Velu Duraisamy, Nasser S. S Awwad, Amal A. Atran, Mohamed S. Hamdy, Shaima M.A. Alshahrani

First published in London, United Kingdom, 2023 by IntechOpen
IntechOpen is the global imprint of INTECHOPEN LIMITED, registered in England and Wales, registration number: 11086078, 5 Princes Gate Court, London, SW7 2QJ, United Kingdom

British Library Cataloguing-in-Publication Data
A catalogue record for this book is available from the British Library

Additional hard and PDF copies can be obtained from orders@intechopen.com

Photocatalysts - New Perspectives
Edited by Nasser S. Awwad, Saleh Saeed Alarfaji and Ahmed Alomary
p. cm.
Print ISBN 978-1-83968-973-4
Online ISBN 978-1-83968-974-1
eBook (PDF) ISBN 978-1-83968-982-6

For EU product safety concerns:
IN TECH d.o.o., Prolaz Marije Krucifikse Kozulić 3, 51000 Rijeka, Croatia, info@intechopen.com or visit our website at intechopen.com.

Meet the editors

Dr. Nasser Awwad obtained a Ph.D. in Inorganic and Radiochemistry from Ain Shams University, Egypt, in 2000, and a postdoctoral degree at Sandia National Labs, USA, in 2004. He has been a professor at King Khalid University (KKU), Saudi Arabia, since 2011. He is the editor of five books and co-editor of two. He has published 251 papers in journals. He has supervised four Ph.D. and eighteen MSc students and participated in thirty-two international conferences. He participated in twelve projects with King Abdulaziz City for Science and Technology (KACST), Saudi Arabia, and Sandia National Labs, USA. He is a member of the Arab Society of Forensic Sciences and Forensic Medicine and the American Chemical Society. He is the rapporteur of the Permanent Committee for Nuclear and Radiological Protection at KKU. He is also a member of the Committee for the Development of International Cooperation Management and the Permanent Committee for the Development of Scientific Research Strategies and Policies at KKU. Dr. Awwad is the head of the Scientific Research and International Cooperation Unit, Faculty of Science, KKU.

Dr. Saleh S. Alarfaji obtained a BSc in Chemistry from King Khalid University (KKU), Saudi Arabia, in 2006, and an MS from Murray State University, USA, in 2012. He obtained a Ph.D. from the University of Nottingham, UK, in 2018. In 2019, he was promoted to Assistant Professor of Computational Physical Chemistry at KKU. In 2020, he became the head of the Department of Chemistry at the same university. Dr. Alarfaji helped to achieve NCAAA accreditation for the BSc and MS in Chemistry at KKU. He specializes in computational chemistry fields such as drug design and discovery, material science, physical chemistry challenges, biomaterials, and optoelectronics. He has authored and co-authored thirty-three research articles in highly reputed journals.

Dr. Ahmed Alomary is the general manager of Aseer Education, Saudi Arabia. He obtained his MSc and Ph.D. in Analytical Chemistry from King Saud University (KSU), Saudi Arabia, in 2013 and 2016, respectively. His research interests include the development and application of novel high-throughput analytical techniques (UPLC, UPLC/MS, LC, LC/MS, GC, GC/MS, spectrophotometer, etc.) for water, medicine, and foodstuffs analysis; developing, establishing, and validating new extraction analytical methodology; and characterising and preparing laboratory reference material for the validation of analytical methodology.

Contents

Preface

Photocatalysis depicts a remarkable class of chemical transformations. It uses the energy delivered by light and causes hard, sometimes even impossible, reactions to conduct in the darkness. This book deals with a variety of photocatalysis topics.

Chapter 1 by Awwad et al. introduces the principles, opportunities, and applications of photocatalysis. Chapter 2 by Muslim and Ahmad discusses the many applications of metal-organic frameworks (MOFs), including in healing metals and heavy metal cations and for antimicrobial applications and photocatalytic indoor environmental remediation. The commercial exploitation of MOFs in using planet-saving solar energy photocatalysis innovations could be more influential with the assistance of industrial partners and various stakeholders. Chapter 3 by Barci and Hao clarifies the need for raw materials and eco-friendly and cost-effective strategies for detecting and extracting materials to satisfy the semiconductor market and its applications. Since the semiconductor market is increasing progressively, driven mainly by the automotive industry, solar cells, smartphone products, the Internet of Things (IOT), and virtual reality (VR) applications, the demand for raw materials, suppliers, and material manufacturing is also increasing. With scientific progress, more materials, alloys, and intelligent materials are being used and thus we need to consider environmental, moral, and ethical aspects when choosing materials for any product, service, or system. Novel techniques such as light detection and ranging (LiDAR), robotics, autonomous haulage systems, and remote operating and monitoring centers assist in maintaining a suitable equilibrium and enable the increasing semiconductor market trend. Chapter 4 by Mittal discusses the usage of self–cleaning coatings or surfaces, as there is a growing demand for self-disinfected and hygienic surfaces. Self-cleaning coatings can be employed in automobile windshields, textiles, antifouling membranes, paints and building construction materials, optoelectronic devices like solar panels, and the medical device and food industries. This chapter presents an outlook on nano titanium dioxide (TiO_2)-based super hydrophilic self-cleaning surfaces by giving a general idea of various systems and approaches that could enhance the self-cleaning manners of TiO_2 in the solar spectrum. The exposed crystal facets in TiO_2, metal and non-metal doping, and dye sensitization are tailored by the hybrid structure of TiO_2, graphene, and graphitic monolayer's TiO_2-semiconductor heterojunctions, attributing to visible-light photocatalytic self-cleaning activity. Chapter 5 by Ezhumalai and Kumaresan focuses on the synthesis methods and applications of g-C_3N_4 and g-C_3N_4-based materials in energy storage. g-C_3N_4-based composites have unique characteristics and a wide range of applications for energy storage and conversion. Chapter 6 by Mohapatra and Park addresses the principal objectives of water purification using a triple-phase catalyst. Under irradiation, charge carriers are formed on the surface of the photocatalyst, and the success of pollutant molecule degradation critically depends on the interaction between the surface and the target molecules. The novel triphasic photocatalytic design with superhydrophobic triphasic interface architecture will allow the immediate delivery of oxygen straight from the air to the reaction interface when compared with state-of-the-art diphasic photocatalytic systems, thus minimizing electron-hole recombination and resulting in remarkably high efficiency. Recently, a solid surface's super wettability (especially underwater superoleophobicity) has attracted much attention. Chapter 7 by Nguyen et al. examines the use of "green" agents to fabricate materials. The chapter shows that integrating a sufficiently large amount of Fe_3O4 will make the composites magnetic enough to separate them from the aqueous solution by an external magnetic field. This characteristic is significant, overcoming the inherent disadvantage of nanomaterials, which disperse too well and are challenging to recover. The summarized research also shows an excellent combination between a magnetic metal oxide and a semiconductor metal oxide, which plays a crucial catalytic role in the photodegradation of pollutant compounds, especially organic pollutants. Chapter 8 by Estrada et al. is an overview of semiconductor photocatalysis, emphasizing carbon and metal sulfide nanocomposites as photocatalysts for wastewater treatment. The coupling of metal sulfides with graphene derivatives platforms has been used synergistically to prevent photo-corrosion of metal sulfide phases and increase the photocatalytic performance of heterostructures. The chapter shows that an increase of graphene or graphene derivative content above a certain value in the hybrid nanostructure decreases the degradation efficiency. Using graphene nanocomposites at a large scale with cost-effective production, high photostability, and high photocatalytic efficiency is challenging. These challenges need to be addressed to guarantee the future commercialization of graphene nanocomposites on a large scale for environmental applications. Chapter 9 by Yoshimoto and Satoshi et al. discusses various types of skin aging, including chronologic aging and photo-aging involving ultraviolet (UV) radiation, visible light, and others. UVA and UVA photocatalysis (involving photosensitizers) contribute to the production of chronically induced skin damage that results in photo-aging, especially wrinkles associated with histopathological actinic elastosis in the dermis. The chapter proposes a screening method to study the effects of antioxidants on UVA photosensitization. It provides new findings for photo-aging prevention by discussing the characteristics of UVA photocatalysts in the skin. Chapter 10 by Dzinun et al. discusses mussel and cockle shells as agricultural wastes for photocatalyst applications. The authors use a solid-state dispersion (SSD) method to prepare a composite photocatalyst in which mussel and cockle are integrated with TiO_2 nanoparticles at a ratio of 9:1. It is interesting to note that the prepared composite photocatalyst particle is nanosized to enhance photocatalytic performance. Chapter 11 by Barraza-Jiménez et al. studies anthocyanidins in the gas phase and under the effects of solvents like water, ethanol, n-hexane, and methanol using density-functional theory (DFT) and time-dependent density-functional theory (TD-DFT) electronic structure calculations for applications as natural dyes in photocatalysis. Conceptual DFT results show that cyanidin, malvidin, and pelargonidin present good charge transfer properties. Cyanidin presents a lower electron reorganization energy (λe) when water is used as the solvent. TD-DFT is used for excited state calculation, and absorption data shows prominent peaks in a wavelength between 479.1 and 536.4 nm. The chapter discusses the UV–Vis absorption spectra generated and the solvent effects in each case. These pigments are good options for photocatalysis applications, and the best choices for dye sensitization are cyanidin, malvidin, and petunidin, after including the more common anthocyanidins in the analysis. In Chapter 12, Mohammed reviews the research on natural fiber (NFr)-reinforced composites and the constraints that have emerged in their development and serviceability. These constraints must be addressed before NFrs can be utilized successfully. Furthermore, the chapter describes how surface-based photocatalyst nanoparticles (PHNPs) could increase NFr adhesion to their matrix and reduce NFr moisture absorption. Therefore, surface treatments with PHNPs can modify the characteristics of NFr. Chapter 13 by Mittal discusses the fabrication of advanced nanocomposites and related challenges, such as low efficiency and selectivity towards hydrogen evolution under the illumination of solar energy. This chapter examines recent developments in photocatalysts, and heterojunction fabrication factors influencing the photocatalytic process for the dynamic production of hydrogen. Finally, Chapter 14 by Venkateshwaran et al. discusses the important parameters affecting the activities of both Co_3O_4 and MnO_2, such as phase and morphology engineering, defects and crystal facets engineering, and strains and mixed metal-oxide formations. The chapter also discusses the fundamentals of water oxidation and reduction reactions. Moreover, it investigates the electrochemistry behind MnO_2 and Co_3O_4 for a better understanding of catalyst reactions and mechanisms. This chapter stresses the importance of MnO_2 and Co_3O_4 as bifunctional catalysts by examining their fundamental electrochemistry, structure-activity relationship, and future directions in the field of metal-air batteries.

This book will inspire readers, researchers, and scientists to further examine newly developed photocatalysts and unlock new potential research tracks for additional novel development. We would like to express our gratitude to the contributing authors for their excellent chapters. We would also like to thank the staff at IntechOpen.

Dr. Nasser S. Awwad
Professor,
Faculty of Science,
Department of Chemistry,
King Khalid University,
Abha, Saudi Arabia

Dr. Saleh Saeed Alarfaji
Faculty of Science,
Department of Chemistry,
King Khalid University,
Abha, Saudi Arabia

Dr. Ahmed Alomary
General Manger,
Aseer Education,
Saudi Arabia

Chapter 1

Introductory Chapter: Photocatalysis – Principles, Opportunities, and Applications

Nasser S. Awwad, Amal A. Atran, Shaima M.A. Alshahrani and Mohamed S. Hamdy

1. Introduction

Photocatalysis is an applicable technology to control the pollutants and contaminations that are released by industrial activity to the nature. The important feature of photocatalysis is that it is possible to replace the high-temperature reactions to eliminate contaminations with reactions that can take place at room temperature, hence, maintain fossil fuel for other purposes. Heterogeneous photocatalysis process implied a material (usually semiconductor) with a certain bandgap that can be activated by light (at certain wavelength). When the semiconductor crystal is subjected to light with a wavelength higher than the bandgap, electrons are normally activated and move from valance band to conduction band creating a positive hole instead, this process is called electron/hole separation. After formation of electron and holes, one of the two following situations might take place: a) recombination between the electron and holes or b) electron and holes reach the surface and activate an organic molecule on the surface of the semiconductor crystal, that is, electrons will participate in reduction process, while the holes participate in oxidation process. Several free radicals were detected as a result of photocatalytic process, notably hydroxyl radicals ($^{\bullet}OH$) and superoxide anions ($O_2^{\bullet-}$). The formed radicals are participating in the continuous reactions with the adjacent organic molecules until total minimization. Therefore, the overall process can be summarized in four main steps: (a) adsorption of organic molecule on the surface of the semiconductor crystal, (b) electron/hole formation, (c) reaction between electron and/or holes with the adsorbed molecule, (d) over-reaction might take place, and (e) desorption of the products (**Figure 1**) [1].

According to Scopus®, in the last 10 years (between 2013 and 2022), the researchers in different fields reported their research results in photocatalysis in more than 66,262 papers (**Figure 2**). Out of these papers, 7850 papers reported the findings in the area of air purification. From these papers, only 1258 papers (only 1.8% of the total photocatalysis publications) discussed the utilization of photocatalysis to purify air from short-chain hydrocarbons. These statistics clearly show that the research in the field of air purification from hydrocarbons needs attention to be developed and improved. Hence, more efforts must be performed in this research area because a clean environment is one of the research strategic plans in most countries.

IntechOpen

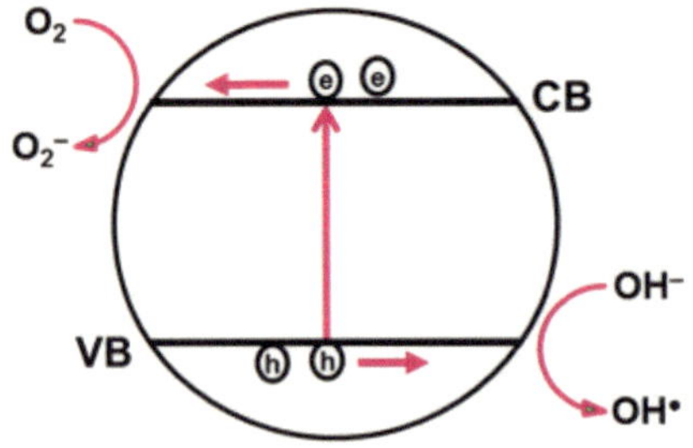

Figure 1.
The photocatalytic process over semiconductor crystals.

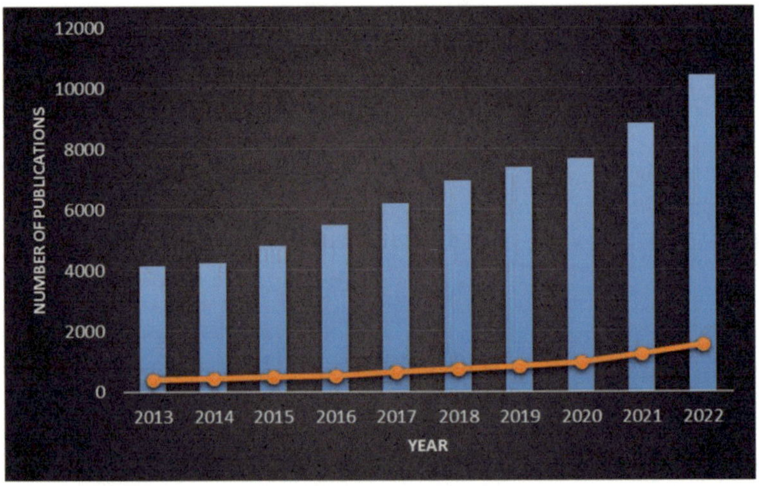

Figure 2.
The number of publications in the last 10 years of the photocatalysis research. The orange line represents the number of photocatalysis in hydrocarbon elimination area.

2. Short-chain hydrocarbons in air

The release of pollutants and contaminants in nature associate with the industrial revolution since the 1800th. Since that, humans were able to release huge amounts of pollutants into air, water, and soil. As the population increases rapidly, the industrial activities have to grow at the same—or even more—rate. As a result, nature is badly affected by releasing such contaminants and severe problems were developing in animals, plants, water, and of course in humans. Few reasons have been identified as main sources of increasing pollutants and contaminations in air, water, and soil. Those reasons include the not-well-planned industrial growth, the continuation of using the old fashion technologies in several simple and manual industries, the presence of huge numbers of small-scale industries with poor facilities, and more importantly, insufficient waste disposal. Scientists realized the fact of fighting/controlling contaminations since 1950s when they realized that contaminations can be a real threat to the nature. Since that, many attempts were reported to control the environmental problems, which have been caused due to the industrial contaminations such as controlling the pollutants in air and in drinking water [2].

The hydrocarbons with a short-chain skeleton, either saturated or unsaturated (also called C1-C3) are representing the threats to the earth. C1-C3 VOC can be released into air from different sources, such as oil plants, during their production and/or in the chemical industries when they are used in them. C1-C3 VOC can be also released as a result of combustion processes, such as waste burning. Moreover, the C1-C3 VOC can also be found in the exhaust fumes of different vehicles, moreover, they can be also found as a result of natural gas combustion in gas power plants. The importance of catalysis for eliminating/reducing human effects on the environment was established several years ago. Generally, almost 95% of the environmental pollution control is carried out by using catalysts. Catalysis considers the major key technology for controlling gas emissions in the different types/scale industries. Catalysis can control gas emissions in two ways, first: by minimizing the number of waste by-products, and second: by treating the gases emitted during the industrial activities. The famous example of automotive catalytic converter is a perfect example of controlling the gas emission by catalysis.

The catalytic reaction of the saturated C1-C3 VOC (e.g. methane, ethane, and propane) is not easy because of the high stability of these compounds. The activation energy needed to oxidize these compounds is high to activate the very stable C-H bonds, while the unsaturated hydrocarbons (e.g. ethylene and propylene) are easy to be oxidized and over oxidized to produce CO_2. Conventional catalysis techniques such as full oxidation by using noble metals (Au, Pt, Rh, or Pd) or the metal oxides of transition elements and doped metal oxides are effective to degrade C1-C3 VOC at elevated temperature (sometimes 800°C). The needed high temperature in addition to the ease poisoning of the catalysts makes the commercialization of such process not feasible because of the extremely high cost of the overall process [3].

3. Semiconductors photocatalysts

Several semiconductors were reported as photocatalysts such as TiO_2, ZnO, CeO_2, ZnSe, ZrO_2, Nb_2O_5, WO_3, SiC, and CdS. The bandgap of each material determine the energy needed from light to be activated, i.e. high bandgap materials need high energy and low wavelength light such as UV, while materials with small bandgap need low energy and higher wavelength light such as visible light. Generally speaking, the photocatalysts should be stable, cost-effective, abundant, non-toxic, active, and operate under different conditions. The photocatalytic process implies the absorption of a photon with a higher energy than the bandgap, hence the electron will be excited from the valence band to the conduction band and electron/hole pairs will be formed. If both reach the surface, electron can participate in reduction reaction and the hole will participate in oxidation reaction. TiO_2 is one of the most interesting materials, and the most studied one in photocatalysis research. This is because TiO_2 is abundant, nontoxic, stable, and very active under UV illumination. However, due to its wide bandgap (3.2 eV), it cannot utilize visible light to be activated. Several attempts have been reported to shift the adsorption band of TiO_2 toward visible light region, such as doping TiO_2 with other metal or metal oxide, creating sub-energy level in TiO_2 lattice, decreasing the crystal size of TiO_2 to nano-level, or forming composite with another material. Several transition elements were reported as dopant for TiO_2, and it showed a shift in the bandgap toward the visible light region such as Cr^{6+}, V^{5+}, and Fe^{3+}. Moreover, nobel metals such as Ag, Au, and Pt were also reported as an electron trap in TiO_2, however, this system is difficult to commercialize due to high cost of the materials. Creating a sub-energy level also attracts several researchers to increase the activity of TiO_2 in visible light. ZnO is a white powder with a bandgap of 3.2 eV, it has been studied as an active photocatalyst, and it exhibited higher photoactivity in several reactions. Moreover, the degradation of several antibiotic compounds is present in water such as amoxicillin, ampicillin, and cloxacillin. The comparison between ZnO and TiO_2 in the fever of ZnO was discussed. It has been shown that the degradation of cellulose bleaching effluent was investigated by using ZnO and TiO_2 as photocatalysts, ZnO showed better activity than TiO_2. ZnO, again, showed better activity than TiO_2 in the degradation of Acid Red 14 dye. Furthermore, high activity is also reported for ZnO than TiO_2 (Degussa P25) in the degradation of Acid Brown 14 dye under different operating conditions. WO_3 is a pale yellow semi-conductor metal oxide with a bandgap of 2.8 eV. Here, WO_3 differs from TiO_2 and ZnO in its light adsorption capacity, it absorbs light up to 500 nm, which indeed gives an advantage over TiO_2 and ZnO. Many authors reported the photocatalytic activity of WO_3 with a certain co-catalyst [4].

4. Oxygen-defected semiconductors

The oxygen-defected semiconductors attracted a lot of interest in the last decade. The idea behind the oxygen-defected semiconductors is modifying the semiconductor lattice through the presence of few oxygen vacancies. This can be achieved *via* either synthesis of the semiconductor in limited oxygen environment (two-step synthesis) or extracting some oxygen from the lattice of the semiconductor (two-step synthesis). The high photocatalytic activity of the oxygen-defected semiconductor can be related to the creation of a sub-energy level below the conduction band of the semiconductor. This sub-energy level can be used for electron relaxing after electron/hole pair formation, therefore, this relaxing minimizes the recombination between the photo-generated electrons and holes. In **Figure 3**, the creation of sub-energy level and its role in photocatalysis process is illustrated.

TiO_2 was the first reported semiconductor that can create oxygen-defected sites in the crystals lattice. Mao et al. reported in Science [5] that reduced titania (TiO_{2-x}), which contains oxygen vacancies sites, V_O-Ti^{3+}, is much more active under the visible light illumination than the equivalent TiO_2. Mao used the hydrogenation technique at elevated temperature to reduce the commercially available TiO_2. Later, it was reported the one-step *in-situ* reduction of TiO_2 by NO and CO as reducing gases to produce blue titania, the produced material showed high photocatalytic performance in water

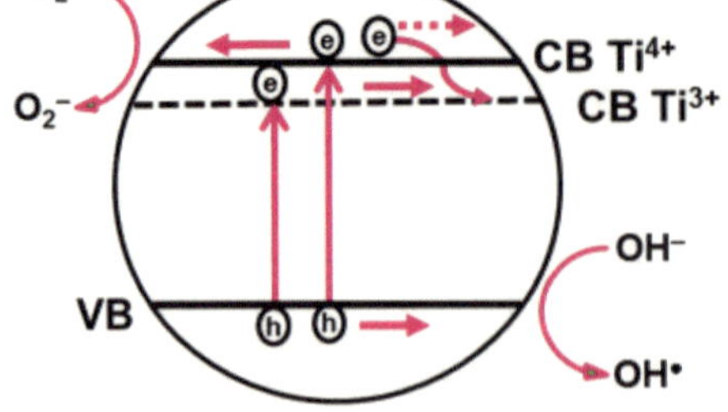

Figure 3.
The oxygen-defected TiO_2 and the sub-energy level created as a result of oxygen vacancies.

splitting reaction than neat titania. It has been pointed out that the high activity of the hydrogenated titania in the decomposition of water contaminants (sulfosalicylic acid and phenol) under the illumination of UV. Several reports were published describing different techniques to create Ti^{3+} in TiO_2 such as the thermal treatment under vacuum or poor oxygen environment and thermal treatment at elevated temperature with reducing agents. More complicated methods were also reported such as laser treatment at elevated temperature > 500 K or bombardment with high-energy particles such as neutrons or γ-ray.

Moreover, the oxygen-defected ZnO was synthesized by several techniques such as the reduction of ZnO thin films by biogenic tactic. The photocatalytic performance of the prepared material was evaluated in the degradation of different dyes and 4-nitrophenol, results showed higher degradation rate than neat ZnO, however, stability of the oxygen-defected ZnO was not discussed. Furthermore, it was presented a computational study about the oxygen-defected sites in ZnO. In a third study, it was discussed the synthesis of oxygen-defected ZnO nanorods by thermal treatment for zinc acetate as a precursor, and although the photocatalytic activity was higher than neat ZnO in the degradation of methylene blue dye, stability was not discussed. It is interesting to mention that sometimes oxygen-defected semiconductors can create colored material such as blue titania black ZnO. The synthesis of black ZnO was achieved by hydrogenation technique. Black ZnO showed higher photocatalytic activity than the corresponding ZnO.

Oxygen-defected WO_3 as a photocatalyst was less explored. Only few studies demonstrated the computational calculations about the presence of oxygen-defected WO_3, stability was not discussed. In a recent report, it has been reported the electronic structure of WO_3 was changed by incorporating different metals, however, the discussion about oxygen-defected and the stability were not discussed.

5. Carbon nitride g-C_3N_4

Graphitic carbon nitride is a very versatile material discovered in 1843 by Berzelius and Liebig [6]. The interesting thing about is that several techniques have been reported for the synthesis of C_3N_4 with a variety of precursors such as the thermal decomposition of melamine. Moreover, it was reported that the synthesis of carbon nitrides can be proceeded through a condensation method with cyanurchloride and calcium cyanamide. In another synthesis procedure, high pressure and high temperature were applied to create carbon nitride from 2-amino-4, 6-dichlorotriazine. In another study, it was reported that the synthesis of carbon nitride can be performed by using cyanurchloride and sodium amide by heating at 200°C in benzene. Recently, carbon nitride has attracted the researchers in photocatalysis area because its small bandgap makes it utilize light in visible light areas of the solar spectrum. Hence, the choice of the precursor and the synthesis conditions are extremely important factors to achieve the demanded structure.

One of the famous composites was g-C_3N_4 with TiO_2. Chang et al. [7] reported a sol-gel technique to create a series of TiO_2/g-C_3N_4 composites. The composite showed an excellent liquid phase photocatalytic decolorization of rhodamine B (RhB) dyed solution. C_3N_4-TiO_2 composites exhibited 2.4 to 7.0 times higher than solo TiO_2 or N-TiO_2. Gu et al. [8] reported the synthesis of anatase TiO_2 nano-sheets composite with (g-C_3N_4). The synthesis procedure was the solvent evaporation method. The composite exhibited superior photocatalytic degradation activity of several organic compounds under the illumination of UV and visible than the parent TiO_2 and C_3N_4. Zhou et al. [9] reported the synthesis of g-C_3N_4/TiO_2 by pyrolysis process of urea and titanium hydroxide. The formed material was evaluated in the gas phase photoreduction of carbon dioxide and water vapor to form CO and CH_4. In this paper, the authors confirmed the formation of nitrogen-doped TiO_2 together with g-C_3N_4 as a separate phase. The photocatalytic behavior of the composite was much higher than the commercial P25. Wang et al. [10] reported the heating of carbon nitride precursor together with TiO_2; however, the formed composite was adjacent to particles C_3N_4 and TiO_2. Although the formed composite was two separate phases, however, the photocatalytic activity was much higher than the parent TiO_2 in H_2 evaluation reaction. Another trial has been reported for the solid state reaction of C_3N_4 precursor and TiO_2 by Boonprakob et al. [11]. The composite of g-C_3N_4/TiO_2 was prepared under Ar flow, and the formed sheets were tested in the degradation of methylene blue under visible light. The composite exhibited also higher photocatalytic activity than the parent TiO_2 and C_3N_4. However, again, the structure formed was adjacent to two separate phases of the mesoporous C_3N_4 and the crystalline TiO_2. Core-shell structure was not feasible. A third trial was to perform the solid state reaction between the carbon nitride precursor and the pre-synthetized TiO_2. Although the authors claimed the formation of a thin layer of carbon nitride around the titania particles, but the HR-TEM images they presented did not show such structure, in addition to, and based on our primary experiments, the ratio of C_3N_4 precursor/TiO_2, which they presented, cannot lead to a core-shell structure.

6. Porous ceria

Recently, our research group succeeded to fabricate porous ceria with high surface area, almost six times higher than commercial ceria. The interesting thing about the porous ceria is the morphological structure. Commercial ceria has bulky plate-like structure, which is always present in agglomerated form. However, the prepared porous ceria exhibited sponge-like structure with rough surface containing plenty of voids, caves, and channels. This structure was achieved by using citric acid as a fuel in a simple flash combustion method (**Figure 4**) [12].

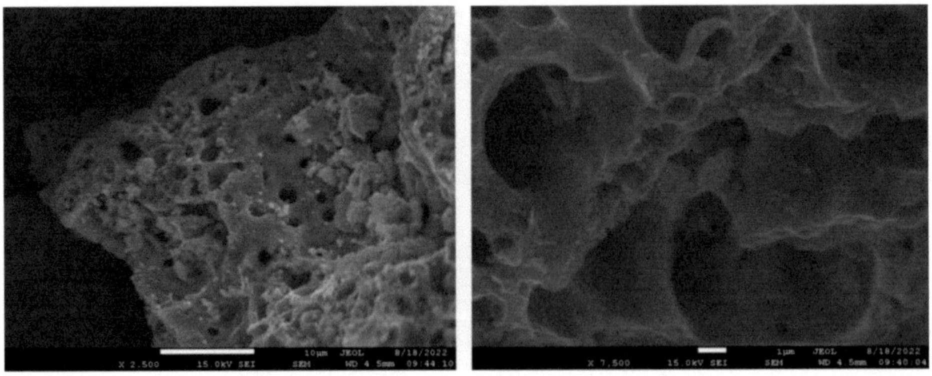

Figure 4.
SEM micrograph of porous ceria.

The prepared porous ceria exhibited high-photocatalytic activity under the illumination of visible light in different applications than commercial ceria. In dye decolorization experiment, methyl green was used as a model compound. It was found that the activity of porous ceria is 3.4 times higher than that of commercial ceria. The second application, the photocatalytic elimination of a gas mixture contains five different short-chain hydrocarbons was applied. Porous ceria was 1.8 times higher than commercial ceria. Currently, our research team is developing the porous ceria by doping with transition elements and the photocatalytic activity is under investigation.

Author details

Nasser S. Awwad*, Amal A. Atran, Shaima M.A. Alshahrani and Mohamed S. Hamdy
Department of Chemistry, College of Science, King Khalid University, Abha, Saudi Arabia

*Address all correspondence to: aawwad@kku.edu.sa

References

[1] Choudhary TV, Banerjee S, Choudhary VR. Catalysts for combustion of methane and lower alkanes. Applied Catalysis, A: General. 2002;**234**:1

[2] Pohanish RP. Sittig's Handbook of Toxic and Hazardous Chemicals and Carcinogens. New York: William Andrews Publishing; 2002

[3] Choudhary VR, Upadhe BS, Pataskar SG, Keshavraja A. Low-temperature complete combustion of methane over Mn-, Co-, and Fe-stabilized ZrO_2. Angewandte Chemie, International Edition. 1996;**35**:2393

[4] Wang S, Yun J, Luo B, Butburee T, Peerakiatkhajohn P, Thaweesak S, et al. Recent progress on visible light responsive heterojunctions for photocatalytic applications. Journal of Materials Science and Technology. 2017;**33**:1-22

[5] Chen X, Liu L, Yu P, Mao S. Increasing solar absorption for photocatalysis with black hydrogenated titanium dioxide nanocrystals. Science. 2011;**331**:746

[6] Acar C, Dincer I. Photoactive materials. Comprehensive Energy Systems. 2018;**2**:524-572

[7] Chang F, Zhang J, Xie Y, Chen J, Li C, Wang J, et al. Fabrication, characterization, and photocatalytic performance of exfoliated g-C_3N_4–TiO_2 hybrids. Applied Surface Science. 2014;**311**:574

[8] Gu L, Wang J, Zou Z, Han X. Graphitic-C_3N_4-hybridized TiO_2 nanosheets with reactive {0 0 1} facets to enhance the UV-and visible-light photocatalytic activity. Journal of Hazardous Materials. 2014;**268**:216

[9] Zhou S, Liu Y, Wang Y, Jiang G, Zhao Z, Wang D, et al. Facile in situ synthesis of graphitic carbon nitride (g-C_3N_4)-N-TiO_2 heterojunction as an efficient photocatalyst for the selective photoreduction of CO_2 to CO. Applied Catalysis B: Environmental. 2014;**158-159**:20

[10] Wang J, Huang J, Xie H, Qu A. Synthesis of g-C_3N_4/TiO_2 with enhanced photocatalytic activity for H_2 evolution by a simple method. International Journal of Hydrogen Energy. 2014;**39**:6354

[11] Boonprakob N, Wetchakun N, Phanichphant S, Waxler D, Sherrell P, Nattestad A, et al. Enhanced visible-light photocatalytic activity of g-C_3N_4/TiO_2 films. Journal of Colloid and Interface Science. 2014;**417**:402

[12] Atran AA, Ibrahim FA, Awwad NS, Shkir M, Hamdy MS. Facial one-pot synthesis, characterization, and photocatalytic performance of porous ceria. Catalysts. 2023;**13**:240

Chapter 2

Historical Developments in Synthesis Approaches and Photocatalytic Perspectives of Metal-Organic Frameworks

Mohd Muslim and Musheer Ahmad

Abstract

Metal–organic frameworks (MOFs) have witnessed fast-growing development in inorganic chemistry as well as material chemistry due to their attractive tunable property, structural specificity, high surface area, and porosity of 3D structures. The conventional semiconductor nature of MOFs is dependent on the photoactive organic ligands and their optimization with incorporated active metal center ion, which have enjoyed these properties in the photocatalytic mechanism via efficient photogenerated charge carriers under the illumination of sunlight (UV-Visible) and other different types of lights. To improve photocatalytic efficiency, a wide range of MOFs could be easily designed to cover and harvest UV irradiation from the sunlight. A wide variety of MOFs have been designed and synthesized as photocatalysts for photocatalytic degradation of organic pollutants, photocatalytic specific redox in organic synthesis, and function in photoelectrodes. In addition, the mechanisms and current challenges for MOFs in photocatalytic degradation of organic pollutants will be thoroughly discussed. This chapter discusses recent research advances in the use of MOFs as emerging photocatalysts.

Keywords: metal–organic frameworks, photocatalytic degradation, photocatalysis mechanism, photoelectrodes, photocatalytic selective redox

1. Introduction

1.1 General introduction

Metal–organic frameworks (MOFs) are crystalline three-dimensional (3D) hybrid materials composed of metal ions and metal clusters linked by polydentate organic ligands [1]. MOF metal centers act as templates, connecting to organic linkers via coordinative metal–ligand interactions and electrostatic attraction. MOFs esthetical chemistry is determined by the interaction of a specific metal secondary building unit with organic ligands [2]. The organic linker may have the same topology but a different metric, creating an isoreticular set of structures that share the same basic net. The importance of coordination bonds and other weak interactions (pi-electron, H-bond, or

IntechOpen

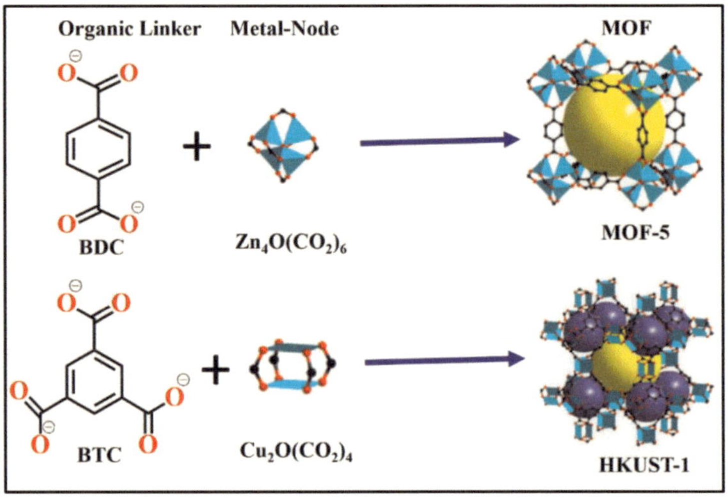

Figure 1.
Schematic representation of the synthesis of MOF-5 and HKUST-1 using different secondary building units (SBUs) and organic linkers. Free spaces in the framework are represented by yellow and blue spheres (reproduced from Ref. [8]).

Van der Waals interaction) in MOF synthesis could be taken for granted. MOFs have high crystallinity, large surface area, high pore volume, and low framework density [3]. They are promising materials for a variety of applications, including clean energy storage (methane and hydrogen), CO_2 capture, absorption, and various separation processes [4–6]. In general, MOFs are made up of two parts: cluster or metal ion nodes and organic linkers that connect the SBUs, resulting in crystalline structures with significant porous texture development. MOFs can also be used as thin-film devices, for biomedical imaging, light harvesting, optical luminescence, catalysis, and other various applications [7]. MOF-5 and HKUST-1 are two well-known MOFs used as a photocatalyst in the synthesis of synthetic organic molecules (**Figure 1**) [9]. The interactions of a specific metal secondary building unit with organic SBUs determine the chemistry of MOFs. The combination of these structures results in an enormous number of possibilities for synthesizing various MOFs with tailored functional properties [10].

The various types of MOFs are produced by using various SBU and organic linkers. As can be seen, different pore shapes of the MOFs framework can be achieved depending on the organic linker. Polytopic organic linkers include carboxylates, phosphonates, sulphides, azoles, and heterocyclic compounds [8]. Several SBUs and organic ligands used in the synthesis of MOFs are depicted in **Figure 2**. When Yaghi et al. synthesized MOF-5 in 1995, they made the first reference to the synthesis of metal–organic frameworks. Since then, a large number of these materials have been studied and classified into various categories in the literature [3, 11, 12]. UIO-66 from Universiteti I Oslo, MIL from Materials of Institute Lavoisier, and ZIF-based MOFs from Zeolite Imidazolate Framework, among many others MOFs have been used in the photocatalytic degradation [13, 14].

1.2 Historical developments

MOF-5 was first synthesized by Yaghi and his colleagues in 1999 as $Zn_4O(BDC)_3{\cdot}(DMF)_8(C_6H_5Cl)$ using zinc nitrate and H_2BDC (1,4-benzenedicarboxylate) as a precursor. Structural transformation of MOFs occurs when exposed to variable

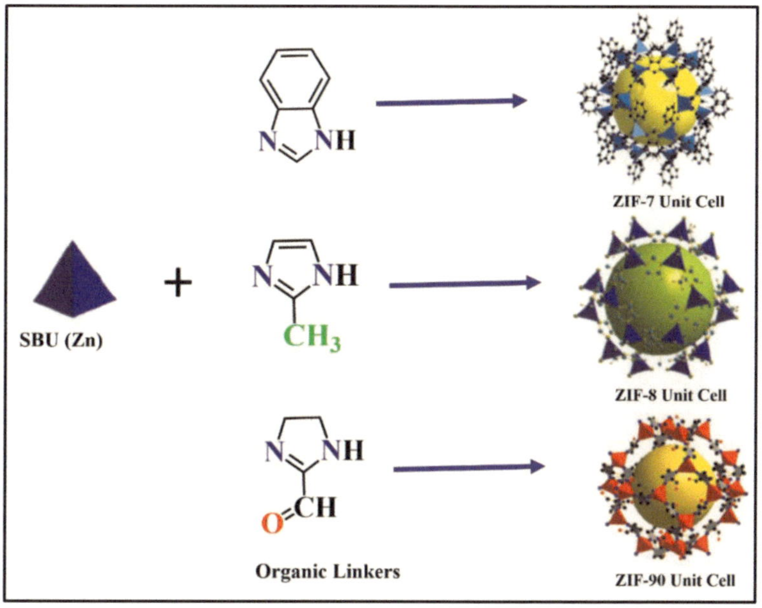

Figure 2.
Synthetic scheme for different zirconium-based metal–organic frameworks (MOFs) were synthesized using the same secondary building units (SBUs) and different organic ligands (reproduced from Ref. [8]).

water concentration environments [15]. Despite the fact that such a structure can be reversed by thermal treatment of the frameworks. Hausdorf and his colleagues studied the photocatalytic activity of zinc carboxylate-based MOFs (MOF-5) in water. Furthermore, Laurier et al. reported in 2013 that when exposed to visible light, iron (III)-based MOFs can photodegrade Rhodamine 6G in an aqueous solution [16]. Meanwhile, Serre and Sanchez synthesized the $Ti_8O_8(OH)_4(O_2C–C_6H_4–CO_2)_6$ (MIL-125(Ti)) in 2009 [17]. When exposed to visible light, iron(III)-based MOFs can photodegrade Rhodamine 6G in an aqueous solution. The Fe-O cluster itself could indeed act as a semiconductor to absorb visible light and then induce electrons from organic ligands and entire photocatalysts' surface [18].

2. Advance features of MOFs

2.1 Ultrahigh porosity of MOFs

Metal–organic frameworks (MOFs) with ultrahigh porosity are useful in a variety of applications, such as gas storage, separation, and catalysis. It is usually vulnerable to conscience because of the large void space inside the crystal framework. Expanding the organic linker chains should lead to increased porosity of MOFs in general [19]. The porous nature of high porosity MOFs was first demonstrated in the 1990s. The reported metal–organic framework (MOF-2010) had a large Langmuir surface area (6240 $m^2\ g^{-1}$) and pore volume (3.60 $cm^3\ g^{-1}$ and 0.89 $cm^3\ cm^{-3}$) (**Figure 3**) [21]. The porous nature of high porosity MOFs was first demonstrated in the 1990s, with no encapsulation of guest molecules in their pores. MOFs frameworks exhibit ultrahigh porous behavior with reversible gas storage properties [22]. They are excellent candidates for use in the creation of novel and valuable MOF materials.

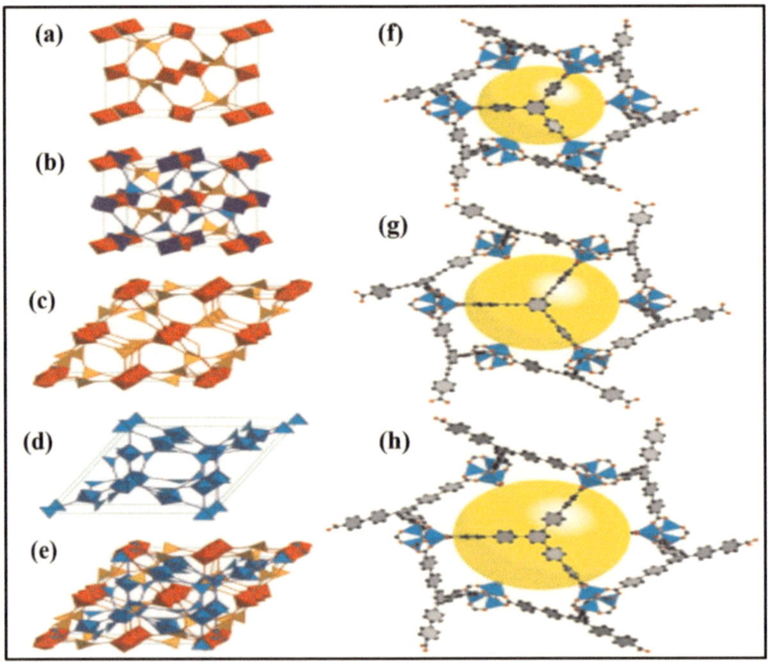

Figure 3.
(a) The connectivity of pyr and qom nets along with [3, 6] coordinates; (b) pairs of pyr nets; (c–e) ***qom*** *is not self-dual; (d)* ***qom*** *connectivity with dual tiling net; (e) different net from the original net of the (c); (f) porous net of* ***MOF-177****; (g) porous net of* ***MOF-180****; and (h) porous net of* ***MOF-200****. Where yellow ball indicates the porous cages,* ***Zn*** *is blue;* ***O*** *is red;* ***C*** *is black, and hydrogen atoms are omitted for clarity (reproduced from Ref. [20]).*

2.2 Ultrahigh surface area of MOFs

The isoreticular expansion premise has made significant progress in the generation of ultrahigh surface area MOFs. This method has been used to summarize some of the appreciable surface areas of MOFs, such as MOF-2105 and NU-1004 [23, 24]. When the solvent is removed, an increase in linker causes MOFs to collapse. Supercritical carbon dioxide activation has proven useful in addressing this issue. Although a long chain of linkers can result in the formation of interpenetrated structures, synthesizing MOFs in topographic networks can mitigate this tendency. Ultrahigh surface area of MOFs has been developed to overcome the problem of water shortage (or high humidity) consistency [22, 25]. This has led to the development of NU-1106 and DUT-327, both of which are based on the ***rht*** and ***umt*** topologies. Large surface area MOFs (NU-1103) have been reported with a larger surface area of 5646 $m^2\ g^{-1}$ (BET area of 6550 m^2g^{-1}) (**Figure 4**) [23, 26, 27].

2.3 MOFs with Lewis acid frameworks

Multicomponent reactions (MCRs) combine three or more reaction partners in one skillet to produce organic products. MCRs have played an important role in drug discovery and pharmaceutical applications. Brønsted and Lewis acids have been used to accelerate multiple MCR reactions at the same time [28]. Metal–organic frameworks (MOFs) have emerged as an important class of crystalline porous materials for

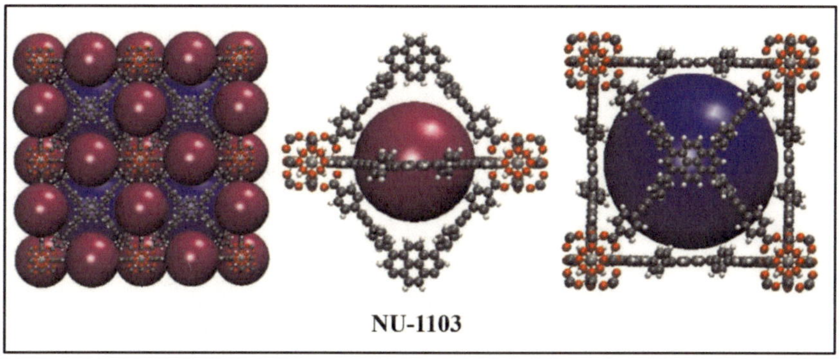

Figure 4.
Representation of large pores (indicated by blue spheres) and small pores (indicated by purple spheres) in the **ftw** *topological networks of NU-1103 (reproduced from Ref. [23]).*

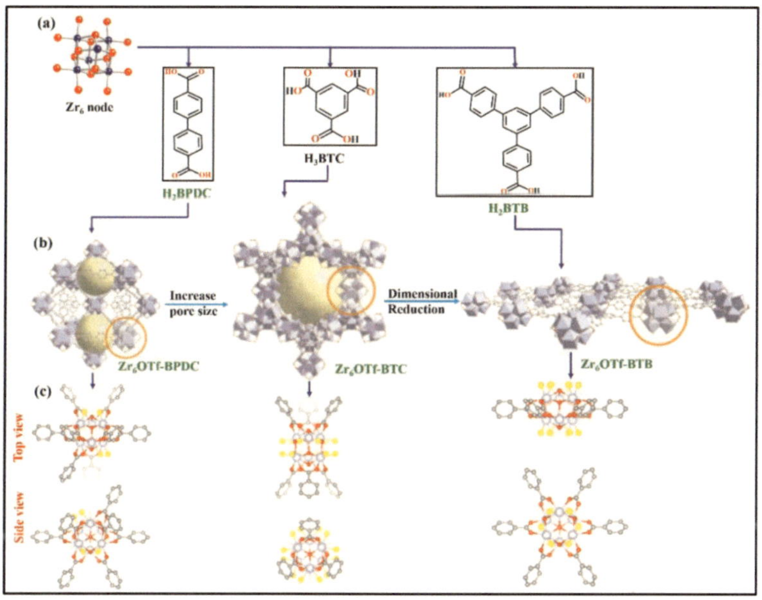

Figure 5.
The generation of strongly Lewis acidic Zr-OTf sites in Zr_6OTf-BPDC, Zr_6OTf-BTC, and Zr_6OTf-BTB is illustrated and compared using (a) MOF nodes and ligands, (b) structures and pore distributions, and (c) coordination defects or capping residues (yellow color) (purple: ***Zr****, red:* ***O****, gray:* ***C****, yellow: Lewis acidic site). Where H atoms are omitted for clarity (reproduced from Ref. [28]).*

the development of high-efficiency single-site solid photocatalysts. MOFs are composed of inorganic metal ions or clusters and organic linkers with organic atoms and molecules (**Figure 5a**) [20, 29]. A set of strict MOFs with acidic sites based on electron-deficient high-valent metallic sources (Zr^{IV}, Hf^{IV}, etc.) have been developed and used to catalyze biologically important transformations. The acidity of Brønsted and Lewis acids was increased by converting immaculate doped or other Zr-capping substituents (**Figure 5b**) [28, 30]. A 2D MOF with self-supporting nanostructure morphological characteristics and freely available Lewis acidic Zr-OTf sites has outperformed two three-dimensional (3D) MOFs for the fabrication of a wide range of synthesized tetrahydroquinoline and aziridine carboxyl group derivative products

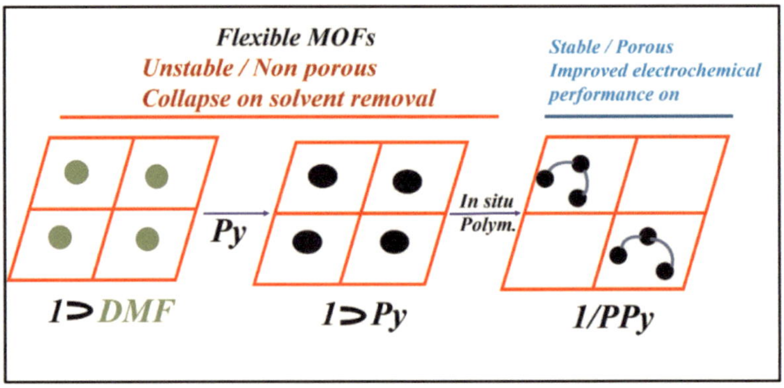

Figure 6.
The reversible phase transformation for stabilization of soft nature MOFs (reproduced from Ref. [37]).

(**Figure 5c**). Zr_6OTf-BTB outperformed the relatively homogeneous standard Sc $(OTf)_3$ in terms of significantly higher turnover numbers and 9–14 times longer catalyst lifetime [31]. It was eventually used to effectively create a few biologically active drug targets via MCRs.

2.4 Flexible and porous MOFs

Porous coordination polymers (CPs) or metal–organic frameworks (MOFs) have received a great deal of attention as smart materials. MOF-based materials, such as MOF composite materials, have piqued the interest of electrochemical energy storage and conversion researchers [32]. In addition to MOFs, there are also soft porous crystals (SPCs), which appear to be reversible or multistable crystalline solids with long-range structural ordering and repairable state transformation [33]. Flexibility frequently comes at the expense of decreased stability, and porosity loss is common. Polymeric guests prevent the framework from collapsing spontaneously, resulting in novel and stable porous phases [34]. This strategy was also used to stabilize highly porous MOFs after activation, preserving porosity. Polymerization of monomer units within MOF pore spaces is widely acknowledged as a simple and convenient method for polymer-porous material interbreeding. Kitagawa and colleagues demonstrated the incorporation of common vinyl polymers, polypyrrole (PPy), and polythiophene (PTh) into appropriate nanochannels of different MOFs [35]. Polypyrrole (PPy) is a polymer that has been shown to be a good electrical conductor for superconductors and has been used to improve MOF electrical properties and encourage their use in energy storage applications and supercapacitors [36] (**Figure 6**). Researchers believe that inserting a conducting polymer guest into such materials could change the porous behavior of the host frameworks.

3. Synthesis approaches of MOFs

The solvothermal synthesis method has been used for the synthesis of MOFs using organic solvents, such as N, N dimethylformamide (DMF), methanol (MeOH), ethanol (EtOH), and acetonitrile [38, 39]. To form and self-assemble MOF crystals, metal components and organic linkers are commonly dissolved in an organic solvent. The heating rate during synthesis is typically less than 220°C, and crystal growth times

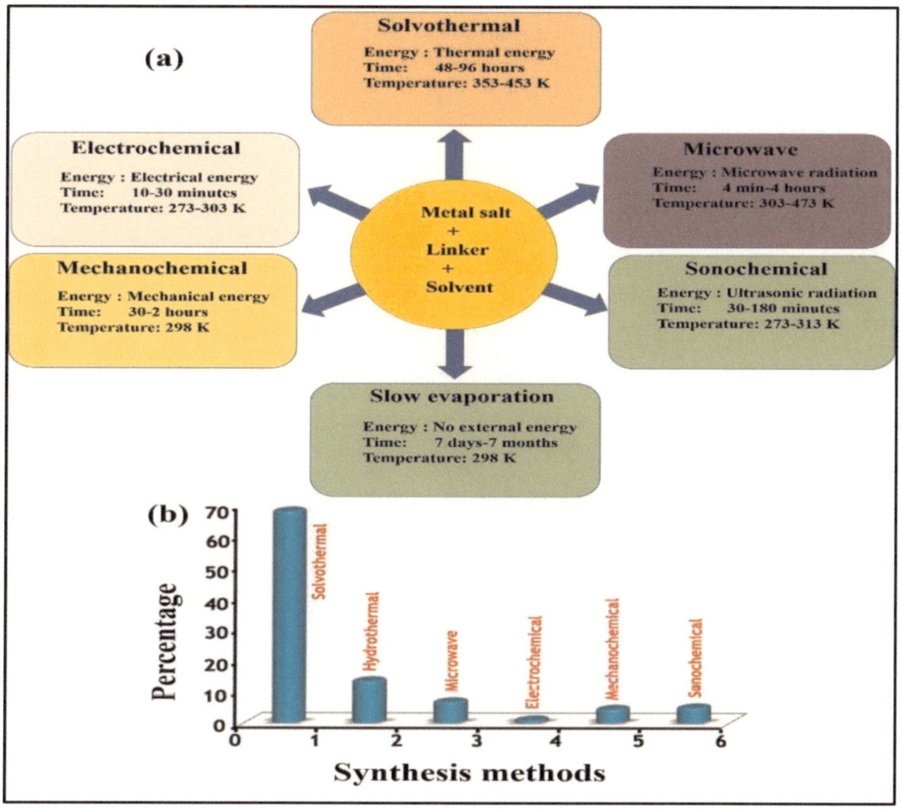

Figure 7.
(a) The popular synthesis methods for synthesis of MOFs; (b) the utility percentages of synthesis methods (reproduced from Ref. [41]).

range from a few hours to several days. Significant advances in the synthesis of MOFs have been made after two decades. Several useful synthesis techniques for MOFs and their utility have been reported, including electrochemical, microwave-assisted, mechanochemical, and sonochemical methods [40] (**Figure 7a**, **b**).

The ability to manipulate and customize the morphology of mesoporous crystals, as well as their synthetic functionalization, is critical in delivering the desired characteristics and outcomes for MOF materials [42–44]. MOF crystal growth has led to the development of more advanced and tunable methods of synthesis for controlling MOF crystal morphology and size, as well as heavily doped to begin creating hybrid MOF crystals [45]. This section of the time frame that follows provides an overview of some of the key advances in this area of research (**Figure 8**).

3.1 Solvothermal method

Metal–organic frameworks can be successfully synthesized using solvothermal (**Figure 9**) and hydrothermal strategies, which appear to be very simple and well-known methods modified from zeolite synthesis [48]. In most cases, the metal component and the carboxylic acid linker are bonded in a suitable solvent. The ability to produce large crystals, less expensive, and use a high bandwidth of heat are just a few advantages of using this method over simpler approaches [49]. The DMF solvent is widely recognized

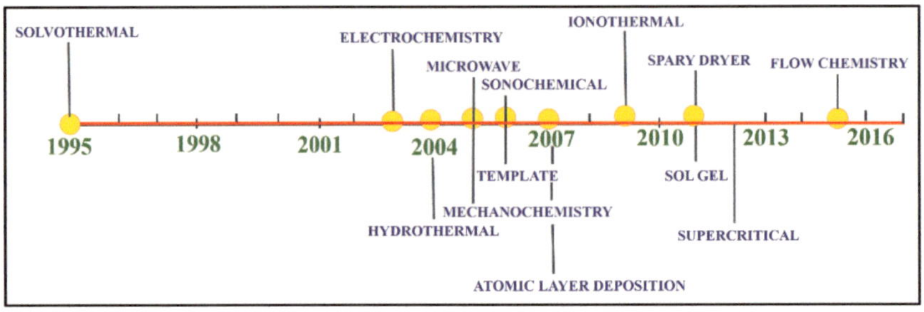

Figure 8.
The time framework for the development of the MOF synthesis methods (reproduced from Ref. [46]).

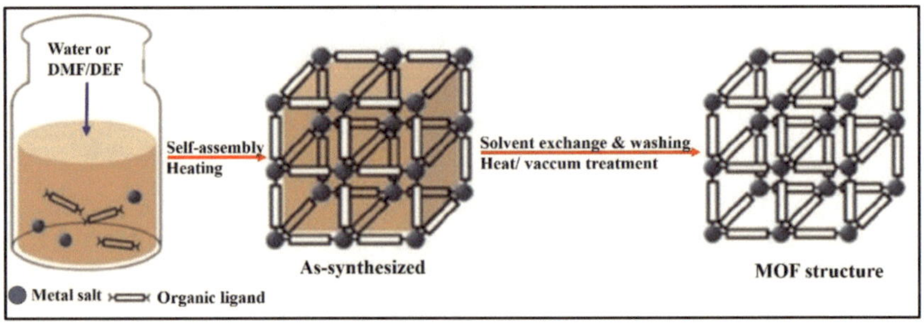

Figure 9.
Schematic representation of the solvothermal synthesis method for synthesis of metalorganic frameworks (reproduced from Ref. [47]).

as the best solvent for this purpose. Microwave-assisted applications can help to improve reaction efficiency but are limited by low product yield and long reaction times, high heat, and the use of toxic organic solvents, such as cadmium iodobenzene [50].

3.2 Microwave-assisted method

Microwave-assisted synthesis methods make use of the interaction of electromagnetic waves with polar solvent ionic species (**Figure 10**). MIL-101(Fe) was synthesized at 150°C using dimethyl formamide, with a yield of 20% and particle sizes similar to 200 nm [51]. Increasing the concentration of water or the pH reduced crystal size. Microwave heating at 210°C was also used to synthesize Cr-Mil-101 MOF. Zr-based MOFs exhibit excellent chemical and thermal stability, resulting in strong coordination interactions of zirconium Zr(IV) ions with organic ligands [52]. The microwave was also used to help in the synthesis of Cr-MIL-101 at 95 °C and for a shorter duration of 9 minutes to generate MOF-5. A solvothermal method for preparing pure phase MIL-140 is likely to result in superior chemical stability with less preparation time than conventional electric heating strategies. In 2013, Ren et al. used this process to produce highly crystalline UiO-66 MOF octahedral-shaped crystals for H_2 storage capacity [53].

3.3 Vapor diffusion method

The vapor diffusion method was the first synthetic route used to create MOFs structure. This method produces high-quality crystals, but it requires high ligand

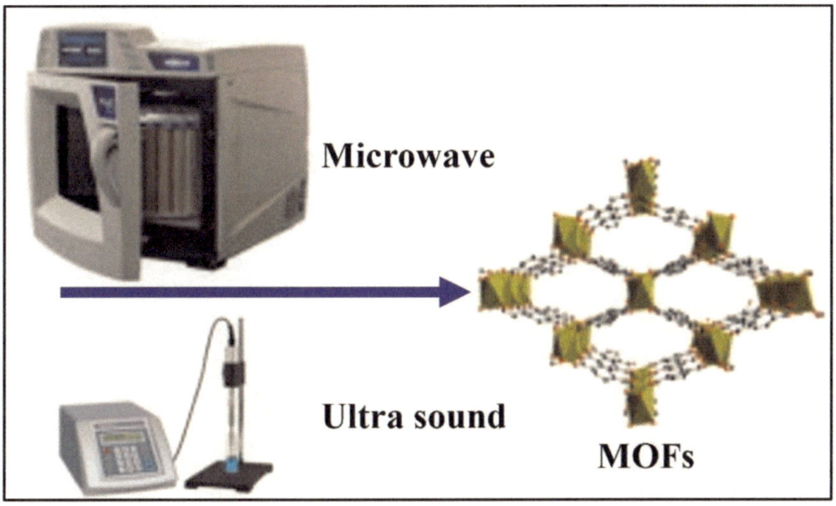

Figure 10.
Schematic representation of the microwave-assisted method for synthesis of metalorganic frameworks (reproduced from Ref. [40]).

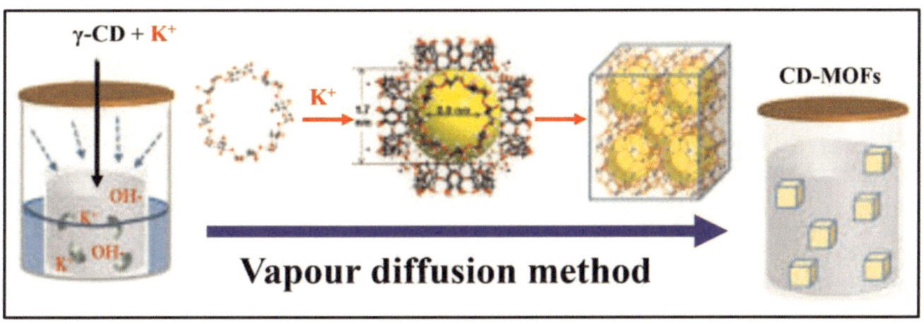

Figure 11.
Schematic representation of the vapor diffusion method for synthesis of metal–organic frameworks (reproduced from Ref. [55]).

solubility. For the first time, Smaldone and his colleagues characterized a series of cyclodextrin-based metal–organic frameworks (CD-MOFs) [54]. In addition, Forgan's group created cyclodextrin-based metal–organic frameworks (-CD-MOFs) by combining -CD with K^+, Rb^+, and Cs^+ in an aqueous medium and then vapor diffusion with MeOH [54]. They obtained single crystals of -CD-MOFs with crystal sizes ranging from 200 to 400 nm in 2–7 days (**Figure 11**). Wu et al. used this method to make [Pb(1,4-NDC)(DMF)] by dissolving $Pb(NO_3)_2$ and H_2-1,4-NDC (naphthalene dicarboxylate) in DMF inside one vessel and triethylamine in the other. The continuous growth of MOF crystals was caused by the sustained diffusion of triethylamine from the outer to the inner container [56].

3.4 Gel crystallization

Gel crystal growth is a useful strategy for MOF synthesis that involves adding an emulsifier to the reaction medium. Das et al. synthesized MOF $[Ba_2(O_3P(CH_2)_3PO_3)]$ $3H_2O$ by first dissolving the metal component, barium(II) chloride in water [57]. The

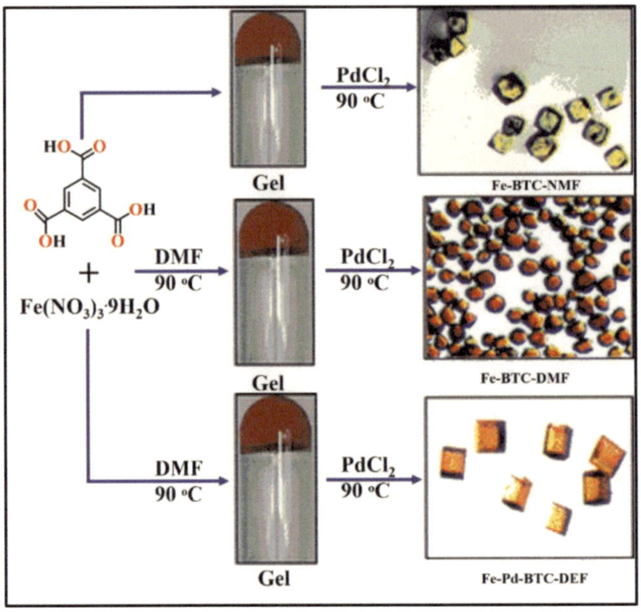

Figure 12.
Synthesis of Fe-MOFs via $PdCl_2$-mediated gel crystallization methods (reproduced from Ref. [58]).

mixture was thoroughly mixed prior to forming a gel, and sheets of ethanediphosphonic aqueous acidic remedy were equipped and nurtured for three weeks. Despite the fact that this procedure is time-consuming, additional separation phases are required to detoxify the product using the gel crystallization method (**Figure 12**).

3.5 Solventless method

Solventless synthesis methods are more advantageous because they allow the fabrication of MOFs without the use of toxic solvents and create a new identity through studies of a few solventless synthesis of MOFs [59]. Mechanosynthesis is a fast, scalable, and nontoxic method for producing MOFs. It involves ball bearings in a stainless-steel vessel with reagents and a stoichiometric amount of solvent [60]. Solventless synthesis reduces solvent toxicity by expelling it from the reaction. The nanocrystalline nature of the material is provided by the finishing, which can be biosynthesized quickly at room temperature. After the vessel has been completely closed and the reagent kits have been pulverized by proper mixing, the metal–organic framework is addressed (**Figure 13**).

3.6 Sonochemical

The sonochemical synthesis of MOFs is a simple and effective method that involves exposing the mixture to sonogram waves with frequencies ranging from 20 kHz to 10 MHz. The advantages of this approach include the lack of additional heat required, the quick reaction time, and the creation of a narrow size distribution crystallization product (**Figure 14**). The sonochemical synthesis method is suitable for the synthesis of nanoscale crystals because particles are generated instantly inside local solvent cavity regions with a short total lifetime (ms) and dimensions in the 10 nm range. MOFs

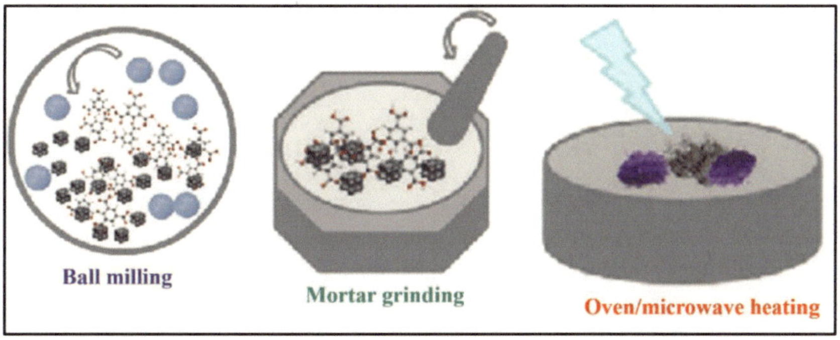

Figure 13.
Schematic representation of solventless synthesis method of metal–organic frameworks (reproduced from Ref. [61]).

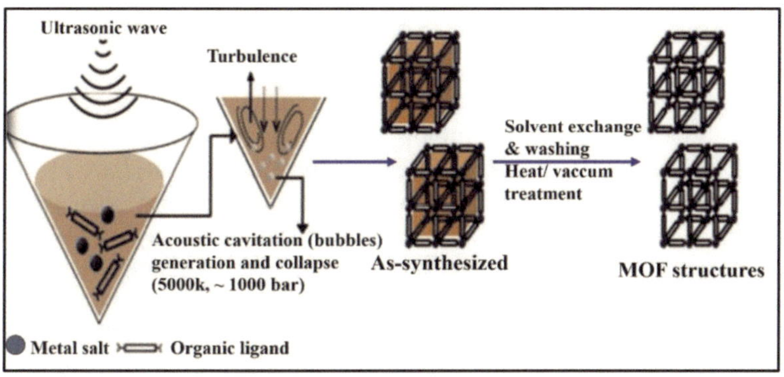

Figure 14.
Sonochemical synthesis of the metal–organic framework (reproduced from Ref. [62]).

produced at room temperature using a sonochemical synthesis method have dimensions in the 10 nm range [63]. Fard et al. demonstrated the creation of a 2D MOF [$Pb_2(N_3)(NO_3)L_2$], (L = 8-hydroxy quinolate) in an aqueous medium [64].

3.7 Electrochemical method

The electrochemical synthesis method provides an alternative approach to the sonochemical procedure that does not require an external source of heat [65] (**Figure 15**). Joaristi et al [67]. reported the synthesis of HKUST-1 on an anode side with only an isopropanol solution and a copper mesh. MIL-100(Fe) has been synthesized for the first time using a simple nucleation method [68]. The method involves dissolving 1,3,5-benzenetricarboxylic acid in ethyl alcohol and Milli-Q moisture in an electrolytic system under high pressure and temperature. MIL-100(Fe) crystals were grown on pure iron substrates using a Fe electrode at temperatures ranging from 110 to 190°C. Heating a solution of BDC: HNO_3: H_2O: AA: DMF contributed to UiO-66 anodic/cathodic film deposition on zirconium foil. The anodic deposition has been shown to improve MOF adhesion on zinc metal ions, in a study published in the Journal of Organic Chemistry and Biomaterials (JICB) by researchers at the University of California, Los Angeles [69].

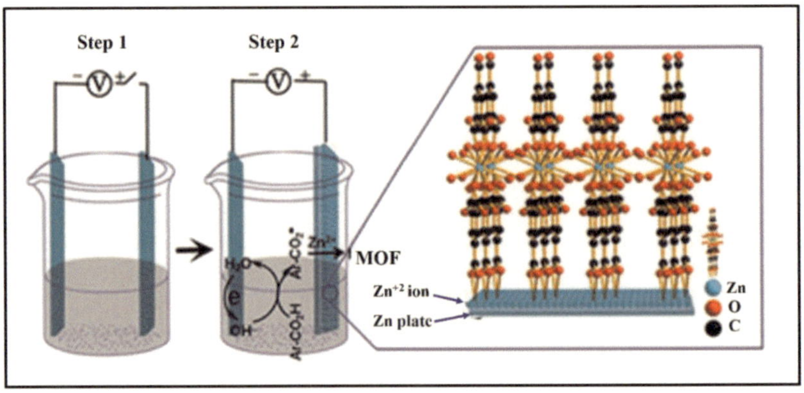

Figure 15.
Schematic representation of electrochemical synthesis method of metal–organic frameworks (reproduced from Ref. [66]).

4. Photocatalytic applications of MOFs

MOFs have recently received a lot of attention in the field of photocatalytic degradation. Photocatalysis is a green technology that converts sunlight into chemical energy [70]. MOFs have a large surface area and high porosity and are ideal candidates for photovoltaics. MOF breakthroughs in organic contaminant degradation, water splitting, and CO_2 photoreduction [71]. TiO_2 has a large bandgap (3.2 eV), which limits its photocatalytic properties in the UV light region. This region accounts for only 5% of the spectral region, resulting in low photocatalytic activity. In the last decade, far more effort has been expended to improve light utilization by changing TiO_2 via anion/cation doping or incorporating it with other metals/semiconductors [72]. These evaluations are limited to delineating the engineering premise and the photocatalysts working principle.

4.1 Fundamental processes

MOF-based photocatalysis, like the vintage photodegradation phenomenon formed for conventional semiconductor photocatalysts, followed four basic steps of processes (**Figure 16**), which operate from their crystalline structure. The four fundamental operating processes in MOF-based heterogeneous catalysis are as follows [74]:

4.1.1 Photoexcitation step

MOF-based photocatalysts could occur via inorganic SBU or organic ligands. MOFs generally have low absorptions in the UV-Vis region, depending on their fluorophore centers [75]. The spectra shown above are frequently associated with various π–π^* transitions of aromatic units. MOF photocatalysts exhibit a few distinct absorption coefficient bands as a result of inorganic SBUs and organic ligands having distinct molecular orbitals, as well as their lowest energy bands. Semiconductor band theory fails to describe light absorption and subsequent transitions observed in MOFs. LMCT transitions are important for understanding the mechanisms and reaction products in photocatalysis involving MOFs. Even though photo-generated holes and electrons

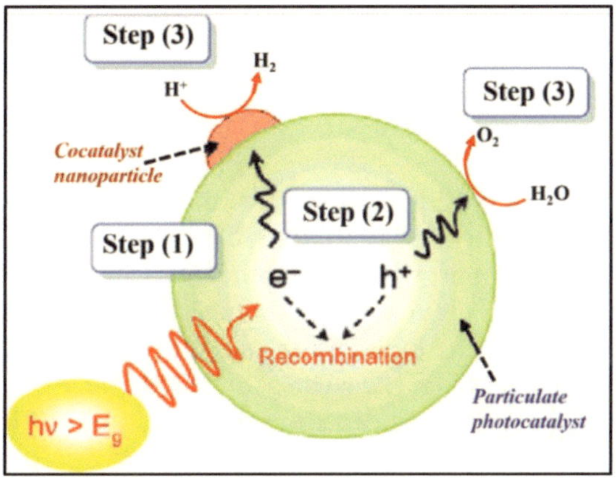

Figure 16.
Fundamental steps in a traditional photocatalysis system (reproduced from Ref. [73]).

have the same energy stages for MOF molecular orbitals (HOMO and LUMO), this is critical for understanding MOF reactions [76].

4.1.2 Charge-transfer step

This procedure begins with the absorption of massive amounts of light energy, which results in the excitation of positive and negative charges from the bulk to the photocatalyst surface. The efficiency of these procedures is determined by the material type, crystallinity, and particle size. High charge mobility, for example, is frequently associated with higher crystalline semiconductor materials that exhibit focused charge mobility. The adsorption and photodegradation activities of various dyes on three Cd-containing MOFs have been investigated. Anionic dyes (e.g., sulphonated) were significantly adsorbed by $Cd_2(4,4'\text{-bpy})_3(S_2O_3)_2$, $Cd_1(4,.4')\cdot 2H_2O$ in the dark, but not by non-cationic dyes. The presence of chromophore centers, which may also serve as recombination centers where charge carriers are trapped and quenched, causes structural deficiency in inorganic photocatalysts, resulting in significant energy loss through realized heat [77]. The charge-transfer performance of semiconductor photocatalysts with small granules is satisfactory. If the particles are too small in size, the replication process may occur due to improved surface defects.

4.1.3 Surface activation step

Charge carriers travel to the photocatalyst surface, where they can be stimulated to perform specific chemical reactions. Photogenerated charge carriers are gathered by photoelectrodes and allowed to open circuit [78]. The first step in a fairly standard photocatalytic reaction is the physical adsorption of adsorbent materials. Anionic and cationic dyes showed different surface adsorption and catalytic properties. Furthermore, dye macromolecular adsorption resulted in non-covalent weak interactions rather than radical reforms in the cadmium thiosulphate-based MOF structures. The hydroxyl radical pathway is critical in the breakdown of anthraquinonic anionic dyes, whereas a surface-controlled N-de-ethylation reaction mechanism was proposed to explain the systematic degradation of cationic dyes via sequential intermediates, where MLCT inferred from Cd metal to HOMO as well as ligand-associated LUMO play dominant roles [79].

4.1.4 Charge-carrier recombination step

Photoluminescence (PL) spectrophotometry could be used to capture and interpret the energy released during recombination. Charge-carrier recombination accounts for the majority of energy loss in photocatalysts and PEC processes, and it remains one of the most difficult challenges to overcome [80]. Charge-transport recombination occurs in both the bulk and the surface of the photocatalyst. Reducing charge-carrier recombination from the surface and bulk phases is critical. Surface metallization with noble metals has been shown to be beneficial in charge-carrier combinations [81].

4.2 MOF-based photocatalytic degradation of organic pollutants

MOFs are composed of metal-containing nodes linked by organic ligands via strong covalent bonds. When exposed to light, a few MOFs begin to behave like semiconductor materials, implying they could be useful as photocatalysts [82]. New research has demonstrated porous MOF materials to be a new type of photocatalyst. MOFs have a promising future, despite the fact that they have not been widely investigated to date. It is simple to synthesize MOFs with tunability in light absorption capacity, thereby activating appealing photocatalytic properties. Research into the application of MOFs in this area has so far been largely unexplored. In this chapter, we highlight the importance of photocatalysis in the degradation of organic pollutants into MOFs. The reaction pathway as well as the impact of external variables on electrocatalytic activity are discussed. The main issues in photocatalytic degradation and potential opportunities have been thoroughly discussed [83].

4.3 Photocatalytic properties of d-block metal-based MOFs

MOFs offer a unique opportunity for the discovery of new catalysts capable of degrading organic pollutants. More effort has been devoted to developing innovative photocatalyst materials based on MOFs. MOFs could also have potential applications in the environmentally friendly removal of organic compounds. Some d-block metal-based MOFs have good photocatalytic efficiency for organic pollutants. The d-block transition metal MOFs are important for their contributions to a variety of fields such as magnetism, catalysis, gas separation, drug delivery, and so on. The transition metal (Zn(II), Cu(II), and Cd(II)) based MOFs that have been studied as photocatalysts for photocatalytic degradation of an organic pollutant under UV, visible, or UV-vis light illumination are summarized in **Table 1** [95].

The MOF-5 is made up of Zn_4O clusters that are orthogonally linked by 1,4-bdc linkers at the corners of a cubic framework structure. This MOF was discovered to have a broad absorption band in the wavelength range of 500–840 nm. MOF-5 is a highly efficient photocatalyst that would most likely succeed due to the light source [96]. MOF-5 may improve overall photocatalytic activity efficiency and photodegradation of phenol, like TiO_2, could occur via a network of reactions, such as the formation of a radical cation by electron transfer from phenol to MOF (**Figure 17a**). It degraded phenol in aqueous solutions in a manner similar to commercial TiO_2 and could improve overall photocatalytic activity efficiency (**Figure 17b**).

MOFs	Light irradiation	Degradation (%)	Time (min)	Organic pollutants	Ref.
Cd-based MOFs					
Cd(4-bpah)(1,3-bdc)(H_2O)	UV	40	240	MB	[84]
Cd $(btec)_{0.5}(bimb)_{0.5}$	Vis	90	540	X_3B	[12]
Cd(npdyda)$(H_2O)_2$	UV	13.7	150	MO	[85]
Ni-based MOFs					
$[Ni_2(4,4'\text{-bpy})_2](4,4'\text{-obb})_2 \cdot H_2O$	UV	50	90	RhB	[86]
$[Ni_2(4,4'\text{-bpy})_2](4,4'\ \text{obb})_2 \cdot H_2O$	UV	80	90	MB	[87]
Co-based MOFs					
$[Co_2(4,4'\text{-bpy})](4,4'\text{-obb})_2$	UV	60	90	RhB	[85]
$Co(btec)_{0.5}(4,4'\text{-bimb})$	Vis	80	540	X3B	[88]
$[Co_2(4,4'\text{-bpy})](4,4'\text{-obb})_2$	UV	90	90	MB	[86]
Zn-based MOFs					
$[Zn_2(4,4'\text{-bpy})](4,4'\text{-obb})_2$	UV	40	90	RhB	[86]
MOF-5	UV	50	180	Phenol	[89]
MOF-5	UV	100	180	DTBP	[90]
(UTSA-38)	UV–vis	100	120	MO	[91]
$[Zn(1,4\text{-biyb})(adtz)] \cdot H_2O$	UV	62	180	MB	[88]
Fe-based MOFs					
Fe_2(bhbdh)	Vis	90	15	RhB(H_2O_2)	[92]
MIL-53(Fe)	Vis	20	20	MB(H_2O_2)	[93]
Fe_2(bhbdh)	Vis	90	15	MO(H_2O_2)	[92]
MIL-53(Fe)	UV–vis	11	40	MB	[93]
Cu-based MOFs					
Cu(ptz)(I)	Vis	98	24	MB(H_2O_2)	[93]
$[Cu_3(3\text{-dpsea})(1,3,5\text{-btc})_2(H_2O)_5]$	UV	56	45	MB	[94]
Cu(ptz)(I)	Vis	100	35	RhB(H_2O_2)	[88]
Cu(dm-bim)	Vis	95	34	MO	[94]
Cu(dm-bim)	Vis	100	40	RhB	[94]

Table 1.
The photocatalytic degradation of organic pollutants in aqueous media using some d-block metal-based MOFs as photocatalysts.

MOF-5 is a highly efficient photocatalyst that would most likely succeed due to the light source. Visible light irradiation (cut-off filter $\lambda > 380$ nm) would significantly degrade TiO_2 and ZnO activity due to a lack of uptake at wavelengths $\lambda > 350$ nm (**Figure 17c**) [98].

MOF-5 exhibited opposite morphology forward into various compounds, including large phenolic molecules that can flexibly disperse into the micro pores of MOF-5 deteriorated significantly faster than small ones can gain access to the inner of MOF-5,

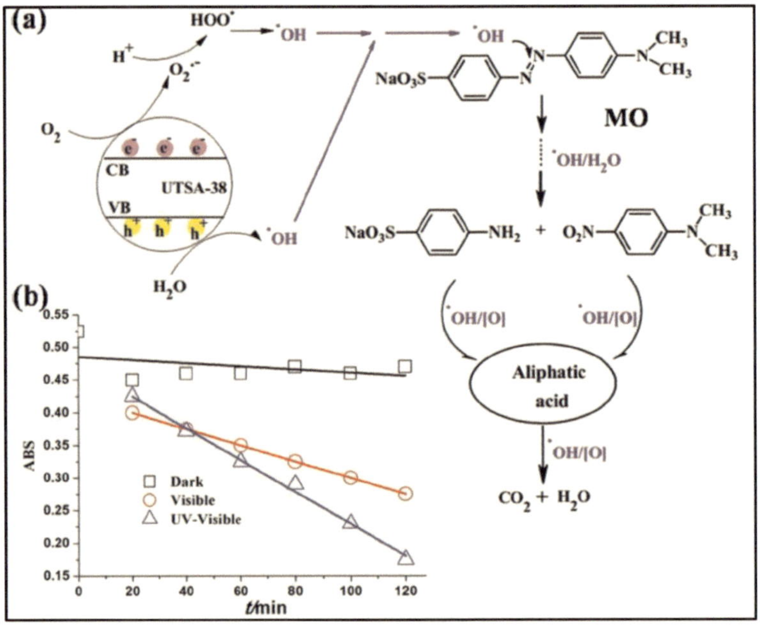

Figure 17.
(a) The comparison of calculated bandgap for TiO_2 and MOF-5; (b) plots for photocatalytic degradation of phenols using TiO_2, ZnO, and MOF-5; (c) the plausible mechanism of photocatalytic degradation using MOF-5 as a photocatalyst (reproduced from Ref. [97]).

as investigated by Garcia and his colleagues. Researchers have studied the photodegradation of DTBP and P, where DTBP is 2,6-di-tert-butylphenol and P is significantly larger. They found that DTBP deteriorated at a similar rate to P in terms of MOF-5 at first (**Figure 18a**) [100]. MOF-5 has demonstrated size-selective photocatalytic activity. When a mixture containing P_{mix} and $DTBP_{mix}$ was exposed to radiation, DTBP deteriorated 4,4'-fold greater in comparison to P after 180 minutes

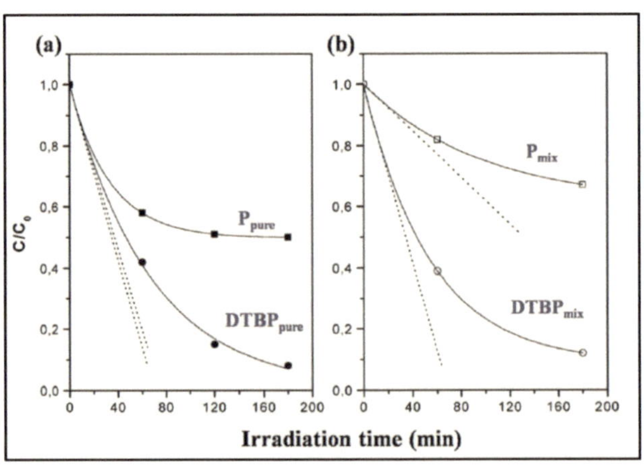

Figure 18.
(a) Photodegradation curves for phenol (P) and 2,6-di-tert-butylphenol (DTBP) of the pure species at 40 mg L^{-1}; (b) irradiation of a mixture of 20 mg L^{-1} of both molecules using MOF-5 as a photocatalyst (reproduced from Ref. [99]).

of irradiation, degrading nearly 50% of the phenol and 100% of the DTBP (**Figure 18b**) [101].

Porous MOFs with 2.85 eV bandgap energy have shown photocatalytic properties for the degradation of methyl orange (MO) in an aqueous solution. The concentration of MO in water must have gradually decreased over time in the presence of light, implying perceptible decay of MO. MO can be degraded completely into colorless molecules in 120 minutes, implying that UV light was far more effective than visible light for this type of photocatalytic activity [102]. Furthermore, the UTSA-38 catalyst was recovered from the reaction mixtures with simple filtration, with no discernible loss of catalytic performance. The main pathways proposed by UTSA-38 for MO photoreduction when exposed to UV or visible light are depicted in **Figure 19a**. Charged particles reduced oxygen (O_2) to oxygen radicals, which then changed into hydroxyl radicals (OH°), which were efficient at decaying MO [104].

The Langmuir–Hinshelwood kinetic has been successfully applied to heterogeneous photodegradation. The relationship between the initial degradation rate and the initial dye concentration of the organic substrate can be written as $r_O = k_O C_O/(1 + K_O C_O)$. The photodegradation of the four dyes in [Co_2(4,4'-bpy), Ni_2, Zn_2, and H_2O has been studied. The majority of these reactions produced very low K_O values, which were discovered. A low value of K_O indicates poor adsorption, despite the fact that K_O is the equilibrium adhesion coefficient. The photocatalysts [Co_2(4,4'-obb)$_2$, [Ni_2('bpy')·$2H_2O$, and [Zn_2 bpy) performed better than commercial TiO_2 catalysts under laboratory conditions (**Figure 19b**) [101]. These MOF catalysts were previously reported, but their kinetic rates and degradation efficiencies have been summarized in **Table 2**.

The photocatalytic decomposition of organic dyes in [Co_2(4,4'-bpy),] a simple mechanism has been proposed further. One electron moves from the HOMO to the LUMO when exposed to UV light and $2H_2O$. The excited M^{2+} center decomposes rapidly to its ground state. If any molecules are within an acceptable distance but have the proper orientation, transitional energetic compounds may form. This results in the

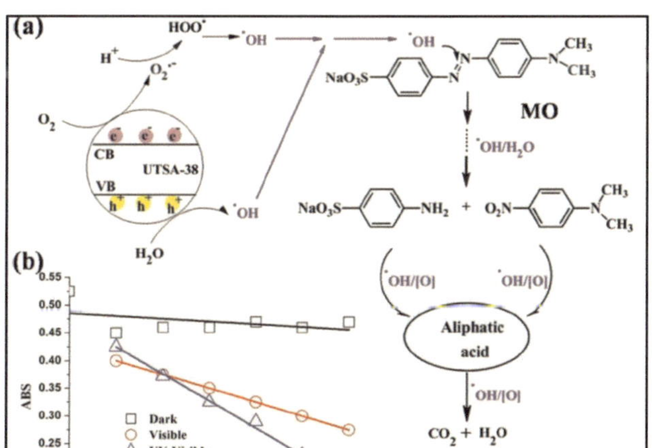

Figure 19.
(a) Photodegradation mechanism for methyl orange by UTSA-38 in the presence of UV-visible or visible light; (b) absorbance plots for degradation of methyl orange solution degraded by UTSA-38 in the presence of different light sources, such as UV-visible, visible, and dark light (reproduced from Ref. [103]).

MOFs	k_O (min^{-1})	Dyes	K_O (mg L^{-1})
$[Ni_2(4,4'\text{-bpy})_2](4,4'\text{-obb})_2{\cdot}H_2O$	0.029	RBBR	0.0015
$[Co_2(4,4'\text{-bpy})](4,4'\text{-obb})_2$	0.013	RHB	0.0035
$[Zn_2(4,4'\text{-bpy})](4,4'\text{-obb})_2$	0.020	OG	0.0029
$[Co_2(4,4'\text{-bpy})](4,4'\text{-obb})_2$	0.032	MB	0.0064
$[Ni_2(4,4'\text{-bpy})_2](4,4'\text{-obb})_2.H_2O$	0.029	OG	0.0049
$[Zn_2(4,4'\text{-bpy})](4,4'\text{-obb})_2$	0.007	RhB	0.0020
$[Ni_2(4,4'\text{-bpy})_2](4,4'\text{-obb})_2{\cdot}H_2O$	0.008	RhB	0.0023
$[Co_2(4,4'\text{-bpy})](4,4'\text{-obb})_2$	0.031	OG	0.0022
$[Zn_2(4,4'\text{-bpy})](4,4'\text{-obb})_2$	0.023	MB	0.0029

Table 2.
The kinetic parameters for dye degradation using $[Co_2(4,4'\text{-bpy})](4,4'\ obb)_2$, $[Ni_2(4,4'\text{-bpy})_2](4,4'\text{-obb})\cdot 2H_2O$, as well as $[Zn_2(4,4'\text{-bpy})](4,4'\text{-obb})_2$ [105].

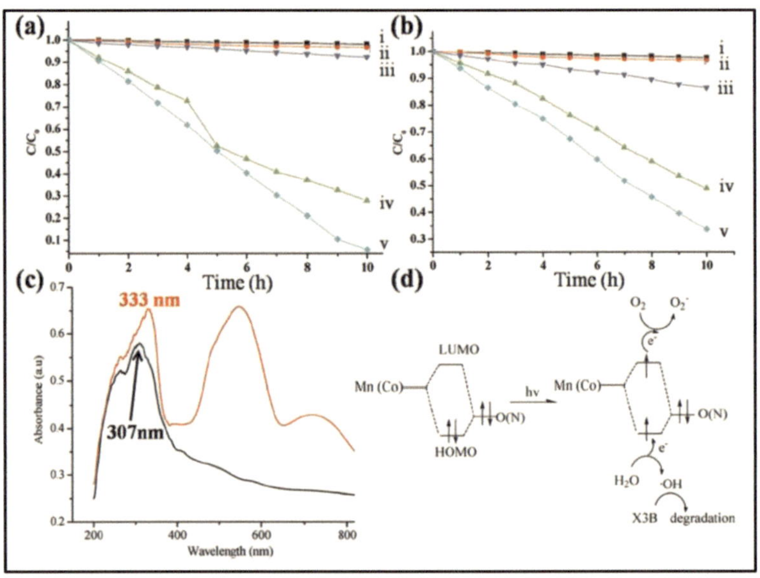

Figure 20.
(a) X_3B photodegradation experiments: (i)$X_3B/[Mn_3(btc)_2(bimb)_2]$ ii) X_3B/UV light (without catalyst); iii) $X_3B/[Mn_3(btc)_2(bimb)_2]$ (iv) $X_3B/[Mn_3(btc)_2(bimb)_2]$; $4H_2O$/visible light; $X_3B/[Mn_3(btc)_2(bimb)_2]$; and (v) $4H_2O$/tert-butyl alcohol/UV light. $4H_2O$/UV light. (b) X_3B photodegradation experiments: (i) $X_3B/[Co_3(btc)_2(bimb)_2]$ (ii) X_3B/UV light (without catalyst); (iii) $X_3B/[Co_3(btc)_2(bimb)_2]$ (iv) $X_3B/[Co_3(btc)_2(bimb)_2]$; $4H_2O$/tert-butyl alcohol/UV light; $4H_2O$/visible light, as well as (v) $X_3B/[Co_3(btc)_2(bimb)_2]$. UV light/$4H_2O$ $[Mn_3(btc)_2(bimb)_2]$ UV/vis diffuse-reflectance spectra $[Co_3(btc)_2(bimb)_2]$ and $4H_2O$ (black line) $4H_2O$ (red line) with a background of $BaSO_4$. (d) A simplified model of X_3B's photocatalytic reaction mechanism with $[Mn_3(btc)_2(bimb)_2]$. $[Co_3(btc)_2(bimb)_2]$ and $4H_2O$ (reproduced from Ref. [106]).

cleavage of the C–N bond and the gradual N-deethylation of RhB. The HOMO and LUMO MOFs have different bandgap sizes (4.04 and 3.72 eV, respectively), resulting in photocatalytic degradation differences (**Figure 20a** and **b**) [107]. Despite the fact that the two MOFs share the same hierarchical architecture, different focal metal ions result in different radioactivity levels. $Mn_3(btc)_2(bimb)_2]\cdot 4H_2O$ could be assigned to ligand-to-metal charge transfer (LMCT), as shown in **Figure 20c**. In the latter MOF, two additional peaks at 547 and 721 nm are detected, which are most likely the result of the spin-allowed transition of d^7 Co^{2+} ion. The photocatalytic properties of $[Mn_3(btc)_2(bimb)_2]$ have been improved. Under UV light, the aforementioned was able to degrade X_3B almost completely in 10 hours [87]. The energy bandgap between $4H_2O$ and Co_3 has also been found to be larger under UV light. $Mn_3(btc)_2(bimb)_2]$. $4H_2O$ could be attributed to their distinct UV/vis absorption properties. The HOMO is primarily attributed by the oxygen and (or) nitrogen 2p bonding orbitals. The LUMO is caused by empty Mn(Co) orbitals (conduction band). Electrons were transferred from oxygen and (or) nitrogen to Mn throughout the photoinduced process. In this case, one electron was extracted from the water molecule and aerated to produce the OH° hydroxyl radicals [101]. Meanwhile, electrons in the LUMO combined with oxygen adsorbed on the MOF surfaces to form O_2, which was then converted to hydroxide (OH) (**Figure 20d**).

4.4 Photocatalytic selective redox in organic synthesis

MOFs can be used to promote photocatalytic oxidations in the absence or presence of another semiconductor. This is important because the oxidation of alcohols to aldehydes and ketones is an important reaction in organic synthesis [71]. **Table 3** summarises the studies that describe the use of MOFs as photocatalysis catalysts. Amine-functionalized UiO-66 has been reported as a high-efficiency and high-selectivity visible-light photocatalyst for the selective aerobic oxygenation of various organic compounds such as alcohols, olefins, and cycloalkanes.

The $-NH_2$ group inside the bdc linker introduces a new absorption edge in the diffuse reflectance UV/Vis spectrum of NH_2-UiO-66 at λ_{max} (450 nm). Exposure to visible light increased the conversion of the studied alkenes steadily over time. MOFs can be used as photocatalysts for H_2 generation, CO_2 reduction, photooxygenation, and nitro reduction [115]. The experimental results show that the solvent used and the reacting precursors now influence the final product selectivity. From b-methylstyrene, styrene, and 1,2-diphenylethylene, epoxides with selectivity values

Catalyst	Stability evidence	Photoactivity	Photolysis source	Ref.
TiO_2@HKUST-1[e]	IR, XRD, reuse,	89% conversion[b]	sunlight	[108]
CdS-NH_2-UiO-66	XRD, XPS, reuse	31% conversion[b]	300W Xe arc lamp	[109]
NH2-MIL-125(Ti)	XRD, reuse	73% conversion[c]	300W Xe lamp	[110]
Au/MIL-125(Ti)	–	36% conversion[b]	300W Xe arc lamp (λ = 320–780 nm)	[111]
multicore Au@ZIF-8	–	51.6% conversion[b]	500W Xe lamp	[112]
MR-MIL-125(Ti)[d]	reuse	86.7 $nmolg^{-1}min^{-1}$ [b]	150W Xe lamp	[113]
NH_2-UiO-66-F	–	53.9% conversion[b]	26 W helical light bulb	[114]
NH_2-UiO-66	–	1.234 h^{-1} (TOF)[a]	300W Xe lamp	[91]

Table 3.
Summary of photooxidation reactions catalyzed using MOFs-based photocatalysts.

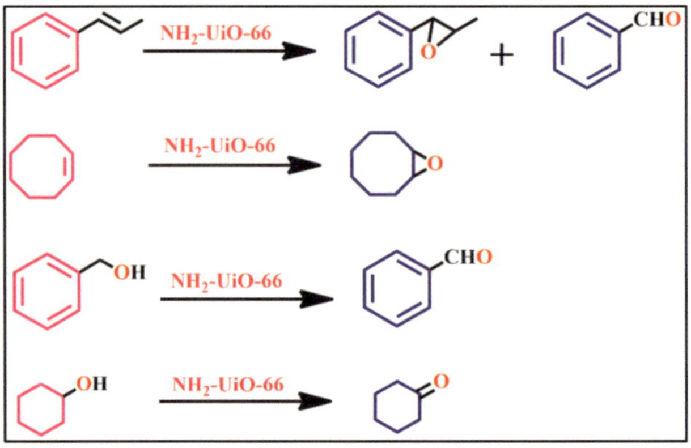

Figure 21.
NH_2-UiO-66 catalyzes photooxidation of various substrates (reproduced from Ref. [91]).

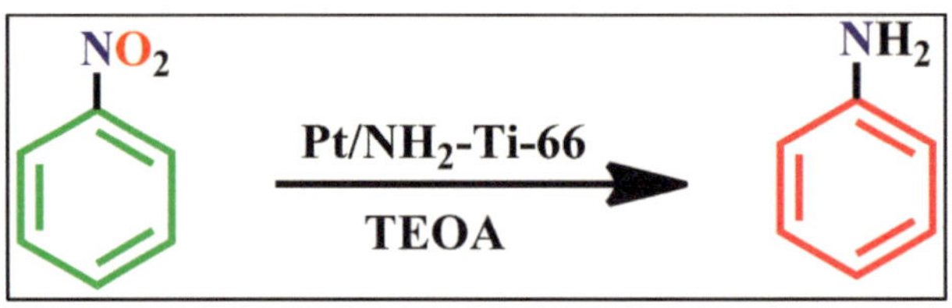

Figure 22.
Nitrobenzene photoreduction catalyzed by Pt(1.5)/NH_2-Ti-MOF (reproduced from Ref. [117]).

ranging from 15 to 65% were obtained. Cyclooctene resulted in low conversion due to its larger kinetic diameter, particularly when compared to the pore diameter of NH_2-UiO-66. An ^{18}O-isotope labeling experiment for such photocatalytic epoxidation of cyclooctene has shown that the product contains oxygen (**Figure 21**). This is consistent with the fact that oxygen can be inferred from molecular sufficient oxygen in the gaseous state [116].

Photocatalysis that results in charge separation may promote both oxidation and reduction (via the reaction with photogenerated holes). All of these methods must occur at the same rate, but depending on the material, either of the two half-reactions could occur. Pt/NH_2-Ti-MOF was discovered to act as a photocatalyst for such nitrobenzene reduction under visible light illumination (500W Xe lamp), resulting in aniline as the final product. Other photocatalytic reductions of aromatic nitro groups are discussed further below. Pt(1.5)/NH_2-Ti-MOF (3.3 mmol^{-1}) demonstrated superior catalytic activity to NH_2-Ti-MoF (2.3 mol^{-1}), suggesting that hoarded Pt also acts as a co-catalyst in this framework (**Figure 22**). The reaction selectivities appear to be nearly identical regardless of the presence of Pt species.

4.5 Functions of MOFs in photoelectrodes

MOFs play critical roles in increasing photoelectrode efficiency and achievability during the fabrication process of photomicrography devices. They improve light-harvesting capability, carrier separation efficiency, carrier potential efficiency, and electrode potential efficiency. MOF photovoltaics can be used to improve light-harvesting capability and accelerate carrier separation efficiency. Light usage efficiency is the most important factor influencing solar energy conversion efficiency in PEC system applications. Improving light resource efficiency as much as possible is critical to improving photoelectrochemical performance [118]. TiO_2 has a low light optimum utilization and a bandgap (3.0–3.2 eV), but it plays an important role in photoelectrode mechanism studies. Because of their adjustable bandgap and absorptivity, MOFs are thought to be effective photosensitizers [119]. They first grew ZIFs in situ on semiconducting ZnO. They sulfurized ZnO@Zn-ZIF, ZnO@Co-ZIF, and ZnO@ZnCo-ZIF to obtain high surface area shells with abundant porosity. They eventually succeeded in fabricating honeycomb ZnO@ZnS, ZnO@CoS, and ZnO@ZnS/CoS heterojunction photoelectrodes (**Figure 23a**). The structure properties after vulcanization provided long photoelectric effect transmitting pathways and an abundance of exposure catalyst surface to achieve effective optical absorption [119]. The sulphide MOFs elevated the photoelectrochemical spectral range to red-shift to varying degrees in the ultraviolet-visible spectral range (UV-vis). MOFs are frequently used as photosensitizers to extend the absorption of visible light in order to improve light usage. Liu et al. used a hydrothermal process to create an ultra-thin MIL-101 (Fe) layer on the surface of Mo: $BiVO_4$ (**Figure 23b**) [121].

MOF photogeneration is a critical step in improving photoelectrode PEC efficiency. The energy levels of MOFs highest occupied molecular orbital and lowest unoccupied molecular orbital can be changed to better match semiconducting levels of energy and charge-transport carriers [122]. A team of researchers in China's Zhejiang Province of Hebei has developed a novel way to improve charge separation at the electrolyte/semiconductor interface. They used a hydrothermal deposition technique to form a binary photoanode from their 3D bimetallic MOFs and $BiVO_4$ (**Figure 24a**). When exposed to visible light, the holes produced by $BiVO_4$ after absorbing photons migrated to CoNi-MOFs, and Co^{2+} and Ni^{2+} were able to capture and oxidize the holes.

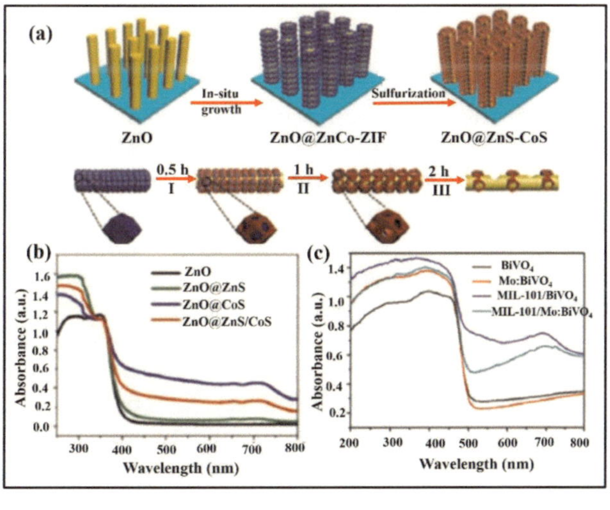

Figure 23.
(a) The fabrication and formation of cellular ZnO@ZnS/CoS are depicted schematically. (b) UV-vis diffuse reflectance of ZnO, ZnO@ZnS, ZnO@CoS, and ZnO@ZnS/CoS. (a-b) are adapted from Ref. Elsevier. All rights reserved. (c) UV-vis spectra of photoanodes $BiVO_4$, MIL-101(Fe)/$BiVO_4$, Mo: $BiVO_4$, and MIL-101(Fe)/Mo: $BiVO_4$ (reproduced from Ref. [120]).

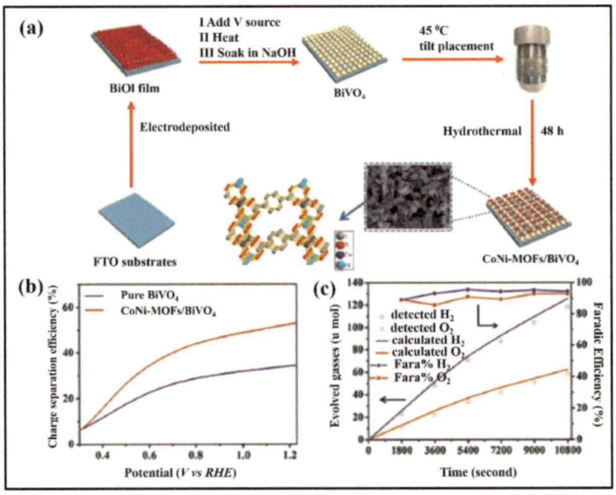

Figure 24.
(a) CoNi-MOFs/BiVO$_4$ schematic illustration; (b) BiVO$_4$ and CoNi-MOFs/BiVO$_4$ charge separation efficiency; (c) The evolution of H_2 and O_2 gases in comparison to the evolution predicted by the current generation and faradaic efficiency. (a–c) (reproduced from Ref. [123]).

The heavy metal ions served as active sites for the interfacial H_2O to O_2 reaction. The photogeneration of CoNi-MOFs/BiVO$_4$ has produced O_2 and H_2 at energies close to theoretical, and the Faraday effectiveness was approximately 90%, demonstrating that the majority of the charges were isolated in time to produce O_2 or H_2 (**Figure 24b** and **c**) [124].

5. Conclusions and perspectives

Emerging metal–organic frameworks (MOFs) have been regarded as the most promising artificial photocatalysts for addressing numerous challenges in the disciplines of energy and environmental remediation due to their exemplary structure and diversity. The enslavement of novel photocatalysts has piqued the interest of numerous research groups. Semiconducting MOFs have a promising future as nascent photocatalysts, but they face significant barriers to widespread adoption. MOFs can be used for a wide range of applications, including healing metals and heavy metal cations, high antimicrobial applications, and photocatalytic indoor environmental remediation. Artificial photosynthesis, such as water splitting and CO_2 photoreduction, is a novel and rapidly expanding application of MOFs. MOF-based photocatalysis equipment could be used for photovoltaic solar cells and detectors. On the surface of MOFs, direct CO_2 capture and photoreduction from the atmosphere may be possible. Adsorption and separate redox active sites allow the constructed MOFs to mimic native plants. MOFs have a promising future as emerging photocatalysts. With the assistance of industrial partners and diverse stakeholders, the commercial exploitation of MOFs in the use of planet-saving solar energy photocatalysis innovations could be more influential. Numerous laboratory studies have shown their potential in the application of MOFs as photovoltaics.

Acknowledgements

The authors are especially grateful to the Department of Applied Chemistry, ZHCET, Faculty of Engineering and Technology, Aligarh Muslim University, Aligarh U.P.-202002, India for providing extensive assistance during the completion of this project.

Conflict of interest

The authors declare no conflict of interest.

Declarations

The authors declare that they have no known competing financial interests or personal relationships that could have appeared to influence the work reported in this chapter.

Author details

Mohd Muslim and Musheer Ahmad*
Faculty of Engineering and Technology, Department of Applied Chemistry, ZHCET, Aligarh Muslim University, Aligarh, UP, India

*Address all correspondence to: amusheer4@gmail.com

References

[1] Park SS, Hendon CH, Fielding AJ, Walsh A, O'Keeffe M, Dincă M. The organic secondary building unit: Strong intermolecular π interactions define topology in MIT-25, a mesoporous MOF with proton-replete channels. Journal of the American Chemical Society. 2017; **139**(10):3619-3622

[2] Joseph J, Iftekhar S, Srivastava V, Fallah Z, Zare EN, Sillanpää M. Iron-based metal-organic framework: Synthesis, structure and current technologies for water reclamation with deep insight into framework integrity. Chemosphere. 2021;**284**:131171

[3] Cook TR, Zheng YR, Stang PJ. Metal-organic frameworks and self-assembled supramolecular coordination complexes: Comparing and contrasting the design, synthesis, and functionality of metal-organic materials. Chemical Reviews. 2013;**113**(1):734-777

[4] Wang X, Song C. Carbon capture from flue gas and the atmosphere: A perspective. Frontiers in Energy Research. 2020;**8**:265

[5] Osman AI, Hefny M, Abdel Maksoud MIA, Elgarahy AM, Rooney DW. Recent advances in carbon capture storage and utilisation technologies: A review. Environmental Chemistry Letters. 2021; **19**(2):797-849

[6] Furukawa H, Yaghi OM. Storage of hydrogen, methane, and carbon dioxide in highly porous covalent organic frameworks for clean energy applications. Journal of the American Chemical Society. 2009;**131**(25):8875-8883

[7] Li X, Yu J, Gosztola DJ, Fry HC, Deria P. Wavelength-dependent energy and charge transfer in MOF: A step toward artificial porous light-harvesting system. Journal of the American Chemical Society. 2019;**141**(42):16849-16857

[8] Bedia J, Muelas-Ramos V, Peñas-Garzón M, Gómez-Avilés A, Rodríguez JJ, Belver C. A review on the synthesis and characterization of metal organic frameworks for photocatalytic water purification. Catalysts. 2019;**9**:52

[9] Soni S, Bajpai PK, Arora C. A review on metal-organic framework: Synthesis, properties and application. Characters in Applied Nanomaterials. 2018;**2**:2

[10] Zhang YF, Zhang ZH, Ritter L, Fang H, Wang Q, Space B, et al. New reticular chemistry of the rod secondary building unit: Synthesis, structure, and natural gas storage of a series of three-way rod amide-functionalized metal-organic frameworks. Journal of the American Chemical Society. 2021;**143**(31):12202-12211

[11] Kalmutzki MJ, Hanikel N, Yaghi OM. Secondary building units as the turning point in the development of the reticular chemistry of MOFs. Scientic Advances. 2018;**4**:10

[12] Metal-organic frameworks history and structural features. 2020;**2020**:1-29. Available from: https://doi.org/10.1142/9781786346735_0001

[13] Kandiah M, Nilsen MH, Usseglio S, Jakobsen S, Olsbye U, Tilset M, et al. Synthesis and stability of tagged UiO-66 Zr-MOFs. Chemistry of Materials. 2010; **22**(24):6632-6640

[14] Liu X. Metal-organic framework UiO-66 membranes. Frontiers of Chemical Science and Engineering. 2020;**14**(2):216-232

[15] Brozek CK, Michaelis VK, Ong TC, Bellarosa L, López N, Griffin RG, et al.

Dynamic DMF binding in MOF-5 enables the formation of metastable cobalt-substituted MOF-5 analogues. ACS Central Science. 2015;**1**(5):252-260

[16] Laurier KGM, Vermoortele F, Ameloot R, De Vos DE, Hofkens J, Roeffaers MBJ. Iron(III)-based metal-organic frameworks as visible light photocatalysts. Journal of the American Chemical Society. 2013;**135**(39): 14488-14491

[17] Yuan S, Qin JS, Lollar CT, Zhou HC. Stable metal-organic frameworks with group 4 metals: Current status and trends. ACS Central Science. 2018;**4**(4):440-450

[18] Pino E, Calderón C, Herrera F, Cifuentes G, Arteaga G. Photocatalytic degradation of aqueous rhodamine 6G using supported TiO2 catalysts. A model for the removal of organic contaminants from aqueous samples. Frontiers in Chemistry. 2020;**8**:365

[19] Yuan S, Zou L, Qin JS, Li J, Huang L, Feng L, et al. Construction of hierarchically porous metal–organic frameworks through linker labilization. Nature Communications. 2017;**8**:15356

[20] Zulkifli ZI, Lim KL, Teh LP. Metal-organic frameworks (MOFs) and their applications in CO_2 adsorption and conversion. Chemistry Selected. 2022;7 (22):e202200572

[21] Furukawa H, Ko N, Go YB, Aratani N, Choi SB, Choi E, et al. Ultrahigh porosity in metal-organic frameworks. Science. 2010;**329**(5990):424-428

[22] Furukawa H, Gándara F, Zhang YB, Jiang J, Queen WL, Hudson MR, et al. Water adsorption in porous metal-organic frameworks and related materials. Journal of the American Chemical Society. 2014;**136**(11): 4369-4381

[23] Wang TC, Bury W, Gómez-Gualdrón DA, Vermeulen NA, Mondloch JE, Deria P, et al. Ultrahigh surface area zirconium MOFs and insights into the applicability of the BET theory. Journal of the American Chemical Society. 2015;**137**(10):3585-3591

[24] Farha OK, Eryazici I, Jeong NC, Hauser BG, Wilmer CE, Sarjeant AA, et al. Metal-organic framework materials with ultrahigh surface areas: Is the sky the limit? Journal of the American Chemical Society. 2012;**134**(36):15016-15021

[25] Liu X, Wang X, Kapteijn F. Water and metal-organic frameworks: From interaction toward utilization. Chemical Reviews. 2020;**120**(16):8303-8377

[26] Gómez-Gualdrón DA, Moghadam PZ, Hupp JT, Farha OK, Snurr RQ. Application of consistency criteria to calculate BET areas of micro- and mesoporous metal-organic frameworks. Journal of the American Chemical Society. 2016;**138**(1):215-224

[27] Archana K, Asif A, Jose D, Sujith R. MOFs for hydrogen storage. Metal Frame Nanomater. 2022:589-607. Available from: https://doi.org/10.1016/B978-0-323-91179-5.00019-X

[28] Feng X, Song Y, Lin W. Dimensional reduction of Lewis acidic metal-organic frameworks for multicomponent reactions. Journal of the American Chemical Society. 2021;**143**(21): 8184-8192

[29] Wen M, Mori K, Kuwahara Y, An T, Yamashita H. Design of single-site photocatalysts by using metal–organic frameworks as a matrix. Chemical: An Asian Journal. 2018;**13**(14):1767-1779

[30] Hu Z, Zhao D. Metal–organic frameworks with Lewis acidity: Synthesis, characterization, and catalytic

applications. CrystEngComm. 2017; **19**(29):4066-4081

[31] Liu YL, Liu XY, Feng L, Shao LX, Li SJ, Tang J, et al. Two-dimensional metal–organic framework nanosheets: Synthesis and applications in electrocatalysis and photocatalysis. ChemSusChem. 2022;**15**(10): e202102603

[32] Lee KJ, Lee JH, Jeoung S, Moon HR. Transformation of metal-organic frameworks/coordination polymers into functional nanostructured materials: Experimental approaches based on mechanistic insights. Accounts of Chemical Research. 2017;**50**(11): 2684-2692

[33] Pallach R, Keupp J, Terlinden K, Frentzel-Beyme L, Kloß M, Machalica A, et al. Frustrated flexibility in metal-organic frameworks. Nature Communications. 2021;**12**:1

[34] Yang S, Karve VV, Justin A, Kochetygov I, Espín J, Asgari M, et al. Enhancing MOF performance through the introduction of polymer guests. Coordination Chemistry Reviews. 2021; **427**:213525

[35] Le Ouay B, Uemura T. Polymer in MOF nanospace: From controlled chain assembly to new functional materials. Israel Journal of Chemistry. 2018;**58**(9): 995-1009

[36] Xie LS, Skorupskii G, Dincă M. Electrically conductive metal-organic frameworks. Chemical Reviews. 2020; **120**(16):8536-8580

[37] Li Z, Fraile J, Viñas C, Teixidor F, Planas JG. Post-synthetic modification of a highly flexible 3D soft porous metal–organic framework by incorporating conducting polypyrrole: Enhanced MOF stability and capacitance as an electrode material. Chemical Communications. 2021;**57**(20):2523-2526

[38] Dai S, Tissot A, Serre C. Metal-organic frameworks: From ambient green synthesis to applications. Bulletin of the Chemical Society Japan. 2021;**94** (11):2623-2636

[39] Sun Y, Zhou HC. Recent progress in the synthesis of metal-organic frameworks. Science and Technology of Advanced Materials. 2015;**25**:16

[40] Khan NA, Jhung SH. Synthesis of metal-organic frameworks (MOFs) with microwave or ultrasound: Rapid reaction, phase-selectivity, and size reduction. Coordination Chemistry Reviews. 2015;**285**:11-23

[41] Dey C, Kundu T, Biswal BP, Mallick A, Banerjee R. Crystalline metal-organic frameworks (MOFs): Synthesis, structure and function. Nano-Micro Letters. 2013;**70**(1):3-10

[42] Doan HV, Amer Hamzah H, Karikkethu Prabhakaran P, Petrillo C, Ting VP. Hierarchical metal–organic frameworks with macroporosity: Synthesis, achievements, and challenges. Nano-Micro Letters. 2019;**11**(1):54

[43] Wu H, Lou XW. Metal-organic frameworks and their derived materials for electrochemical energy storage and conversion: Promises and challenges. Scientific Advances. 2017;**3**(12):AAP9252

[44] Kalaj M, Cohen SM. Postsynthetic modification: An enabling technology for the advancement of metal-organic frameworks. ACS Central Science. 2020; **6**(7):1046-1057

[45] Cubillas P, Anderson MW, Attfield MP. Crystal growth mechanisms and morphological control of the

prototypical metal-organic framework MOF-5 revealed by atomic force microscopy. Chemistry - A European Journal. 2012;**18**(48):15406-15415

[46] Rubio-Martinez M, Avci-Camur C, Thornton AW, Imaz I, Maspoch D, Hill MR. New synthetic routes towards MOF production at scale. Chemical Society Reviews. 2017;**46**(11):3453-3480

[47] Zhao Y, Song Z, Li X, Sun Q, Cheng N, Lawes S, et al. Metal organic frameworks for energy storage and conversion. Energy Storage Materials. 2016;**2**:35-62

[48] Han Y, Yang H, Guo X. Synthesis Methods and Crystallization of MOFs. Synth Methods Cryst. 2020. Available from: undefined/state.item.id

[49] Lohse MS, Bein T. Covalent organic frameworks: Structures, synthesis, and applications. Advanced Functional Materials. 2018;**28**(33):1705553

[50] Shah J, Mohanraj K. Comparison of conventional and microwave-assisted synthesis of benzotriazole derivatives. Indian Journal of Pharmaceutical Sciences. 2014;**76**(1):46

[51] Arenas-Vivo A, Avila D, Horcajada P. Phase-selective microwave assisted synthesis of iron(III) aminoterephthalate MOFs. Materials. 2020;**13**:6

[52] Jhung SH, Lee JH, Yoon JW, Serre C, Férey G, Chang JS. Microwave synthesis of chromium terephthalate MIL-101 and its benzene sorption ability. Advanced Materials. 2007;**19**(1):121-124

[53] Zhao Y, Zhang Q, Li Y, Zhang R, Lu G. Large-scale synthesis of monodisperse UiO-66 crystals with tunable sizes and missing linker defects via acid/base co-modulation. ACS Applied Materials & Interfaces. 2017;**9**(17):15079-15085

[54] Han Y, Liu W, Huang J, Qiu S, Zhong H, Liu D, et al. Cyclodextrin-based metal-organic frameworks (CD-MOFs) in pharmaceutics and biomedicine. Pharmaceutics. 2018;**10**:271

[55] Yuan J, Du Y, Zhang H. Recent advances on visible-light-driven CO2 reduction: Strategies for boosting solar energy transformation. APL Materials. 2020;**8**(6):060904

[56] Gangu KK, Maddila S, Jonnalagadda SB. A review on synthesis, crystal structure and functionality of naphthalenedicarboxylate ligated metal-organic frameworks. Inorganica Chimica Acta. 2017;**466**:308-323

[57] Das AK, Vemuri RS, Kutnyakov I, McGrail BP, Motkuri RK. An efficient synthesis strategy for metal-organic frameworks: Dry-gel synthesis of MOF-74 framework with high yield and improved performance. Scientific Reports. 2016;**6**:28050

[58] Fumanal M, Corminboeuf C, Smit B, Tavernelli I. Optical absorption properties of metal–organic frameworks: Solid state versus molecular perspective. Physical Chemistry Chemical Physics. 2020;**22**(35):19512-19521

[59] Campanelli M, Del Giacco T, De Angelis F, Mosconi E, Taddei M, Marmottini F, et al. Solvent-free synthetic route for cerium(IV) metal-organic frameworks with UiO-66 architecture and their photocatalytic applications. ACS Applied Materials & Interfaces. 2019;**11**(48):45031-45037

[60] Titi HM, Do JL, Howarth AJ, Nagapudi K, Friščić T. Simple, scalable mechanosynthesis of metal–organic frameworks using liquid-assisted resonant acoustic mixing (LA-RAM). Chemical Science. 2020;**11**(29): 7578-7584

[61] Mostaghimi M, Rêgo CRC, Haldar R, Wöll C, Wenzel W, Kozlowska M. Automated virtual design of organic semiconductors based on metal-organic frameworks. Front Materials. 2022;**9**:82

[62] Wen Y, Feng M, Zhang P, Zhou H-C, Sharma VK, Ma X. Metal organic frameworks (MOFs) as photocatalysts for the degradation of agricultural pollutants in water. ACS ES&T Engineering. 2021;**1**(5):804-826

[63] Vaitsis C, Sourkouni G, Argirusis C. Sonochemical synthesis of MOFs. Metal Frameworks for Biomedical Applications. 2020:223-244. Available from: https://doi.org/10.1016/B978-0-12-816984-1.00013-5

[64] Fard MJS, Rastaghi F, Ghanbari N. Sonochemical synthesis of new nano-two-dimensional lead(II) coordination polymer: As precursor for preparation of PbO nano-structure. Journal of Molecular Structure. 2013;**1032**:133-137

[65] Li Z, Zhuang T, Dong J, Wang L, Xia J, Wang H, et al. Sonochemical fabrication of inorganic nanoparticles for applications in catalysis. Ultrasonics Sonochemistry. 2021;**71**:105384

[66] Li WJ, Lü J, Gao SY, Li QH, Cao R. Electrochemical preparation of metal–organic framework films for fast detection of nitro explosives. Journal of Materials Chemistry A. 2014;**2**(45): 19473-19478

[67] Martinez Joaristi A, Juan-Alcañiz J, Serra-Crespo P, Kapteijn F, Gascon J. Electrochemical synthesis of some archetypical Zn 2+, Cu 2+, and Al 3+ metal-organic frameworks. Crystal Growth & Design. 2012;**12**:3489-3498. DOI: 10.1021/CG300552W

[68] Campagnol N, Van Assche T, Boudewijns T, Denayer J, Binnemans K, De Vos D, et al. High pressure, high temperature electrochemical synthesis of metal–organic frameworks: Films of MIL-100 (Fe) and HKUST-1 in different morphologies. Journal of Materials Chemistry A. 2013;**1**(19):5827-5830

[69] Chen X, Zhang Y, Kong X, Guo Z, Xu W, Fang Z, et al. Controlling crystal growth of MIL-100(Fe) on Ag nanowire surface for optimizing catalytic performance. RSC Advances. 2020; **10**(42):25260-25265

[70] Zhang Y, Liu H, Gao F, Tan X, Cai Y, Hu B, et al. Application of MOFs and COFs for photocatalysis in CO2 reduction, H2 generation, and environmental treatment. EnergyChem. 2022;**2022**:100078

[71] Gao J, Huang Q, Wu Y, Lan Y-Q, Chen B. Metal–organic frameworks for photo/electrocatalysis. AdvancedEnergy Sustainable Reseaarch. 2021;**2**(8):2100033

[72] Etacheri V, Di Valentin C, Schneider J, Bahnemann D, Pillai SC. Visible-light activation of TiO2 photocatalysts: Advances in theory and experiments. Journal of Photochemistry and Photobiology C Photochemistry Reviews. 2015;**25**:1-29

[73] Fung CM, Er CC, Tan LL, Mohamed AR, Chai SP. Red phosphorus: An up-and-coming photocatalyst on the horizon for sustainable energy development and environmental remediation. Chemical Reviews. 2022; **122**(3):3879-3965

[74] Alshammari A, Jiang Z, Cordova KE. Metal organic frameworks as emerging photocatalysts. Semiconductors in Photocatalysis. 2016. Available from: https://doi.org/10.5772/63489

[75] Nasalevich MA, Van Der Veen M, Kapteijn F, Gascon J. Metal–organic frameworks as heterogeneous

photocatalysts: Advantages and challenges. CrystEngComm. 2014;**16** (23):4919-4926

[76] Santaclara JG, Kapteijn F, Gascon J, Van Der Veen MA. Understanding metal–organic frameworks for photocatalytic solar fuel production. CrystEngComm. 2017;**19**(29):4118-4125

[77] Tansell AJ, Jones CL, Easun TL. MOF the beaten track: Unusual structures and uncommon applications of metal-organic frameworks. Chemical in Central Journal. 2017;**11**(1):100

[78] Hesari M, Mao X, Chen P. Charge carrier activity on single-particle photo (electro)catalysts: Toward function in solar energy conversion. Journal of the American Chemical Society. 2018;**140** (22):6729-6740

[79] Fan Y, Liu HJ, Zhang Y, Chen Y. Adsorption of anionic MO or cationic MB from MO/MB mixture using polyacrylonitrile fiber hydrothermally treated with hyperbranched polyethylenimine. Journal of Hazardous Materials. 2015;**283**:321-328

[80] Liqiang J, Yichun Q, Baiqi W, Shudan L, Baojiang J, Libin Y, et al. Review of photoluminescence performance of nano-sized semiconductor materials and its relationships with photocatalytic activity. Solar Energy Materials & Solar Cells. 2006;**90**(12):1773-1787

[81] Zhang L, Mohamed HH, Dillert R, Bahnemann D. Kinetics and mechanisms of charge transfer processes in photocatalytic systems: A review. Journal of Photochemistry and Photobiology C Photochemistry Reviews. 2012;**13**(4):263-276

[82] Kong XJ, Li JR. An overview of metal–organic frameworks for green chemical engineering. Engineering. 2021;7(8):1115-1139

[83] Wang Y, Niu J, Zhang Z, Long X. Photocatalytic degradation of organic pollutants in water. Progress in Chemistry. 2013;**20**(10): 1621-1627

[84] Ahmad R, Mondal PK. Adsorption and photodegradation of methylene blue by using PAni/TiO_2 nanocomposite. Environmental Chemistry Letters. 2012; **33**(3):380-386

[85] Tan YX, He YP, Zhang J. Tuning MOF stability and porosity via adding rigid pillars. Inorganic Chemistry. 2012; **51**(18):9649-9654

[86] Zhang J, Zivic N, Dumur F, Guo C, Li Y, Xiao P, et al. Panchromatic photoinitiators for radical, cationic and thiol-ene polymerization reactions: A search in the diketopyrrolopyrrole or indigo dye series. Mater Today Communication. 2015;**4**:101-108

[87] Wang F, Wang C, Yu Z, Xu K, Li X, Fu Y. Two multifunctional Mn(II) metal–organic frameworks: Synthesis, structures and applications as photocatalysis and luminescent sensor. Polyhedron. 2016;**105**:49-55

[88] Song Z, Dong X, Wang N, Zhu L, Luo Z, Fang J, et al. Efficient photocatalytic defluorination of perfluorooctanoic acid over BiOCl nanosheets via a hole direct oxidation mechanism. Chemical Engineering Journal. 2017;**317**:925-934

[89] Li Y, Wen M, Wang Y, Tian G, Wang C, Zhao J. Plasmonic hot electrons from oxygen vacancies for infrared light-driven catalytic CO_2 reduction on Bi_2O_{3-x}. Angewandte Chemie. 2021; **133**(2):923-929

[90] Porcheddu A, Colacino E, Cravotto G, Delogu F, De Luca L. Mechanically induced oxidation of alcohols to aldehydes and ketones in ambient air: Revisiting TEMPO-assisted oxidations. Beilstein Journal of Organic Chemistry. 2017;**13**(1):2049-2055

[91] Long J, Wang S, Ding Z, Wang S, Zhou Y, Huang L, et al. Amine-functionalized zirconium metal–organic framework as efficient visible-light photocatalyst for aerobic organic transformations. Chemical Communications. 2012;**48**(95):11656-11658

[92] Dadashi-Silab S, Doran S, Yagci Y. Photoinduced electron transfer reactions for macromolecular syntheses. Chemical Reviews. 2016;**116**(17):10212-10275

[93] Chen X, Wu Z, Liu D, Gao Z. Preparation of ZnO photocatalyst for the efficient and rapid photocatalytic degradation of azo dyes. Nanoscale Research Letters. 2017;**12**(1):1-10

[94] Habisreutinger SN, Schmidt-Mende L, Stolarczyk JK, Schmidt-Mende L, Stolarczyk JK, Habisreutinger SN. Photocatalytic reduction of CO_2 on TiO_2 and other semiconductors. Angewandte Chemie, International Edition. 2013;**52**(29):7372-7408

[95] López YC, Viltres H, Gupta NK, Acevedo-Peña P, Leyva C, Ghaffari Y, et al. Transition metal-based metal–organic frameworks for environmental applications: A review. Environmental Chemistry Letters. 2021;**19**(2):1295-1334

[96] Ahnert SE, Grant WP, Pickard CJ. Revealing and exploiting hierarchical material structure through complex atomic networks. NPJ Computing Materials. 2017;**3**(1):35

[97] Kou J, Lu C, Wang J, Chen Y, Xu Z, Varma RS. Selectivity enhancement in heterogeneous photocatalytic transformations. Chemical Reviews. 2017;**117**(3):1445-1514

[98] Chiou CH, Juang RS. Photocatalytic degradation of phenol in aqueous solutions by Pr-doped TiO2 nanoparticles. Journal of Hazardous Materials. 2007;**149**(1):1-7

[99] Yu X, Cohen SM. Photocatalytic metal–organic frameworks for the aerobic oxidation of arylboronic acids. Chemical Communications. 2015;**51**(48):9880-9883

[100] Zhao F, Wang P, Lucardi RD, Su Z, Li S. Natural sources and bioactivities of 2,4-di-tert-butylphenol and its analogs. Toxins. 2020;**12**(1):35

[101] Wang C-C, Li J-R, Lv X-L, Zhang Y-Q, Guo G. Photocatalytic organic pollutants degradation in metal-organic frameworks. 2014. Available from: www.rsc.org/ees

[102] Panda J, Sahu SN, Tripathy RR, Sahoo T, Sahu JR, Pattanayak SK, et al. Metal–organic frameworks for heterogeneous photocatalysis of organic dyes. Photocatalytic Degradation of Dyes: Current Trends and Futur Perspectives. 2021:489-508. Available from: https://doi.org/10.1016/B978-0-12-823876-9.00001-9

[103] Subudhi S, Swain G, Tripathy SP, Parida K. UiO-66-NH2Metal-organic frameworks with embedded MoS2Nanoflakes for visible-light-mediated H2 and O2 evolution. Inorganic Chemistry. 2020;**59**(14):9824-9837

[104] Miceli M, Frontera P, Macario A, Malara A. Recovery/reuse of heterogeneous supported spent catalysts. Catalysis. 2021;**11**(5):591

[105] Wang CC, Li JR, Lv XL, Zhang YQ, Guo G. Photocatalytic organic pollutants

degradation in metal-organic frameworks. Energy & Environmental Science. 2014;7(9):2831-2867

[106] Pandey RK. Bimetallic metal-organic frameworks (BMOFs) and their potential applications. ACS Symposium Series. 2021;**1393**:3-15

[107] An Y, Xu B, Liu Y, Wang Z, Wang P, Dai Y, et al. Photocatalytic overall water splitting over MIL-125(Ti) upon CoPi and Pt Co-catalyst deposition. ChemistryOpen. 2017;**6**(6):701-705

[108] Abedi S, Morsali A. Ordered mesoporous metal-organic frameworks incorporated with amorphous TiO2 as photocatalyst for selective aerobic oxidation in sunlight irradiation. ACS Catalysis. 2014;**4**(5):1398-1403. 10, 16439-16447

[109] Fang B, Xing Z, Du F, Kong W, Li Z, Zhou W. CdS nanocages@defective-CoNi-LDH with bilayer porous hollow frameworks toward optimized sono-photocatalytic performance. Journal of Materials Chemistry A. 2022

[110] Sun D, Ye L, Li Z. Visible-light-assisted aerobic photocatalytic oxidation of amines to imines over NH2-MIL-125 (Ti). Applied Catalysis B: Environmental. 2015;**164**:428-432

[111] Liang R, Huang R, Ying S, Wang X, Yan G, Wu L. Facile in situ growth of highly dispersed palladium on phosphotungstic-acid-encapsulated MIL-100(Fe) for the degradation of pharmaceuticals and personal care products under visible light. Nano Research. 2017;**11**(2):1109-1123

[112] Zeng M, Chai Z, Deng X, Li Q, Feng S, Wang J, et al. Core–shell CdS@ZIF-8 structures for improved selectivity in photocatalytic H2 generation from formic acid. Nano Research. 2016;**9**(9): 2729-2734

[113] Nasalevich MA, Goesten MG, Savenije TJ, Kapteijn F, Gascon J. Enhancing optical absorption of metal–organic frameworks for improved visible light photocatalysis. Chemical Communications. 2013;**49**(90): 10575-10577

[114] Goh TW, Xiao C, Maligal-Ganesh RV, Li X, Huang W. Utilizing mixed-linker zirconium based metal-organic frameworks to enhance the visible light photocatalytic oxidation of alcohol. Chemical Engineering Science. 2015;**124**: 45-51

[115] Hosseini M-S, Abbasi A, Masteri-Farahani M. Extending the visible light absorption of NH2-UiO-66 through diazotization reaction for photocatalytic chromium (VI) reduction. Environmental Science and Pollution Research. 2022. DOI: 10.21203/rs.3.rs-1398473/v1

[116] Jin K, Maalouf JH, Lazouski N, Corbin N, Yang D, Manthiram K. Epoxidation of cyclooctene using water as the oxygen atom source at manganese oxide electrocatalysts. Journal of the American Chemical Society. 2019;**141** (15):6413-6418

[117] Zhong Y, Wang R, Wang X, Lin Z, Jiang G, Yang M, et al. A Ti-MOF decorated with a Pt nanoparticle cocatalyst for efficient photocatalytic H2 evolution: A Theoretical Study. Frontiers in Chemistry. 2020;**8**:660

[118] Jiang C, Moniz SJA, Wang A, Zhang T, Tang J. Photoelectrochemical devices for solar water splitting – materials and challenges. Chemical Society Reviews. 2017;**46**(15):4645-4660

[119] Khan MM, Ansari SA, Pradhan D, Ansari MO, Han DH, Lee J, et al. Band gap engineered TiO2 nanoparticles for visible light induced

photoelectrochemical and photocatalytic studies. Journal of Materials Chemistry A. 2014;**2**(3):637-644

[120] Ali M, Pervaiz E, Noor T, Rabi O, Zahra R, Yang M. Recent advancements in MOF-based catalysts for applications in electrochemical and photoelectrochemical water splitting: A review. International Journal of Energy Research. 2014;**45**(2):1190-1226

[121] Gao L, Long X, Wei S, Wang C, Wang T, Li F, et al. Facile growth of AgVO3 nanoparticles on Mo-doped BiVO4 film for enhanced photoelectrochemical water oxidation. Chemical Engineering Journal. 2019;**378**: 122193

[122] Zhang L, Cui P, Yang H, Chen J, Xiao F, Guo Y, et al. Metal–organic frameworks as promising photosensitizers for photoelectrochemical water splitting. Advancement of Science. 2016;**3**(1): 1500243

[123] Zhou S, Chen K, Huang J, Wang L, Zhang M, Bai B, et al. Preparation of heterometallic CoNi-MOFs-modified BiVO4: A steady photoanode for improved performance in photoelectrochemical water splitting. Applied Catalysis B: Environmental. 2020;**266**:118513

[124] Gao L, Li F, Hu H, Long X, Xu N, Hu Y, et al. Dual modification of a BiVO4 photoanode for enhanced photoelectrochemical performance. ChemSusChem. 2018;**11**(15):2502-2509

Chapter 3

Understanding the Need of Raw Materials, and Eco-Friendly and Cost-Effective Methods for Detection and Extraction of Materials to Satisfy Semiconductor Market and Its Applications

Marinela Barci and Wu Hao

Abstract

This chapter describes the need of raw materials, and eco-friendly and cost-effective methods for detection and extraction of materials to satisfy semiconductor market and its applications. Since semiconductor market is increasing progressively mainly driven by automotive, solar cells, smartphone products and IOT, VR applications, the need of raw material also increases. To keep this trend, strategies to increase R&D and expertise in mobility and sustainability, European and intercontinental policies for supplier and material manufacturing are needed. Critical raw materials are explained and since there are few countries who supply them, other countries should invest in finding new ways to recycle and manufacture them in-house. With the scientific progress, more materials, alloys, and smart materials are being used, so we need to consider environmental, moral, and ethical aspects when we choose materials in any product, service, or system. Novel techniques such as LiDAR, robotics, autonomous haulage systems, remote operating, and monitoring centers help to maintain a good equilibrium and enable the increasing semiconductor market trend. Photocatalysis, especially more advanced artificial systems used to directly convert solar energy into storable energy, might be of interest to solve environmental pollution and energy storage issues. The idea is to develop photocatalyst with high efficiency, stability, and low cost using a single-material component.

Keywords: nanotechnology, semiconductors, raw and critical materials, photocatalysis, robotics, LiDAR

1. Introduction

Nanotechnology is a wide discipline, which includes science, engineering, and technology conducted at the nanoscale level of 1 to 100 nanometers according to the

Figure 1.
Estimated materials in one iPhone require 46 different elements reported according to Stanford Advanced Materials (SAM) Corporation [2].

Noble Prize scientist Richard Feynman. At such low scale, even gold has different properties in color, and electrical and chemical compositions, that the ones used to characterize the typical gold in large scale. Nanotechnology includes many sectors such as automotive, medical and healthcare, aerospace, photovoltaics, communication, railways and semiconductors, etc. [1], and each of them uses a wide range of materials and alloys. Our focus on this chapter addresses semiconductor's need for raw materials and how to identify more cost-efficient and environmentally friendly ways of extracting and storing them. The most used semiconductor materials are silicon, germanium, and gallium arsenide, but for each sector the list of more raw materials increases. To better understand the concept, we use as an example a smartphone, a typical electronic device each of us possess. Inside an iPhone are used more than 46 different elements according to Stanford Advanced Materials as reported in **Figure 1** adapted from [2]. The detailed list of materials divided by each electronic component is shown below [3]:

1. The phone's **processor chip** is made of silicon, phosphorus, antimony, arsenic, boron, indium, and gallium

2. **The electrical connections** within iPhone use metals such as copper, gold, silver, and tungsten

3. For **micro-capacitors** component used mainly is tantalum

4. The materials used for **soldering** are tin, copper, and silver

Most of these materials are not abundant in nature and are supplied by few countries. Considering the market of mobile phones, more materials are used daily and to keep up with the demand it is important to recycle. Apple announced that their product iPhone 12 is made of 100% recycled raw materials, and especially, rare earth materials are used for magnetic components [4]. Many other companies follow the same trend as they valorize green energy and carbon-free products.

Industry is collaborating more with research and academia to identify what is the cost of bringing the materials from lab to fab [5] and identifying new fields to help technology facilitate the extraction, storage, and utility of the materials [6].

Many countries are increasing the attention and research in the entire raw material chain from exploration, mining, and mineral processing to substitution, recycling, and circular economy [7, 8]. This opens new possibilities related to R&D and expertise in mobility, sustainability, and European and intercontinental policies for supplier and material manufacturing. The growth is driven by both increasing demand and applications in semiconductor industry and by the investment of big semiconductor companies in the field.

This chapter is divided into the following sections: market study for semiconductor raw materials, suppliers and main semiconductor's companies, geopolitical and social impact, characterization techniques used for the evaluation of material properties, traditional and novel technologies, and semiconductor and new applications for photocatalysis and the last part is dedicated to conclusions.

2. Market study for semiconductor raw materials

The semiconductor material market is expected to register a CAGR of 4.17% over the forecast period (2021–2026) stated by Mordor Intelligence Pvt. Ltd. [9].

Market experts project that the semiconductor market (including smartphones, tablet, PC, smartwatch) is about to increase exponentially in the upcoming years, mainly driven by novel applications such as artificial intelligence, internet of things, nano-robotics, high power computing, virtual and augmented reality.

The Semiconductor Industry Association (SIA)-announced global semiconductor industry sales showed an increase of 21.1% over the year 2021 to 2022 with an increase of a total of $42.0 billion [9]. The most growth rate is expected in Asia, while the least impact is predicted for South America and Africa as shown in **Figure 2**.

Other sources such as Techcet [10] divide the global semiconductor market according to process materials, packaging materials, consumable equipment components, and silicon wafers and they project an increase in the market in the upcoming years up to 30% as reported in **Figure 3**.

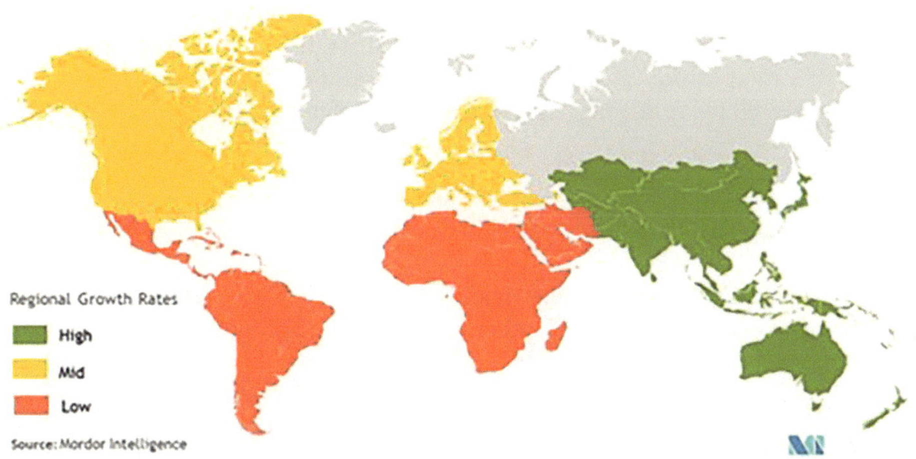

Figure 2.
Global semiconductor material market. Growth rate by region for 2021–2026 according to Mordor Intelligence Pvt. Ltd. market study [9].

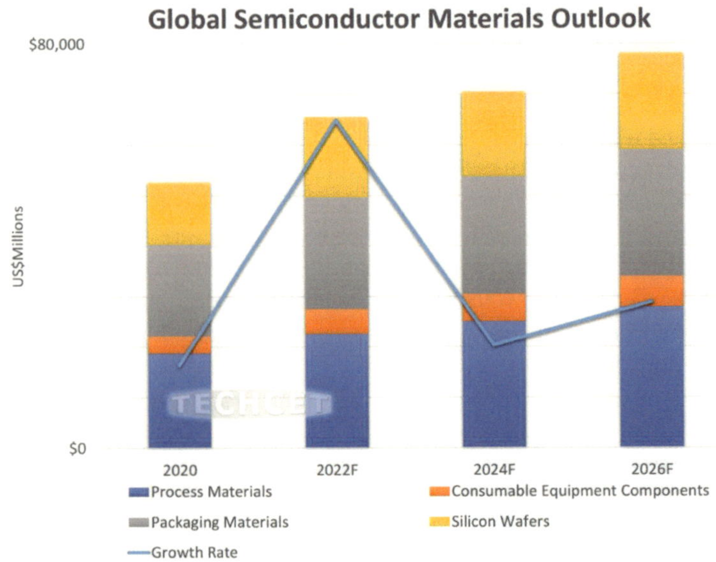

Figure 3.
Global semiconductor material projection by Techcet [10].

Moreover with the new policies for green and carbon-free energy, according to the International Energy Association, the rise of low-carbon power generation is projected to triple mineral demand from this sector by 2040 [11].

3. Suppliers and main semiconductor's companies

Semiconductor devices are made on a single wafer or an epitaxial wafer with a single wafer as the substrate. The Czochralski is a typical melting growth method to create single crystals. The next steps in the semiconductor process are crystal orientation, barrel grinding, reference surface, slicing, grinding, chamfering, polishing, etching, cleaning, inspection, packaging, etc., to produce the final wafer. The growth of a single-crystal thin film into a single-crystal substrate can be made by epitaxy, chemical vapor deposition, magnetron sputtering, and other methods [12, 13].

The modern global semiconductor industry according to [14] is generally divided into two models:

1. Integrated Device Manufacturers that design, produce, and sell their own chips like Intel and Samsung

2. Fabless foundry companies like Nvidia and Arm that design their semiconductors and manufacture them by other foundries like TSMC and SMIC

Semiconductor manufacturing is an integrated, multi-stage process that occupies a big part of the industry chain, such as semiconductor designers and manufacturing suppliers. It includes silicon wafers, photoresists, wet electronic chemicals, electronic gases, reagents, CMP polishing materials, and target materials. Chip packaging materials include package substrates, lead frames, resins, bonding wires, solder balls, and plating solutions [12].

A good explanation of the overall chain is described in **Figure 4**, where it includes as input: manufacturing equipment to chemical and materials, and electronic design automation, while combined with design and manufacture explained before, provides wafers to assemble, test, and package the final product reported to the consumer [14].

Big companies in semiconductors, such as TSMC, Samsung, Broadcom, NXP, Intel, Micron, SK Hynix, Qualcomm [15], announced that they are going to increase their presence worldwide by investing in new plants of production. This news overcomes the issues of chip shortage (definition: sociological or physical change that prevented certain chips from being produced in enough large numbers to satisfy demand) and post-pandemic [16].

Some of the big supplier companies in the materials are Basf SE, Indium Corporation, Kyocera Corporation, Showa Denko Materials Co, Ltd., and Intel Corporation reported in **Figure 5**. More details are given in [9].

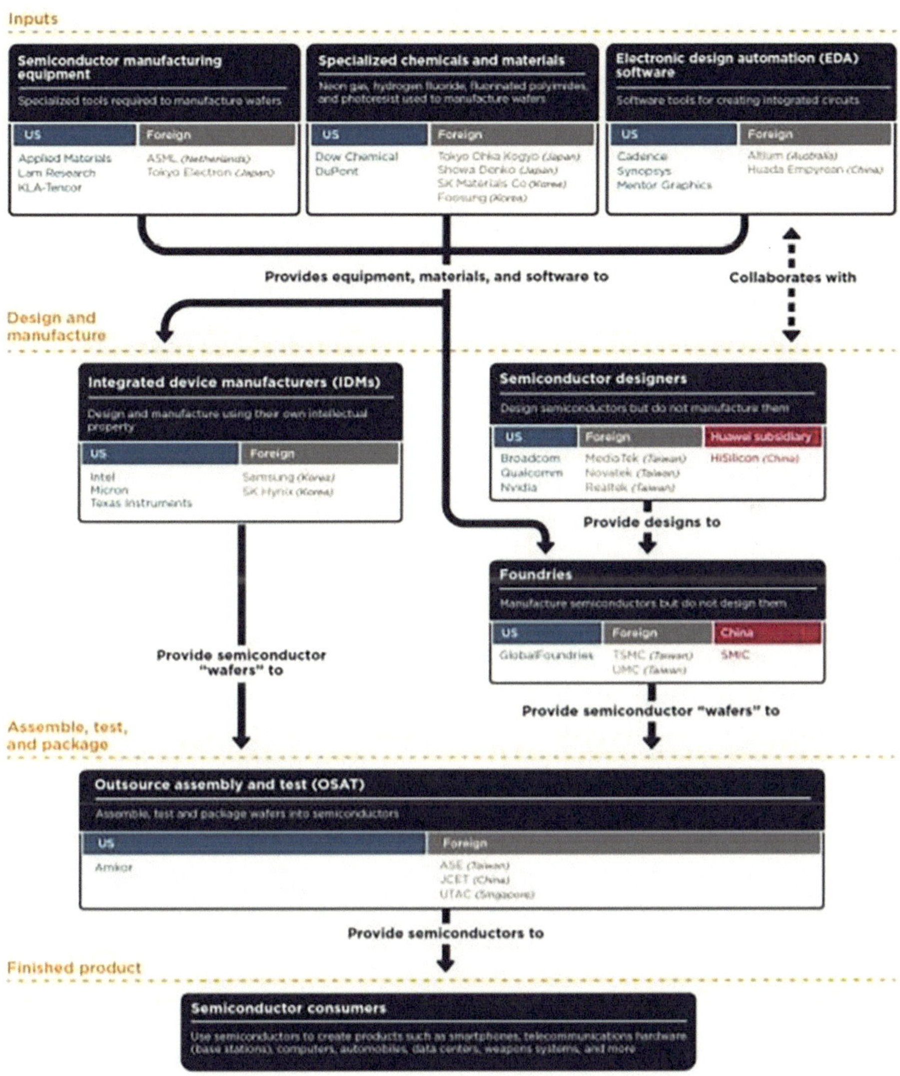

Figure 4.
Semiconductor chain from input to finished product according to source PIIE [14].

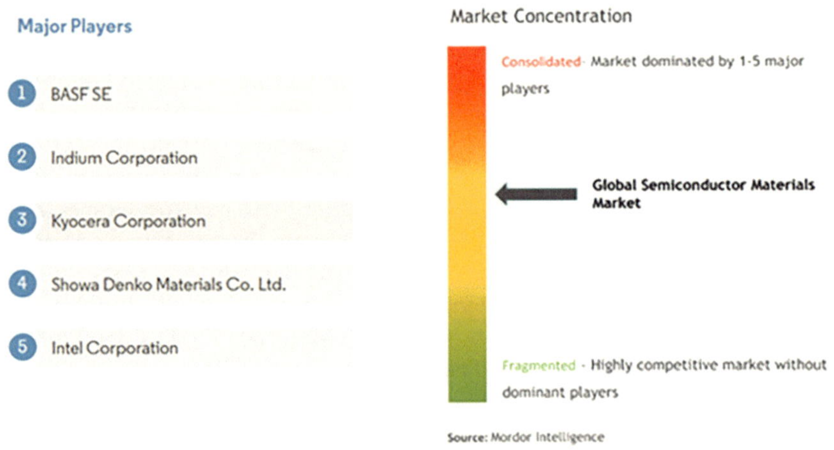

Figure 5.
Global semiconductor material major players—Source Mordor Intelligence Pvt Ltd [9].

4. Geopolitical and social impact

Nowadays, China dominates the manufacturing industry, also due to its abundance in raw materials and low cost-effective ways of producing them [17–21]. China refines nearly 90% of the world's rare earths, which mostly are part of critical raw material list. The raw materials are used in various products such as EVs, advanced ceramics, computers, smartphones, wind turbines, monitors, and fiber optics [21].

Many countries such as Europe, USA, and Canada import most of the critical raw materials from China. A critical material is defined as vital for the world's economy, whose supply may be at risk due to geological scarcity, geopolitical issues, trade policy, or other factors.

In the last years, we have faced coronavirus pandemic and COVID-19 quarantine rules and political tensions [16, 22–24], so it is important that each country develops its internal strategy and has less dependence on other countries. This can be done by increasing the attention in the entire raw material chain from exploration, mining, and mineral processing to substitution, recycling, and circular economy as reported in **Figure 6** [7, 25, 26]. This section will address European Union strategy to overcome the shortage problem. Their experts suggest that the key solution is to boost mineral recycling, identify alternative materials, and promote greater exploration and trade better with World Trade Organization (WTO) to ensure the supply of important raw materials to European industry is not interrupted [26].

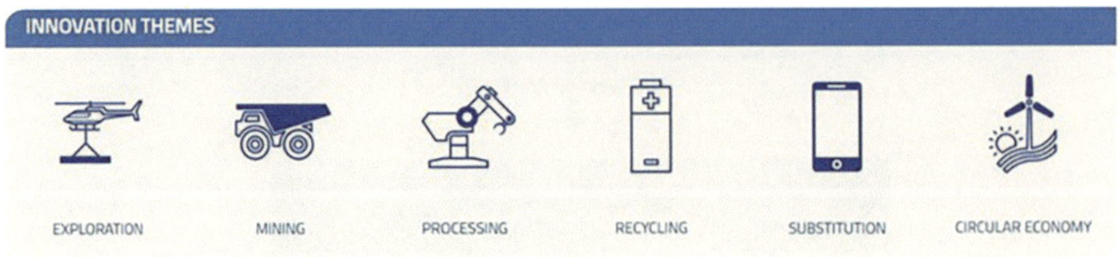

Figure 6.
Raw materials of EU project with the purpose of developing new materials into a major strength for Europe [7].

5. Characterization techniques used for the evaluation of material properties

The increase of scientific discoveries and new technologies has impact in material science widening the choice of materials or smart materials (such as shape memory polymer, shape memory alloy, and hydrogels) to be used in semiconductors. It is important as pointed out by Reference 27 to consider environmental, moral, and ethical aspects when we choose materials in any product, service, or system [27]. The main properties to characterize them are as follows:

- Physical properties such as mass, volume, weight, density
- Electrical resistivity measures the ability of a material to conduct electricity and thermal conductivity how fast the heat is dissipated when there is a gradient of temperature in the material
- Hardness defines the resistance of the material to scratching
- Mechanical properties such as tensile and compressive strength, stiffness, toughness, brittleness, and ductility
- Stress as the ability of a material to withstand pulling force and strain as the ability of a material to withstand being pushed away
- Plasticity is the ability of a material to be changed in shape permanently
- Esthetic characteristics related to taste, texture, smell, and appearance
- Piezoelectricity, shape memory alloys, photochromicity, electro-and magneto-rheostatic
- Thermoelectricity is when electricity is produced directly from heat

To evaluate which element gives the best properties or the desired properties mentioned above for a given product, there are several characterization techniques. The main parameters that differ one method from the other are resolution or detection limit (of the order of nm), physical basis (scattering, emission of electrons...), material sensitivity (linked to refractive index or atomic number), environment of testing (vacuum, air, liquid), and measured parameters (size, shape, morphology, composition...) that we want to extract and analyze. Some good reference to each method is given in [27, 28]. In this short chapter, we will limit to mention some of the most used characterization techniques and basic utility and working mechanism:

1. Scanning electron microscopy (SEM) produces images using electrons instead of visible light
2. Transmission electron microscopy (TEM) uses electrons instead of light and is used for the evaluation of nanostructures such as particles, fibers, thin films, and imaging of atoms

3. Atomic force microscopy (AFM) is used to study the samples at a nanoscale. It takes an image in a three-dimensional topography and provides surface measurements

4. Energy-dispersive X-ray spectroscopy (EDS) is a non-destructive analytical technique. It analyzes near-surface elements and estimates the elemental proportion at different positions

5. Ultraviolet-visible spectrometry (UV-vis) consists of tungsten lamp for the ultraviolet and visible region wavelengths, respectively, a monochromator and sample—reference beams and a detector

6. Fourier transform infrared spectroscopy (FTIR) uses infrared light for scanning the organic, inorganic, and polymeric materials

7. X-ray diffraction (XRD) mainly used to detect the crystallinity of the sample

6. Traditional and novel technologies

In the previous sections, we described the semiconductor material market, main players, applications, challenges, and prospective and we understood that the raw material for semiconductor applications is a booming market. The mining sector that provides the raw materials will need to grow tremendously to enable the required technological demands. Technological innovation should satisfy the growth market need, but need to be also eco-friendly and at low cost. Some people refer to sustainable technology as the society economic equilibrium between raw materials, land, industries, sites, and industrial resources like workers, technology, equipment, machinery, fuels. This model comes more into focus as we are threatened by climate change and environmental degradation [29].

The most innovative technologies predicted to transform mining industry include robotics, autonomous haulage systems, remote operating, and monitoring centers as described in [30]. We will focus in particular in light detection and ranging (LiDAR) technology. LiDAR blocking diagram is composed of laser, light detector, signal and data processor, CPU, and display.

LiDAR is a remote sensing technology based on time-of-flight principle: A target is illuminated with a laser, the laser pulses travel at the speed of light to the object, and

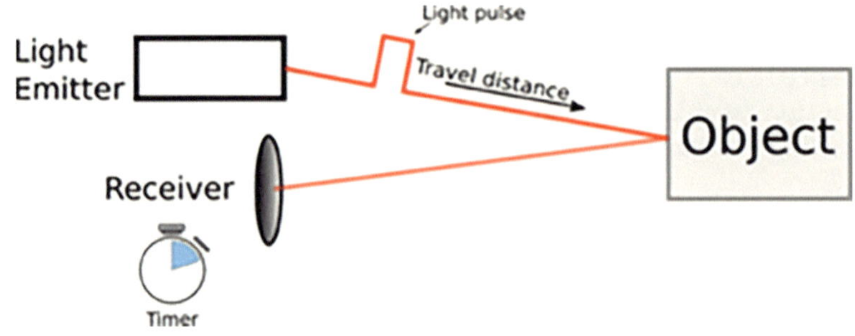

Figure 7.
LiDAR working principle adapted from Reference [31].

the laser is then reflected back to the sensors as illustrated in **Figure 7**. The time taken for the pulses to hit the object and be reflected back to the sensor is calculated and then used to calculate the distance to the object [31].

LiDAR can be used in many fields such as oil and gas exploration, geology, mining, and robotic mining. The information gathered by LiDAR can identify the areas with oil or gas deposits to help geologists know exactly where they need to mine. It can also help to determine the soil structure where the crops or species of plans can grow in a particular region.

Advanced LiDAR technology can detect what minerals exist in a given area and can analyze the mine's structure to prevent them from collapsing after mines have been drilled for minerals.

In robotic mining, LiDAR sensors are used in obstacle detection during mining by the robotic mining vehicles, to prevent the human risk [32].

7. Semiconductor and new applications for photocatalysis

The term photocatalyst is a combination of two words such as **Photo,** which is related to photon, and **Catalyst**, which are substances that alter the reaction rate of materials, mainly semiconductors in the presence of light [33]. This is done through chemical reactions known as photocatalythic reaction or photocatalysis that generate an electron-hole pair, when a semiconducting material is exposed to light. The photocatalysis can be categorized into two types depending on the physical state of reactant and semiconductor.

If both (the semiconductor and reactant) are in the same phase such as gas, solid, or liquid, we refer to homogeneous photocatalysis. Otherwise, if semiconductor and reactant are in different phases, they are called heterogeneous photocatalysis [33, 34].

Photocatalysis is a branch of chemistry and it is referred to as a green chemical pathway, as the main applications studied are environmental applications, biotechnology, and health care such as solar cells, water splitting, and purification, wastewater treatment, pollutant degradation, cancer treatment, antibacterial, air purifying, antifogging, self-cleaning, batteries and conservation and storage of energy, etc. [35].

When considering the main materials compatible for photocatalysis, some parameters need to be taken into account such as photocatalyst loading, pH, surface area and morphology, reaction temperatures, contaminant concentration, and calcination temperature of photocatalysts [36]. Many materials are used for these applications, and among them TiO_2, TiN, Pt, Si, MoS_2, MoS_2/TiO_2 are the most basic materials used as photocatalyst and they are at low-cost, abundance in nature, chemically stable, and transparent to visible light [33, 36–39].

TiO_2 seems to be an excellent photocatalyst and the most used one. The concept to use TiO_2 as photocatalytic for water purification was introduced in 1977 by Frank and Bard. They notice that if you put catalytically active TiO_2 powder into a pool filled with polluted water and then illuminate it with sunlight, the water will gradually become purified. This concept is very interesting and still used by researchers nowadays for purifying water and then extended to purifying air, cancer treatment, etc. [40].

As reported in many scientific works and in the previous paragraphs, environmental pollution and energy shortage are becoming the big challenge for the developed and in developing countries. Photocatalysis, especially more advanced artificial systems used to directly convert solar energy into storable energy, might be of interest to solve these issues. The idea is to develop photocatalyst with high efficiency, stability, and low cost using a single-material component.

8. Conclusion

This chapter clearly describes the need for materials in nanotechnology and its sectors, especially semiconductors, and also brings the attention of experts and researchers in the field. Industry is collaborating more with research and academia to identify what is the cost of bringing the materials from lab to fab and identifying new fields to help technology facilitate the extraction, storage, and utility of the materials.

In the last years, we have faced coronavirus pandemic and COVID-19 quarantine rules and political tensions, so it is important that each country develops its internal strategy and has less dependence on other countries. This can be done by increasing the attention in the entire raw material chain from exploration, mining and mineral processing to substitution, recycling, and circular economy. Novel technologies such as LiDAR, robotics, autonomous haulage systems, remote operating, and monitoring centers can help the raw material industry to enable semiconductor market which is projected to increase mainly by increasing smartphone, tablets, VR, IOT, robotics, photovoltaic, and automotive industry.

Photocatalysis, especially more advanced artificial systems used to directly convert solar energy into storable energy, might be of interest to solve environmental pollution and energy shortage, which remains of particular interest for developed and developing countries.

Acknowledgements

I want to thank the author and the publisher of the book for the invitation and decision that my work is worth to present for publication. I want to thank my company Huawei and my managers for giving me the time to dedicate to this chapter. I want to thank my family for the support and motivation they give me each day. I want to thank my colleagues and professors for the solid technical and practical knowledge I gained during my studies and work experience, which turned to be very useful in writing this chapter.

Author details

Marinela Barci* and Wu Hao
HUAWEI Technologies R&D, Leuven, Belgium

*Address all correspondence to: marinela.barci@huawei.com

References

[1] Wilson M, Kannangara K, Smith G, Simmons M, Raguse B. Nanotechnology: Basic Science and Emerging Technologies. Boca Raton: Chapman & Hall/CRC; 2002

[2] Application of Tantalum and Molybdenum in Mobile Phones. Available from: https://www.samaterials.com/content/how-much-gold-can-an-iphone-refine.html

[3] The Extraordinary Raw Materials in an iPhone 6s. Available from: http://www.visualcapitalist.com

[4] [iPhone 12 Product Environmental Report. Available from: http://www.apple.com

[5] What Will It Take To Bring New Material from Lab To Manufacturing, VLSI 2022 evening panel discussion. Available from: https://www.vlsisymposium.org

[6] Raw Materials Needs for Future Technologies. Available from: https://www.isi.fraunhofer.de

[7] Raw Materials- European Institute of Innovation & Technology – EIT, europa.eu. Available from: https://eit.europa.eu/our-communities/eit-rawmaterials

[8] The raw-materials challenges. Available from: https://www.mckinsey.com

[9] Mordor Intelligence Pvt Ltd. Available from: https://www.mordorintelligence.com/industry-reports

[10] TECHCET Critical Materials Reports. Available from: http://www.blog.baldengineering.com/

[11] The Rising Demand for Critical Minerals. Available from: https://www.mining.com/web/the-50-minerals-critical-to-us-security/#:~:text=According%20to%20the%20International%20Energy%20Association%2C%20the%20rise,minerals%20necessary%20for%20it%20is%20just%20as%20vital

[12] Sze SM. Semiconductor Devices: Physics and Technology. India: John Wiley & Sons; 2008

[13] Semiconductor Materials: Types, Properties and Production Process. Available from: https://www.utmel.com/blog/categories/semiconductor/semiconductor-materials-types-properties-and-production-process

[14] PIIE Working Paper 20-16 "How the United States marched the semiconductor industry into its trade war with China", by Chad P. Bown, originally published by the Peterson Institute for International Economics in December 2020. Available from: https://www.piie.com/sites/default/files/documents/wp20-16.pdf

[15] TSMC, Samsung, Broadcom, NXP, Intel, Micron, SK Hynix, Qualcomm-company website press release

[16] Materials shortage is the latest setback for components industry. Available from: https://www.sourcengine.com

[17] Semiconductor's main raw material suppliers. Available from: https://www.semiconductor-today.com/suppliers.shtml

[18] Why China Is "The World's Factory". Available from: https://www.investopedia.com

[19] China's Rise to Manufacturing Dominance. Available from: https://

www.statista.com/chart/20859/chinese-and-us-share-of-global-manufacturing-output

[20] How China's mines rule the market of critical raw materials. Available from: https://www.dw.com/en/how-chinas-mines-rule-the-market-of-critical-raw-materials/a-57148375

[21] Where is China's manufacturing industry going? Available from: https://www2.deloitte.com/content/dam/Deloitte/global/Documents/Manufacturing/gx_mfg_2011MFGreport_281211.pdf

[22] The 50 Minerals Critical to U.S. Security. Available from: https://www.visualcapitalist.com/the-50-minerals-critical-to-u-s-security

[23] Industrialization, Imperialism and Raw Materials. Available from: https://www.lucianashworth.com

[24] The EU's risky dependency on critical Chinese metals. Available from: https://www.dw.com/en/the-eus-risky-dependency-on-critical-chinese-metals/a-61462687

[25] Investing opportunities in the-semiconductor industry. Available from: https://www.drwealth.com/investing-opportunities-in-the-semiconductor-industry

[26] Europe worries over raw materials – Business, Economy and Finance news from a German perspective. Available from: https://www.dw.com/en/europe-worries-over-raw-materials/a-5731662

[27] Properties of Materials, Design Technology. Available from: https://www.ruthtrumpold.id.au

[28] Eaton P et al. Comparing AFM, SEM and TEM (afmworkshop.com). Ultramicroscopy. 2017;**182**:179-190

[29] What is Sustainable Technology? How will it Revolutionize the Industrial World? (se.com). Available from: https://www.mckinsey.com/industries/metals-and-mining/our-insights/the-raw-materials-challenge-how-the-metals-and-mining-sector-will-be-at-the-core-of-enabling-the-energy-transition

[30] Ten technologies with the power to transform mining. Available from: https://www.mining-technology.com

[31] LiDAR and ToF Cameras – Technologies explained. Available from: https://tof-insights.com/time-of-flight-lidar-and-scanners-technologies-explained

[32] 100 Applications or Uses of LiDAR Technology. Available from: https://www.lidarradar.com/100-applications

[33] Ameta R, Solanki MS, Benjamin S, Ameta SC. Chapter 6 – Photocatalysis. In: Advanced Oxidation Processes for Waste Water Treatment. Academic Press; 2018. pp. 135-175. DOI: 10.1016/B978-0-12-810499-6.00006-1

[34] Qu Y, Duan X. Progress, challenge and perspective of heterogeneous photocatalysts. Chemical Society Reviews. 2013;**42**(7):2568-2580 Available from: https://pubs.rsc.org/en/content/articlelanding/2013/CS/C2CS35355E

[35] Schneider J, Matsuoka M, Takeuchi M, Zhang J, Horiuchi Y, Anpo M, et al. Fundamentals of TiO2 Photocatalysis: Concepts, mechanisms, and challenges. Chemical Reviews;**114**(19):9919-9986. DOI: 10.1002/adma.201901997

[36] Zhang J, Tian B, Wang L, Xing M, Lei J. Mechanism of photocatalysis. In: Photocatalysis. Lecturer Notes in Chemistry. Singapore: Springer; 2018

Available from: https://link.springer.com/chapter/10.1007/978-981-13-2113-9_1

[37] Chen B, Meng Y, Sha J, Zhong C, Hu W, Zhao N. Preparation of MoS2/TiO2 based nanocomposites for photocatalysis and rechargeable batteries: Progress, challenges, and perspective. Nanoscale. 2018;**10**(1):34-68. DOI: 10.1002/0471238961.1608152019051816.a01.pub3

[38] Bard AJ. Photoelectrochemistry and heterogeneous photo-catalysis at semiconductors. Journal of Photochemistry. 1979;**10**(1):59-75. DOI: 10.1016/0047-2670(79)80037-4

[39] Akira Fujishima, Tata N. Rao, Donald A. Tryk, Titanium dioxide photocatalysis, Journal of Photochemistry and Photobiology C: Photochemistry Reviews. 2000;**1**(1):1-21. DOI:10.1016/S1389-5567(00)00002-2

[40] Nakata K, Fujishima A. TiO2 photocatalysis: Design and applications. Journal of Photochemistry and Photobiology C: Photochemistry Reviews. 2012;**13**(3):169-189. DOI: 10.1016/j.jphotochemrev.2012.06.001

Chapter 4

Nano TiO_2-Based Smart Superhydrophilic Self-Cleaning Surfaces

Tanu Mittal

Abstract

In this chapter, we have focused on the use of self-cleaning coating or surfaces which have more concerning the field of research due to the rising demand for self-disinfected and hygienic surfaces. Self-cleaning coatings can be used in automobile windshields, textiles, antifouling membranes, paints, building construction material, optoelectronic devices like solar panels, and in the medical aids and food industry. This chapter gives an outlook on nano TiO2-based superhydrophilic self-cleaning surfaces. The mechanism of superhydrophilicity of nano TiO_2-based systems and photocatalytic properties are cautiously investigated. The photocatalytic behavior and superhydrophilicity of TiO_2 are based on the photogenerated electron–hole pair. This chapter gives a general idea of a variety of systems and methods that could improve the self-cleaning behavior of TiO_2 in the solar spectrum in view of the fact that TiO_2 is only ultraviolet responsive. Evidences have indicated that the superhydrophilic wetting and antifogging performance are affected by porosity and surface hydroxyl (-OH) contents. In this chapter, the effect of multilayer assembly and the role of cumulative concentration of –OH groups on photocatalytic behavior is also discussed.

Keywords: TiO_2, semiconductors, heterojunctions, self-cleaning, hydroxyl group, photocatalytic activity

1. Introduction

The discovery of photoinduced superhydrophilic behavior of Titanium dioxide (TiO_2) so that it is considered as most promising photo-responsive wetting material (Wang et al. in 1997) [1]. TiO_2 is a semiconductor, on excitation of titanium with suitable light energy, the photocatalysis process takes place on the surface so that organic dirt molecules adsorbed on the surface and decomposed with the process of photocatalysis. Photocatalysis is followed by the washing of surface with flow of water and finally surface will be clean and sterilized. TiO_2 an extensively acceptable applicant for superhydrophilic self-cleaning surfaces and coatings due to its superior optoelectronic properties, photostability, low-cost production, nontoxicity, and environmental friendliness. The structural and surface properties and also intrinsic electronic characteristics of titanium dioxide are responsible for its photocatalytic efficiency and photo-induced wettability and consequently its self-cleaning property. In recent times, the research has shown the most interest in field of the development and commercialization of self-cleaning coatings and surfaces with the properties of antireflective, antifouling, and antifogging. Self-cleaning coatings and surfaces have potential applications in major sectors such as the medical industry (especially medical aids with anti-microbial surfaces), food industry, optical industry (lenses, camera, and sensors), marine industry (corrosion protection), aircraft industry, construction industry (cement & paint), automobile industry, etc. In this review article, we give a significant evaluation of the superhydrophilic self-cleaning materials and surfaces, environmental applications, and sustainability. In this chapter, we introduce the concept of superhydrophilicity and self-cleaning action of TiO_2 as a photocatalyst and later water treatment mechanism of such systems. This review emphasizes the potential applications and practical challenges of self-cleaning surfaces and coatings in real-life situations.

2. Self-cleaning action of TiO_2 as photocatalyst

Once a metal oxide semiconductor is irradiated with an energy source that is higher than its band gap (3.2 eV) [2] leads to the absorption of photons and excitation of an electron (e−) in which way generate an electron–hole pair, with the hole (h+) left behind in the valence band. These electron–hole pair in turn undergoes recombination and non-radiatively emits the excess energy in the form of heat or light [3, 4]. This recombination process or charge obliteration reduces the efficiency of TiO_2 photocatalytic activity. Due to crystal defects and impurities present on the surface of TiO_2 this reaction would take place on the surface or within the bulk of TiO_2. Over the surface of TiO_2 excitons (electron–hole pair) were undergoes recombination, this electron–hole pair can combine with the adsorbed molecules and leads to the decomposition of organic volatile compounds. Evidently, electrons in the CB reacts with the oxygen molecules present on the surface and reduces it into superoxide radicals which can react with water to give hydroxyl radicals (% OH); on the other hand holes (h+) in the valence band reacts with water to form %OH radicals. These oxygen-generated species decompose the volatile organic compounds into CO_2 and H_2O on the catalytic surface by the process of free radical mechanism [5–8].

Thus the mechanism of superhydrophilicity on the catalytic surface can be explained by the combination of two processes. The first mechanism of superhydrophilicity includes the surface hydroxylation upon photoexcitation. This step consists of two steps (a) oxygen vacancy generation and (b) reconstruction of photoinduced Ti-OH bonds. The mechanism of superhydrophilicity propagates by the formation of photoinduced electrons in the system by the reduction of metal centers. For example, TiO2 (IV) gets reduced into TiO_2 (III) by the movement of conduction band electrons, however, the holes formed in the valence band oxidized into O_2^- anions, and thus vacancies are formed by the ejection of oxygen atoms. The hydroxyl anions are formed at these vacancies during the photocatalytic mechanism and water molecules get absorbed, leading to the hydrophilic nature of the surface [9]. Sakai et al. proposed that for the superhydrophilic characteristics of a surface photoinduced holes are more important than electrons, these holes can diffuse on the surface of catalyst and get trapped at the oxygen lattice because these photo-induced holes can diffuse on the surface and get trapped at the lattice oxygen sites, this leads to the formation of new hydroxyl bonds with the adsorbed water molecules [10].

2.1 Photoinduced hydrophilicity

Furthermore, it is evident that the hydrophilic conversion rate of TiO_2 can be increased by applying high positive electrode potential, and can be decreased by the usage of hole-scavengers in the photocatalytic reaction [11]. This interpretation suggests that for the hydrophilic conversion of the photocatalytic material, the diffusion of holes to the catalytic surface is very essential. The formation of Ti-OH bands on the TiO_2 surface is attributed to photoinduced hydrophilicity [12]. The diffusion of holes to the catalytic surface can weaken the Ti-O bonds, and water molecules adsorption to this surface can then break the weakened bond. This will lead to the charge separation and the formation of a new OH bond on the surface. Therefore one OH group which is doubly coordinated to Ti atoms converted into two OH groups. It is revealed that the highly amphiphilic catalytic surface of TiO_2 can leads to the reconversion of the surface wettability, which is due to the deposition of a hydroxyl group from the surface. The second mechanism of photocatalytic activity on the surface initiates the decomposition of the adsorbed pollutants on the surface which initiates by the absorption of UV light and thus reduced the organic pollutants on the surface. It has been recommended that in addition to the photocatalytic degradation of organic pollutants on a TiO_2 surface, desorption of water molecules on the surface is also initiated by UV irradiation. The subsequent breakdown in the H-bonded network is supposed to be essential for the hydrophilic conversion [13]. The results reveal that the photocatalytic activity on the surface is not an exclusive requirement for any surface to be hydrophilic or superhydrophjilic. It has been reported that many metal oxides such as vanadium oxide and tungsten oxide have superhydrophilic nature under UV light illumination without and photocatalytic activity mechanism on their surface. In consequence, it has been suggested that a combination of various mechanisms is required to explain the photoinduced hydrophilicity or photocatalytic activity on the surface. The combination of photoinduced hydrophilicity and photocatalytic activity are the major requirements to degrade the organic pollutants, and light-induced superhydrophilicity to sheet water for cleaning the surface by washing off the degraded pollutants, good optical transparency, photostability, and durability.

3. Water treatment

In recent times due to the ever-increasing urbanization and industrialization, the world is facing freshwater disaster. Heavy metal ions and organic pollutants cause foremost severe threats to aquatic life. Industrial and conventional treatment methods such as oxidation, biodegradation, absorption, coagulation, etc. have been used for water purification. Membrane technology is the most commonly used technique for water treatment due to its green and environment-friendly approach. However, membrane technology has faced a major problem of fouling of membranes due to prolonged use of technique. In this line of research, increasing the hydrophilicity of the membrane is considered an effective way to reduce the fouling behavior of the membrane which is caused by pore blocking and fouling material adhesion [14, 15]. By using this method of increasing hydrophilicity of membrane, the hydrophilic material also imparts water permeability stiffness and strength to the polymeric matric used for the preparation of these membranes. The introduction of TiO_2 in the membranes as hydrophilic photocatalytic materials can add self-cleaning behavior to the membrane, which upon photoexcitation degrades the natural organic volatile compounds adhered at the surface of the membrane or pores. It is also commonly seen that during the process of phase separation, TiO_2 particles have been settled down into the bottom of the polymeric bulk due to its high density. However, when TiO_2 nanoparticles hybridized with magnetic nanoparticles, the photocatalytic TiO_2 nanoparticles can be adhered to the surface of the polymer matrix with the help of an external magnetic field [16]. Oil–water separation is an effective technique to demonstrate the interplay of membranes between antifouling and wettability. The oil–water separation technique is very effective to clean offshores oil spills and discharging oil effluents which are very harmful for aquatic life and biotic environment. Another very helpful technique for oil–water separation of molecular sieving technique which allows only water molecules hence block the larger oil droplets, this technique has been worked under a specific applied pressure, hydrophilicity, and selective wettability. Porous materials such as metal meshes, fibers foams, and textiles can be used in addition to polymers and ceramics for oil–water separation. The major difficulty faced in the oil–water separation technique is rapid decline of water permeation due to the clogging of pores by oil droplets, therefore, lowering the filtration flux [17]. Also, significant fouling will occur over the surface of membrane due to the presence of hydroxide groups. A thin layer of TiO_2 has been used for the removal of fouling over the surface of the membrane [18, 19]. In recent times, cellulose acetate nanofiber membrane was fabricated by electro-spinning which is oleophobic in water, superhydrophjilic in oil, and superamphiphilic in air [13]. This membrane can be working for oil/water separation in wastewater treatment; also it has been used for oil/water emulsions and oil/corrosive medium. An interesting Janus membrane is fabricated by in situ growth of TiO_2 on poly (phenylene sulfide) membrane with F-TiO_2@poly (phenylene sulfide) followed by water–oil interfacial grafting in presence of perfluorodecyltriethoxysilane [20].

In addition to environmental remediation such as offshore spills, such multifunctional membranes find great applications in food industry, textile factories, and other chemical plants.

4. Antireflective coatings

High transmittance characteristics of antireflective coatings have great significance in high-performance devices along with the property of transmittance such as flat panel display, solar panels, lenses, telescope, etc. TiO_2 causes a high refractive index and refraction when it is applied to the transparent glass substrate. A very low concentration of TiO_2 in the self-cleaning coatings reduces the photocatalytic activity but at the same time minimized the surface refraction. Therefore for the practical application of self-cleaning coating there should be between balance antifogging and anti-reflective properties of the self-cleaning coating. Preparation of TiO_2 composite with SiO_2 adopted to have self-cleaning anti-reflective properties with a high refractive index. A hierarchical macro-mesoporous SiO_2 thin film with very high porosity showed a significant broadband anti-reflection with an average reflectance of 3.45%

in the wavelength range from 350 nm to 1200 nm [21] The macroporoporous template is superior to the conventional flat SiO_2/TiO_2 composite film.

The sandpaper abrasion test results show that the porous composite film was superhydrophilic with a water contact angle of 2.4° and this contact angle was retained even after 50 cycles of the abrasion test. Raspberry-like core-shell nanoparticles of SiO_2/TiO_2 were synthesized by a sol–gel single pot method. These synthesized SiO_2/TiO_2 core-shell nanoparticles were then fabricated onto glass by layer-by-layer self-assembly dip coating technique. The prepared coating exhibited superhydrophilicity both in the presence and absence of UV irradiation [22]. The SiO_2 particle with the size in the range of sub-micron provides a porous structure and anti-reflective property to the self-cleaning coating. However, the larger surface area of nanoparticles and higher surface roughness in the coating are the main factors in imparting its superhydrophilic property. An example of a high antireflective self-cleaning coating is $Ag/TiO_2/Si$ forest-like hierarchical nano/microstructures like a moth-eye. This type of coating shows solar-weighted reflectance values of 3.5% and 3.3%, respectively, over the wavelength range of 300 – 1000 nm [23]. The estimated contact angle over the surface was <5° which exhibits outstanding anti-fogging and self-cleaning properties by the use of plasmonic silver nanoparticles [24]. The self-cleaning coating also shows photocatalytic dye degradation and self-cleaning under UV–vis light irradiation. Along with these combinations, several researchers have developed double-layered and multilayered TiO_2-SiO_2 films with reduced reflectance [25, 26]. Three-dimensional TiO_2 nanostructures such as nanopores, nanorods, and hierarchical structures with larger surface area have been reported with reduced surface reflectance losses but with enhanced photocatalytic activity [27, 28]. TiO_2 can also be used as a protective layer for thermochromic self-cleaning coating in Vanadium (IV) oxide (VO_2). Vanadium oxide VO_2 is a unique material practically used for thermochromic applications due to its large optical and electronic behavior and because first-order phase transitions from monoclinic to tetragonal geometry [29, 30]. A multifunctional transparent $VO_2/SiO_2/TiO_2$ thin film showed considerable improvement in visible light transmittance which is accompanied by thermochromic and self-cleaning properties. Therefore, these coatings can be applied for energy-efficient, intelligent window applications [31]. TiO_2/VO_2 bilayer composite coating is also capable of supporting the TiO_2 for increasing air purification by absorbing Infrared rays.

4.1 Coatings for building materials

Self-cleaning coatings are the probable substitute for high energy consumption. TiO_2-coated self-cleaning surfaces has been used in construction material such as in cement, tiles, limestone, glass can behold their artistic appearance without contamination or getting dirt throughout their lifetime [32–38]. A number of esthetic images and buildings reported preserving with the help of TiO_2 as a photocatalytic material [39–42]. Sedimentation methods and spray coating methods are usually used for the building materials. According to the report by n-Tech Research in 2015 the self-cleaning market is expected to reach a market of US $3.3 billion by 2020. Pilkington has been marked as the first commercially available coating over the glass. Additionally, these self-cleaning coatings should withstanding varying climatic changes to the outdoor environments. During the development of self-cleaning coatings, a number of weather variables like light, temperature, humidity, etc. should be considered [39–42]. Humidity plays a very important role in stain removal, by influencing the hydration state of TiO_2 this will increase the cumulative hydroxyl ion concentration that will lead to sustainable self-cleaning materials for buildings [43–50].

5. Future outlook and conclusion

This chapter gives emphasis on the specific idea of TiO2-based self-cleaning surfaces. Self-cleaning surfaces cover a significant research area in this era of fast urbanization and industrialization. Also, self-cleaning surfaces are an answer for cleaning contaminated surfaces to compromise large energy consumption. The chapter gives a brief about the outstanding examples of TiO_2 composites for water purification techniques such as TiO_2 hybrid composite with grapheme, Heterojunction of TiO_2 with some other metals, dye-sensitized TiO_2, etc. Applications of TiO_2 and TiO_2 composite materials or hybrids as a superhydrophilic photocatalysts in water purifying membranes, antibacterial surfaces, fabrics, oil–water separation techniques, and in antireflective coatings are discussed in detail. Superhydrophilicity mechanism in semiconductor metal oxides is investigated from different models. Although in this field, there is great progress in the development and fabrication of self-cleaning coatings and surfaces, however, there are still some technical problems and challenges with self-cleaning coatings that will cause bridging the gap between industry and fundamental research.

Mechanical durability and chemical stability are the major concern for the fabrication of self-cleaning coatings. Exposure of these coatings to solvents and outdoor temperature differences can lead to the detachment of the material from the surface of the coating due to which there will be a gradual decline in the performance of coatings. However, the surfaces which are responsive to indoor light sources are more durable and persistent with superhydrophilicity because photoinduced wettability recovers the hydrophobic nature of coating in the dark within minutes to hours. As a result, self-cleaning coatings should be designed in such a way that they should have anti-bacterial, antifogging, and antireflective properties and have good adherence to different surfaces like mirror, ceramic, and furniture. A significant emerging class of self–cleaning coatings and surfaces responds to various external stimuli such as changes in temperature, pH, chemical environment, stress, and humidity. The most recent study also predicts that this field has a promising trend in developing self-cleaning surfaces which has a combination of self-cleaning and photocatalytic properties in this way a single surface find a wide variety of application. Coatings with a combination of anti-bacterial, anti-fogging, and anti-reflective properties can find application in endoscopic surgery examples. The self he antireflective and antifogging can make them more appealing to building materials, solar panels, automobile windshields, etc. under inconsiderate environmental conditions.

Author details

Tanu Mittal
GNA University, Phgwara, India

*Address all correspondence to: drmittaltanu@gmail.com

References

[1] Wang R, Hashimoto K, Fujishima A, Chikuni M, Kojima E, Kitamura A, et al. Light-induced amphiphilic surfaces. Nature. 1997;**388**:431

[2] Pelaez M, Nolan N, Pillai SC, Seery MK, Falaras P, Kontos AG, et al. A review onthe visible light active titanium dioxide photocatalysts for environmental applications. Applied Catalysis B: Environmental. 2012;**125**:331-349

[3] Chen X, Mao SS. Titanium dioxide nanomaterials: Synthesis, properties, modifications, and applications. Chemical Reviews. 2007;**107**:2891-2959

[4] Liqiang J, Yichun Q, Baiqi W, Shudan L, Baojiang J, Libin Y, et al. Review of photoluminescence performance of nanosized semiconductor materials and its relationships with photocatalytic activity. Solar Energy Materials & Solar Cells. 2006;**90**:1773-1787

[5] Fujishima A, Zhang X, Tryk DA. Surface Science Reports TiO2 photocatalysis and related surface phenomena. Surface Science Reports. 2008;**63**:515-582

[6] Nursam NM, Wang X, Caruso RA. High-throughput synthesis and screening of titania-based photocatalysts. ACS Combinatorial Science. 2015;**17**:548-569

[7] Li Z, Luan Y, Qu Y, Jing L. Modification strategies with inorganic acids for efficient photocatalysts by promoting the adsorption of O2. ACS Applied Materials & Interfaces. 2015;7:22727-22740

[8] Banerjee S, Dionysiou DD, Pillai SC. Self-cleaning applications of TiO2 by photo-induced hydrophilicity and photocatalysis. Applied Catalysis B: Environmental. 2015;**176-177**:396-428

[9] Sakai N, Fujishima A, Watanabe T, Hashimoto K. Quantitative evaluation of the photoinduced hydrophilic conversion properties of TiO2 thin film surfaces by the reciprocal of contact angle. The Journal of Physical Chemistry. B. 2003;**107**:1028-1035

[10] Sakai N, Fujishima A, Watanabe T, Hashimoto K. Enhancement of the photoinduced hydrophilic conversion rate of TiO2 film electrode surfaces by anodic polarization. The Journal of Physical Chemistry. B. 2001;**105**:3023-3026

[11] Xu Y, Guo D, Li T, Xiao Y, Shen L, Li R, et al. Manipulating the mussel-inspired co-deposition of tannic acid and amine for fabrication of nanofiltrationmembranes with an enhanced separation performance. Journal of Colloid and Interface Science. 2020;**565**:23-34

[12] Sun T, Liu Y, Shen L, Xu Y, Li R, Huang L, et al. Magnetic field assisted arrangement of photocatalytic TiO2 particles on membrane surface to enhance membrane antifouling performance for water treatment. Journal of Colloid and Interface Science. 2020;**570**:273-285

[13] Dudem B, Bharat LK, Leem JW, Kim DH, Yu JS. Hierarchical Ag/TiO2/Si forest-like Nano/Micro-Architectures as antireflective, plasmonic photocatalytic, and self-cleaning coatings. ACS Sustainable Chemistry & Engineering. 2018;**6**:1580-1591

[14] Zhang F, Bin Zhang W, Shi Z, Wang D, Jin J, Jiang L. Nanowire-haired

inorganic membranes with superhydrophilicity and underwater ultralow adhesive superoleophobicity for high-efficiency oil/water separation. Advanced Materials. 2013;**25**:4192-4198

[15] Gao C, Sun Z, Li K, Chen Y, Cao Y, Zhang S, et al. Integrated oil separation and water purification by a double-layer TiO2-based mesh. Energy & Environmental Science. 2013;**6**:1147-1151

[16] Wang W, Lin J, Cheng J, Cui Z, Si J, Wang Q, et al. Dual superamphiphilic modified cellulose acetate nanofiber membranes with highly efficient oil/water separation and excellent antifouling properties. Journal of Hazardous Materials. 2020;**385**:121582

[17] Yang C, Han N, Han C, Wang M, Zhang W, Wang W, et al. Design of a Janus f-tio2@pps porous membrane with asymmetric wettability for switchable Oil/Water separation. ACS Applied Materials & Interfaces. 2019;**11**:22408-22418

[18] Dai J, Tian Q, Sun Q, Wei W, Zhuang J, Liu M, et al. TiO2- alginate composite aerogels as novel oil/water separation and wastewater remediation filters. Composites. Part B, Engineering. 2019;**160**:480-487

[19] Li X, He J. Synthesis of raspberry-like SiO2 – TiO2 nanoparticles toward antireflective and self-cleaning coatings. ACS Applied Materials & Interfaces. 2013;**5**:5282-5290

[20] Khan SB, Zhang Z, Lee SL. Single component: bilayer TiO2 as a durable antireflective coating. Journal of Alloys and Compounds. 2020;**834**:155137

[21] Guldin S, Kohn P, Stefik M, Song J, Divitini G, Ecarla F, et al. Self-cleaning antireflective optical coatings. Nano Letters. 2013;**13**:5329-5335

[22] Zhang X, Fujishima A, Jin M, Emeline AV, Murakami T. Double-layered TiO2-SiO2 nanostructured films with self-cleaning and antireflective properties. The Journal of Physical Chemistry. B. 2006;**110**:25142-25148

[23] Nakata K, Sakai M, Ochiai T, Murakami T, Takagi K. Antireflection and selfcleaning properties of a moth-eye-Like surface coated with TiO2 particles. Lagmuir. 2011;**27**:3275-3278

[24] Sanz R, Romano L, Zimbone M, Buccheri MA, Scuderi V, Impellizzeri G, et al. UV-black rutile TiO2: An antireflective photocatalytic nanostructure. Journal of Applied Physics. 2015;**117**:74903

[25] Wu J, Wang H, Bao L, Zhong J, Chen R, Sun L. Novel raspberry-like hollow SiO2@TiO2 nanocomposites with improved photocatalytic self-cleaning properties: Toward antireflective coatings. Thin Solid Films. 2018;**651**:48-55

[26] Lin W, Zheng J, Yan L, Zhang X. Sol–gel preparation of self-cleaning SiO2-TiO2/SiO2-TiO2 double-layer antireflective coating for solar glass. Results Physics. 2018;**8**:532-536

[27] Ren Y, Li W, Cao Z, Jiao Y, Xu J, Liu P, et al. Robust TiO2 nanorods-SiO2 core-shell coating with high-performance self-cleaning properties under visible light. Applied Surface Science. 2020;**509**:145377

[28] Salvaggio MG, Passalacqua R, Abate S, Perathoner S, Centi G, Lanza M, et al. Functional nano-textured titania-coatings with self-cleaning and antireflective properties for photovoltaic surfaces. Solar Energy. 2016;**125**:227-242

[29] Mazur M, Wojcieszak D, Domaradzki J, Kaczmarek D, Song S,

Placido F. TiO2/SiO2 multilayer as an antireflective and protective coating deposited by microwave assisted magnetron sputtering. Opto-Electronics Review. 2013;**21**:233-238

[30] Goodenough JB. Anomalous properties of the vanadium oxides. Annual Review of Materials Science. 1971;**1**:101-138

[31] Powell MJ, Quesada-Cabrera R, Taylor A, Teixeira D, Papakonstantinou I, Palgrave RG, et al. Intelligent multifunctional VO2/SiO2/TiO2 coatings for self-cleaning, energy-saving window panels. Chemistry of Materials. 2016;**28**:1369-1376

[32] Matsunaga T, Tomoda R, Nakajima T, Wake H. Photoelectrochemical sterilization of microbial cells by semiconductor powders. FEMS Microbiology Letters. 1985;**29**:211-214

[33] Jiang X, Tian X, Gu J, Huang D, Yang Y. Cotton fabric coated with nano TiO2- acrylate copolymer for photocatalytic self-cleaning by in-situ suspension polymerization. Applied Surface Science. 2011;**257**:8451-8456

[34] Montazer M, Pakdel E, Behzadnia A. Novel feature of nano-titanium dioxide on textiles: Antifelting and antibacterial wool. Journal of Applied Polymer Science. 2011;**121**:3407-3413

[35] Lao L, Fu L, Qi G, Giannelis EP, Fan J. Superhydrophilic wrinkle-free cotton fabrics via plasma and nanofluid treatment. ACS Applied Materials & Interfaces. 2017;**9**:38109-38116

[36] Hu J, Gao Q, Xu L, Wang M, Zhang M, Zhang K, et al. Functionalization of cotton fabrics with highly durable polysiloxane–TiO2 hybrid layers: Potential applications for photo-induced water–oil separation, UV shielding, and self-cleaning. Journal of Materials Chemistry A. 2018;**6**:6085-6095

[37] Shami Z, Holakooei P. Durable light-driven three-dimensional smart switchable superwetting nanotextile as a green scaled-up oil–Water separation technology. ACS Omega. 2020;**5**:4962-4972

[38] Solovyeva M, Selishchev D, Cherepanova S, Stepanov G, Zhuravlev E, Richter V, et al. Self-cleaning photoactive cotton fabric modified with nanocrystalline TiO2 for efficient degradation of volatile organic compounds and DNA contaminants. Chemical Engineering Journal. 2020;**388**:124167

[39] Jiang C, Liu W, Yang M, Liu C, He S, Xie Y, et al. Robust multifunctional superhydrophobic fabric with UV induced reversible wettability, photocatalytic self-cleaning property, and oil–water separation via thiol-ene click chemistry. Applied Surface Science. 2019;**463**:34-44

[40] Stan MS, Nica IC, Popa M, Chifiriuc MC, Iordache O, Dumitrescu I, et al. Reduced graphene oxide/TiO2 nanocomposites coating of cotton fabrics with antibacterial and self-cleaning properties. Journal of Industrial Textiles. 2018;**49**:277-293

[41] Khan MZ, Baheti V, Militky J, Wiener J, Ali A. Self-cleaning properties of polyester fabrics coated with flower-like TiO2 particles and trimethoxy (octadecyl) silane. Journal of Industrial Textiles. 2019;**50**:543-565

[42] Asadnajafi S, Shahidi S, Dorranian D. In situ synthesis and exhaustion of nano TiO2 on fabric samples using laser ablation method. Journal of the Textile Institute. 2020;**111**:122-128

[43] Zhao J, Zhu W, Wang X, Liu L, Yu J, Ding B. Environmentally benign modification of breathable nanofibrous membranes exhibiting superior waterproof and photocatalytic self-cleaning properties. Nanoscale Horiz. 2019;**4**:867-873

[44] Mills A, Lepre A, Elliott N, Bhopal S, Parkin IP, O'Neill SA. Characterization of the photocatalyst Pilkington Activ™: A reference film photocatalyst? Journal of Photochemistry and Photobiology A: Chemistry. 2003;**160**:213-224

[45] Li Q, Liu Q, Peng B, Chai L, Liu H. Self-cleaning performance of TiO2-coating cement materials prepared based on solidification/stabilization of electrolytic manganese residue. Construction and Building Materials. 2016;**106**:236-242

[46] Chabas A, Gentaz L, Lombardo T, Sinegre R, Falcone R, Verità M, et al. Wet and dry atmospheric deposition on TiO2 coated glass. Environmental Pollution. 2010;**158**:3507-3512

[47] da Silva AL, Dondi M, Raimondo M, Hotza D. Photocatalytic ceramic tiles: Challenges and technological solutions. Journal of the European Ceramic Society. 2018;**38**:1002-1017

[48] La Russa MF, Rovella N, Alvarez de Buergo M, Belfiore CM, Pezzino A, Crisci GM, et al. Nano-TiO2 coatings for cultural heritage protection: Therole of the binder on hydrophobic and self-cleaning efficacy. Progress in Organic Coating. 2016;**91**:1-8

[49] Cedillo-González EI, Hernández-López JM, Ruiz-Valdés JJ, Barbieri V, Siligardi C. Self-cleaning TiO_2 coatings for building materials: The influence of morphology and humidity in the stain removal performance. Construction and Building Materials. 2020;**237**:117692

[50] Sun C, Dai J, Zhang H, Zhang F, Zhang N. Preparation and characterization ofspherical titania and their influence on self-cleaning and anticorrosion properties of acrylic resin. Progress in Organic Coating. 2019;**128**:21-31

Chapter 5

Graphite Carbon Nitride

Yamuna Ezhumalai, Prabakaran Kumaresan and Tirupathy Jayapalan

Abstract

Graphitic carbon nitride (g-C_3N_4), is a synthetic polymer made up of carbon, nitrogen, and some impurity hydrogen that is linked together using tris-triazine-based patterns. Because of the inclusion of N and H atoms, it has electron-rich characteristics, basic surface functions, and H-bonding motifs, compared to the bulk of carbon materials. Consequently, it's seen as a possible replacement for carbon in material applications. A brief introduction to g-C_3N_4 is included in this chapter, as are the methods for synthesizing this material with various textural structures and surface morphologies, as well as its physicochemical properties. Furthermore, four parts of g-C_3N_4 applications are discussed. We anticipate that this work will motivate readers to look for new applications for this material in catalysis and other domains.

Keywords: metal-free catalyst, functional material, graphitic carbon nitride, synthesis, applications

1. Introduction

Carbon nitrides are a type of a polymeric substance that is mostly made up of carbon and nitrogen [1, 2]. They may be prepared from carbon materials by replacing carbon atoms with nitrogen atoms, making them intriguing possibilities for a range of uses. Due to the chemical inertness of carbon nitride, solubility problem in acidic, neutral, and basic solvents. Hence, the structure of the material was not completely appreciated until recent decades [3]. Because of the presence of basic surface sites, graphitic carbon nitride (g-C_3N_4) is not only the most stable allotrope of carbon nitrides in the ambient atmosphere, but it also exhibits rich surface characteristics that are appealing for various applications, including catalysis [4]. In the state (this material is a -conjugated polymer), the ideal g-C_3N_4 consists only of an assembly of CN bonds with no electron localization [5, 6].

As shown in **Figure 1**, real materials, such as those made by polycondensation of cyanamide, contain a minor amount of hydrogen, which is present as primary and/or secondary amine groups on the terminal edges. The presence of hydrogen indicates that the real g-C_3N_4 is incompletely condensed and that a number of surface defects exist, which can be useful in catalysis and are thought to promote electron relocalization on the surface, inducing Lewis-base character toward metal-free coordination chemistry and catalysis.

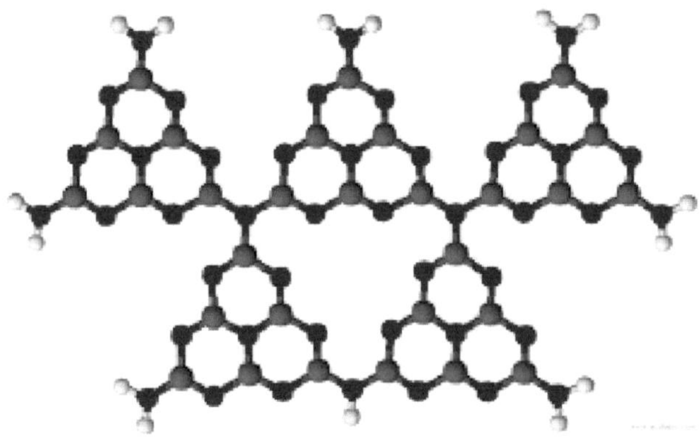

Figure 1.
Multiple surface functionalities reproduced from [7] with permission from the Royal Society of Chemistry.

The energy positions of the conduction band (CB) and valence band (VB) versus the normal hydrogen electrode (NHE) are −1.1 and 1.6 eV, respectively, in g-C_3N_4 (2.7 eV bandgap). Furthermore, g-C_3N_4 is extremely resistant to heat, strong acids, and strong alkaline solutions. The only elements in g-C_3N_4 are carbon and nitrogen, and it can be made by pyrolyzing nitrogen-rich precursors such as melamine, urea, thiourea, and cyanamide. It has been observed that the choice of precursor and pyrolysis temperatures have a significant impact on the electrical structure and bandgap of g-C_3N_4, which will have implications for its prospective uses in a variety of disciplines.

Recently, tremendous progress has been made in the field of g-C_3N_4 research. As a result, a paper summarizing the synthesis of g-C_3N_4-based materials and their prospective energy storage applications is required. The characteristics, production, and possible applications of g-C_3N_4 and g-C_3N_4-based nanocomposites in energy storage and conversion, such as photocatalytic hydrogen evolution, oxygen reduction reaction (ORR), and Li-based battery, are discussed in this book chapter.

2. Synthesis of g-C3N4 materials

It is a stable polymer semiconductor with only C and N components. Many nitrogen-rich organic precursors are now utilized to make g-C_3N_4, including urea, melamine, dicyandiamide, thiourea, and cyanamide. Carbon nitride materials made from direct condensation of these precursors, on the other hand, have bulk structures with low specific surface areas. It's required to introduce well-controlled porous structures in bulk g-C_3N_4 for practical usage as catalysts. Several strategies for obtaining porous g-C_3N_4 have been devised by scientists.

2.1 Synthesis of g-C3N4 materials

2.1.1 Template method

Under the structural inversion, "templating" entails the copying of one structure into another. In its broadest sense, a template is a structure-directing agent. Because the size and shape of the resulting pore structures may be easily controlled by the proper template, the template method is a versatile methodology for the fabrication of nanostructured or porous materials. In g-C_3N_4, hard templates [7] are employed to build porous materials and create hierarchical pore topologies. To regulate nanostructures, silica templates are commonly utilized as a structure directing agent. Groenewolt et al. used multiple pore size mesoporous silica matrices to synthesize unique g-C3N4 nanoparticles of different sizes. Many studies on mesoporous g-C_3N_4 materials made with silica-based hard templates have been published since then. Using aminoguanidine hydrochloride as a precursor and SBA-15 as rigid templates, highly ordered mesoporous g-C_3N_4 with variable pore sizes was produced. To make ordered porous g-C_3N_4, Fukasawa et al. employed uniform-sized silica nanospheres as templates. Using guanidinium chloride as a precursor used a nanocasting process to make mesoporous g-C_3N_4. Two types of pores and high specific surface areas were found in the as-prepared g-C_3N_4 samples. Park et al. used the incipient wetness method to create 2-dimensional (2D) and 3-dimensional (3D) mesostructured g-C_3N_4 using mesoporous silica as a hard template. Open pores and a large specific surface area characterize these materials. The hard-template technique, in particular, necessitates the removal of the template in order to get the requisite g-C_3N_4 structure. This procedure typically uses aqueous NH_4HF_2 or HF, which can harm the environment (**Figure 2**).

2.1.2 Template-free method

A simple thermal treatment of dicyandiamide is used to make porous g-C_3N_4. The synthesized porous g-C_3N_4 has a large pore volume and a high BET surface area (0.50–0.52 m3 g^{-1}). A simple template-free technique to make g-C_3N_4 nanofibers (GCNNFs). Melamine was first reacted with ethanol, then annealed at 450°C for 2 hours to produce GCNNFs, which had a 1D structure and a large specific surface area. Thermal calcinations were also used to make graphene-modified porous g-C_3N_4 (porous g-C_3N_4/graphene). The polymerization process was carried out at various temperatures in this approach, with high calcination temperatures yielding porous g-C_3N_4.

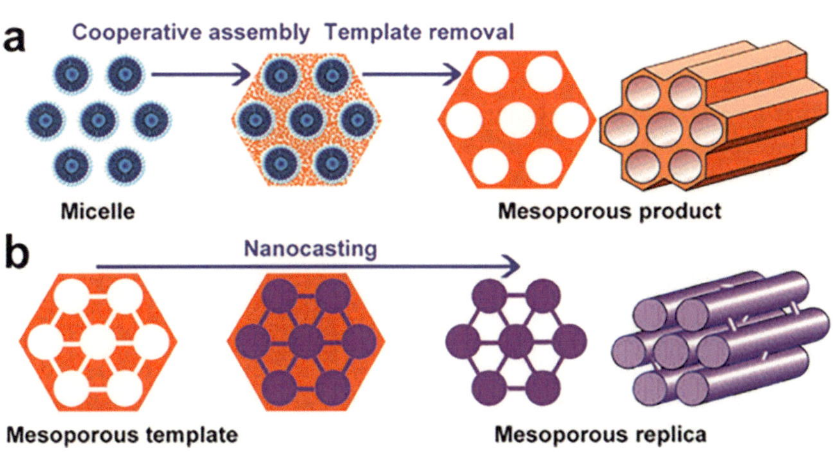

Figure 2.
Scheme of two representative synthesis routes for ordered mesoporous materials: (a) soft templating method and (b) hard templating (nanocasting) method. Reproduced from [8] with permission from the Royal Society of Chemistry.

2.1.3 Fractional thermal polymerization method

Using melamine, guanidine carbonate, and dicyandiamide as starting ingredients, a fractional thermal polymerization process was used to create g-C_3N_4 particles with a large surface area. Melamine, guanidine carbonate, and dicyandiamide were polymerized to create g-C3N4 at 515, 550, and 515 degrees Celsius, respectively. At the temperatures given, no residual component of precursors could be established. Around 200–240°C, all of the products for these three precursor materials had the structure of $C_3H_6N_6$, which converted to tri-s-trizines at 350–500°C. The dense packing between the conjugated aromatic system of g-C_3N_4 became stable using this fractional thermal polymerization approach. After 120 minutes of irradiation, the MO of g-C_3N_4-M (600°C) could reach 54.67 %, whereas that of g-C_3N_4-G (550°C) and g-C_3N_4-D (590°C) could reach 23.46 % and 22.16 %, respectively.

2.1.4 Simple pyrolysis method

Simple pyrolysis of affordable, environmentally friendly, active oxygen-evolving urea in a covered crucible yielded porous g-C_3N_4 with a band gap of 2.87 eV. The photocatalytic hydrogen evolution activity of g-C_3N_4 produced from urea as a precursor was higher than that of thiourea or melamine in the presence of methanol as a sacrificial reagent and Pt as a co-catalyst. This is due to the structure's porous nature and large surface area. The g-C_3N_4 from urea has a somewhat lower degree of polymerization, resulting in more structural flaws acting as active photocatalytic sites for the Pt nanoparticle co-catalyst photodeposition as well as hydrogen production, according to XRD, TGA, XPS, and NMR data.

2.1.5 Ionothermal method

The synthesis of highly crystalline graphitic carbon nitride by dicyandiamide self-condensation in a salt melt of lithium chloride and potassium chloride has been demonstrated, and the resulting g-C_3N_4 has been compared to LiebigLs melon made using the typical bulk condensation technique. The product's FTIR and elemental analysis point to a structure with few flaws and unreacted end groups, indicating a highly condensed framework. Powder XRD analysis and high-resolution TEM reveal pronounced in-plane ordering with a repeat distance of d = 7.30 K, which corresponds to the separation of co-planar, covalently linked heptazine units, and a planar graphitic interlayer distance of d = 3.36 K, which corresponds to the separation of co-planar, covalently linked heptazine units.

3. Applications of g-C3N4 materials

3.1 Friedal: crafts reaction

Friedel–Crafts reactions are a type of aromatic C–H activation reaction that is known to be one of the least environmentally friendly industrial processes, creating roughly 88 percent waste. $AlCl_3$ boosted the standard version. According to Goettmann et al. [9] that meso-g-C_3N_4 is an important Lewis base catalyst allowed for some quite strange aromatic substitution reactions to take place. A Friedel–Crafts type that has been generalized. This metal-free catalyst is not just good for the environment. Only environmentally friendly alkylation agents, such as alcohols or acids, but they also showed unexpected reactivity in the direction of urea and quaternary ammonium compounds.

3.2 Selective oxidation reaction

The selective oxidation of hydrocarbons using pure oxidants is an important step in the synthesis of a wide range of products, from commodity chemicals to specialty medications. Chen et al. [10] proved that the Fe/g-C_3N_4 catalyst was capable of converting benzene to phenol without the use of hydrogen peroxide. The yield of phenol synthesis might be significantly increased by utilizing the photocatalytic capabilities of g-C_3N_4. Su et al. [11] demonstrated that under visible light irradiation, meso-g-C_3N_4 can operate as a photocatalyst to activate O_2 for the selective oxidation of benzyl alcohols.

3.3 Oxygen reduction reaction in fuel cells

Fuel cells have sparked a lot of interest since they provide cleaner, more sustainable energy. The high cost of Pt catalyst and the slow kinetics of ORR now limit the practical applications of fuel cells. Carbon compounds containing nitrogen, such as g-C_3N_4, are worth exploring because they give enough active sites for ORR. However, the low electron transport of g-C_3N_4 limits its electrocatalytic effectiveness. One option for addressing this issue is to use conductive carbon materials as a support to boost electron accumulation and consequently electrocatalytic performance. Lyth et al. employed g-C_3N_4 as an oxygen reduction catalyst and discovered that, while the electrocatalytic activity of g-C_3N_4 was higher than pure carbon, the current densities were low, presumably due to its low surface area. It was discovered that combining C_3N_4 with carbon black boosted current densities. Yang et al. used nanocasting to make graphene-based C_3N_4 (G-CN) nanosheets. The G-CN nanosheets had a high nitrogen content and a large specific surface area, and their electrical conductivities were improved.

4. Advantages of g-C3N4 materials in photocatalyst

The advantages of photocatalysts with varied morphologies are obvious. Two-dimensional photocatalysts made in a specific way, for example, have a bigger specific surface area and more reactive sites. Other decorative components introduced throughout the construction process can also be used. However, there are still certain issues, such as a tight reaction temperature requirement and uneven particle distribution.

4.1 Doping of elements

The use of element doping to change the band structure of photocatalysts and enlarge the light absorption zone is an effective method [9]. Non-metallic (B, O, S, I) doped g-C_3N_4 has been widely utilized to alter the optical absorption of internal band-gap structure and to adjust the photocatalytic performance of g-C_3N_4 to promote the redox potential in visible light [10–14]. The light absorption range of nanosheets, on the other hand, reduces due to the quantum confinement effect, which increases the band gap [15, 16]. As a result, we can apply elemental doping to change the band structure of g-C_3N_4, which improves both light absorption and charge separation [17]. Oxygen doping, boron doping, phosphorus doping, and halogen doping will all be investigated.

4.2 Oxygen doping

By doping oxygen into g-C_3N_4, Wei *et al.* [18] significantly increased photocatalytic performance. **Figure 3** shows the process of making oxygen-doped g-C_3N_4 nanospheres. The copolymerization of the precursor can be used to adjust the degree of oxygen doping. As a result, oxygen-doped g-C_3N_4 demonstrated exceptional photocatalytic performance, with a Rhodamine B visible light degradation rate of 0.249 min^{-1} which was greater than traditional g-C_3N_4, respectively. Zhang et al. [19] also doped varied oxygen element concentrations in typical carbon nitride materials by introducing different quantities of oxalic acid. Photocatalysis efficiency has been significantly enhanced. The addition of oxygen to the energy band structure can change the energy band structure and boost visible light absorption.

4.3 Boron doping

Thaweesak et al. [20] used one-pot thermal condensation to create a novel form of boron-doped graphitic carbon nitride nanoscale material. The best time catalytic hydrogen evolution activity is reported to be 1880 mol g^{-1} h^{-1} (> 400 nm), which is more than 12 times greater than bulk g-C_3N_4. The combination of band structure development and morphological control is credited with the strong photocatalytic performance. He et al. [21] successfully doped boron into a line-shaped carbon nitride photocatalytic material using sodium borohydride corrosion. The photocatalytic performance was successfully improved. Boron doping not only narrows the band gap and absorbs more visible light, but it also has a larger surface area than bulk g-C_3N_4, improving photocatalytic activity significantly. These works show two different preparation tactics that work together.

4.4 Phosphorous doping

Kumar et al. [22] used solid thermal condensation of urea, citric acid, and 1-butyl-3-methylimidazolium hexafluorophosphate (BMIM-PF_6) in a reactor at 200°C to make phosphorus-doped carbon nitride quantum dots (CNPQDs). **Figure 4a** depicts the fabrication of carbon nitride-doped quantum dots (CNPQDs). By condensation polymerization, urea and citric acid produce a basic carbonitride skeleton, and

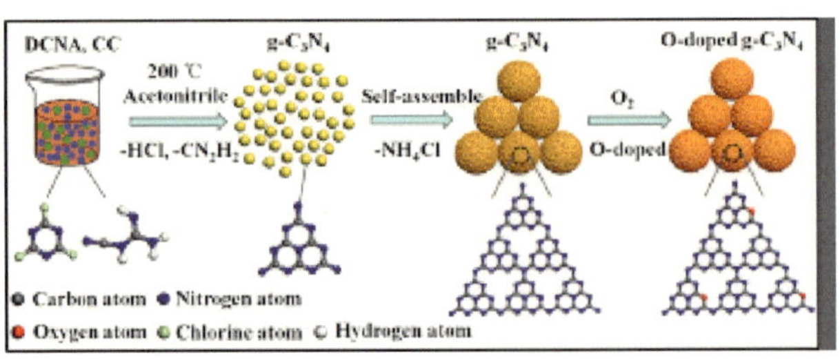

Figure 3.
The preparation of oxygen-doped g-C3N4 nanospheres. Reproduced with permission [18]. Copyright 2018, Royal Society of Chemistry.

BMIM-PF_6 serves as a phosphorus doping source in this synthesis. **Figure 4b** depicts the steady-state photoluminescence (PL) spectra of phosphorus-doped carbon nitride quantum dots (CNPQDs) and carbon nitride quantum dots (CNQDs) in water. 15 N and 31P are two of the most common elements found in nature. **Figure 4c** shows the CPMAS NMR spectra of CNPQDs. **Figures 4d** and **e** depict the process of (CNPQDs) alteration in a titanium dioxide array.

4.5 Iodine doping

Iqbal et al. [23] used a straightforward one-step technique to make a nano-mesoporous g-C_3N_4 doped with iodine and N vacancy at the same time. Synergistic iodine-doped mesoporous g-C_3N_4 epidermoid shedding and the introduction of N vacancy due to the release of nitrogen-rich gas has been observed when glucose and NH_4I are heated simultaneously. Under simulated solar light, the produced iodine-doped mesoporous g-C_3N_4 has an exceptional photocatalytic hydrogen generation performance of 7819.2 mol g^{-1} h^{-1}, approximately 6.5 times greater than bulk g-C_3N_4 and other types of iodine doped g-C_3N_4 photo-catalysts. This paper demonstrates a straightforward and practical method for synthesizing high-performance metal-free g-C_3N_4 photo-catalysts by altering the electronic transition via a well-designed band structure (**Figure 5**).

4.6 Bromine doping

A bromide doped graphite phase carbon nitride technique was proposed by Lan et al. [24]. While preserving the core structure of Triazine as the main component of the material, bromine alteration can increase the optical, conductance, and photocatalytic capabilities of g-C_3N_4. **Figure 6** depicts the photocatalytic mechanism of the CN-Br photocatalyst. The process can be used to modify a variety of g-C_3N_4 precursors, including urea, dicyandiamide, ammonium thiocyanate, and thiourea. This work also demonstrates a viable method for rationally designing and synthesizing g-C_3N_4-based photocatalysts.

Element doping can significantly improve catalytic performance, particularly for visible light. However, there are some drawbacks that cannot be overlooked, such as the lengthy preparation procedure, increased production costs, and decreased industrial production.

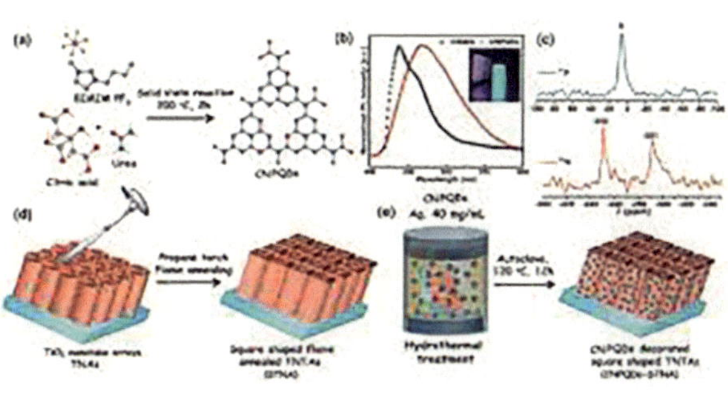

Figure 4.
(a) Schematic diagram illustrating the synthesis of phosphorus-doped carbon nitride quantum dots (CNPQDs). (b) Steady-state PL spectrum of CNPQDs and carbon nitride quantum dots (CNQDs) in water. Inset showing fluorescence of CNPQDs sample under UV light irradiation. (c) N and P CPMAS NMR spectra of CNPQDs. (d-e) the preparation of CNPQDs decorated square-shaped TNTAs. Reproduced with permission [21]. Copyright 2019, Wiley-VCH.

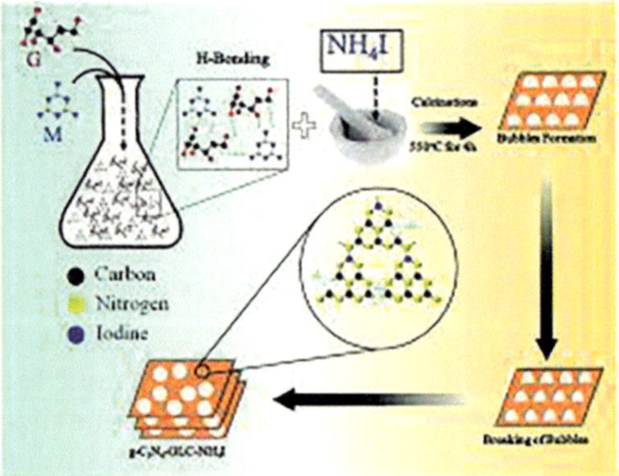

Figure 5.
Schematic of synthesizing N-vacant iodine doped mesoporous g-C3N4 nanosheets. Reproduced with permission [23]. Copyright 2019, Royal Society of Chemistry.

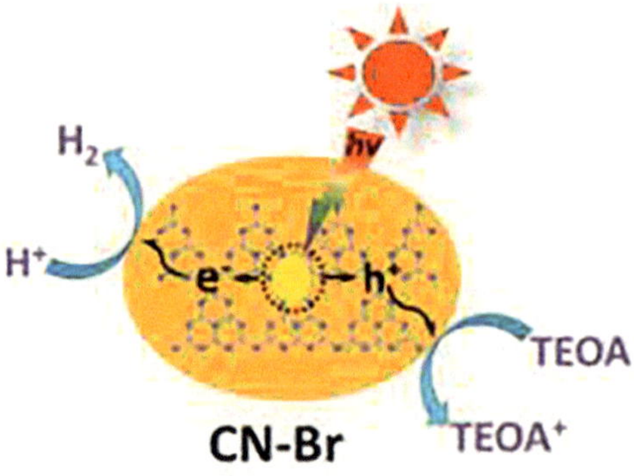

Figure 6.
Photocatalytic mechanism of CN-Br photocatalyst. Reproduced with permission from [24]. Copyright 2016, Elsevier.

5. Conclusions

In summary, this chapter focuses on the synthesis methods and applications of g-C_3N_4 and g-C_3N_4-based materials in energy storage. Because of its unique characteristics and wide range of applications, g-C_3N_4 is a good candidate for effective energy storage. As a result, research into g-C_3N_4-based composites for energy storage and conversion is expected to pick up speed in the near future. Despite significant progress over the years, there are still numerous hurdles in the use of this material in catalysis.

Acknowledgements

I would like to acknowledge the extraordinary debt I owe to the writers who have told me such wise things about writing over the years: I would not be able to get my work done without the continual support and vision of author manager Elena Vracaric. I thank god to give this wonderful time to write this chapter.

Author details

Yamuna Ezhumalai[1*], Prabakaran Kumaresan[2] and Tirupathy Jayapalan[3]

1 Department of Chemistry, Avinashilingam Institute for Home Science and Higher Education for Women, Coimbatore, India

2 Department of Chemistry, PSG College of Arts and Science, Coimbatore, India

3 Department of Physics, JCT College of Engineering and Technology, Pichanur, Coimbatore, India

*Address all correspondence to: yamuchemist@gmail.com

References

[1] Cohen ML. Calculation of bulk moduli of diamond and zincblende s. Physical Review B. 1985;**32**:7988-7991

[2] Liu AY, Cohen ML. Prediction of new low compressibility solids. Science. 1989;**245**:841-842

[3] Liebig J. Uber Einige Stickstoff—Verbindungen. The Annals of Pharmacotherapy. 1834;**10**:1-47

[4] Wang X, Blechert S, Antonietti M. Polymeric graphitic carbon nitride for heterogeneous photocatalysis. ACS Catalysis. 2012;**2**:1596-1606

[5] Thomas A, Fischer A, Goettmann F, Antonietti M, Muller JO, Schlogl R, et al. Graphitic carbon nitride materials: Variation of structure and morphology and their use as metal-free catalysts. Journal of Materials Chemistry. 2008;**18**:4893-4908

[6] Su F, Antonietti M, Wang X. Mpg-C_3n_4 as a Solid Base Catalyst for Knoevenagel Condensations and Transesterification. Sci. Technol., 2012;**2**:1005-1009

[7] Zheng Y, Liu J, Liang J, Jaroniec M, Qiao SZ. Graphitic carbon nitride materials: Controllable synthesis and applications in fuel cells and photocatalysis. Energy & Environmental Science. 2012;**5**(5):6717-6731

[8] Shi, Y, Wan, Y, Zhao, D. Ordered mesoporous non-oxide materials. Chem. Soc. Rev. 2011;**40**:3854-3878

[9] Wang W, Yu JC, Shen Z, Chan DKL, Gu T. g-C3N4 quantum dots: Direct synthesis, upconversion properties and photocatalytuc applications. Chemical Communications. 2014;**50**:10148-10150

[10] Ding K, Wen L, Huang M, et al. Phys. How does the B, F-monodoping and B/F-codoping affect the photocatalytic water-splitting performance of g-C3N4? Chem. Chemical Physics. 2016;**18**:19217-19226

[11] Rahman MZ, Davey K, Qiao SZ. Carbon, nitrogen and phosphorus containing metal-free photocatalysts for hydrogen production: Progress and challenges. Journal of Materials Chemistry A. 2018;**6**:1305-1322

[12] Su F, Xu C, Yu Y, Zhang W. Carbon self-doping induced activation of n* electronic transitions of g-C3N4 nanosheets for efficient photocatalytic H2 evolution. ChemCatChem. 2016;**8**:3527-3535

[13] Zhang H, Han X, Yu H, Zou Y, Dong X. Enhanced photocatalytic performance of boron and phosphorous co-doped graphitic carbon nitride nanosheets for removal of organic pollutants. Separation and Purification Technology. 2019;**226**:128-137

[14] Che W, Cheng W, Yao T, et al. Fast photoelectron transfer in (Cring)-C3N4 plane heterostructural nanosheets for overall water splitting. Journal of the American Chemical Society. 2017;**139**:3021-3026

[15] Hu K, Yao M, Yang Z, et al. pressure tuned photoluminescence and band gap in two-dimensional layered g-C3N4: The effect of interlayer interactions. Nanoscale. 2020;**12**:12300-12307

[16] Xu G, Wang Z, Liu Q, et al. Carbon-bridged g-C3N4 nanosheets for hydrogen evolution rate by a two-step gaseous treatment. Chemistry Select. 2019;**4**:13064-13070

[17] Wan Q, Wei F, Ma Z, Anpo M, Lin S. Novel porous boron nitride nanosheet with carbon dopin: Potential metal-free photocatalyst for visible-light-driven overall water splitting. Advanced Theory and Simulations. 2019;**2**:1800174

[18] Wei F, Liu Y, Zhao H, et al. Oxygen self-doped g-C3N4 with tunable electronic band structure for unprecedentedly enhanced photocatalytic performance. Nanoscale. 2018;**10**:4515-4522

[19] Zhang G, He C, Zhang P, Mi H. Oxygen-doped crystalline carbon nitride with greatly extended visible-light-responsive range for photocatalytic H2 generation. Applied Catalysis B: Environmental. 2020;**283**:119636

[20] Thaweesak S, Wang S, Lyu M, et al. Boron-doped graphitic carbon nitride nanosheets for enhanced visible light photocatalytic water splitting. Dalton Transactions. 2017;**46**:10714-10720

[21] He F, Wang S, Zhao H, et al. Applied Surface Science. 2019;**485**:70-80

[22] Kumar P, Kar P, Manuel AP, et al. Advanced Optical Materials. 2020;**8**:1901275

[23] Iqbal W, Yang B, Zhao X, et al. Facile one-pot synthesis of mesoporous gC3N4 nanosheets with simultaneous iodine doping and N-vacancies for efficient visible-light-driven H_2 evolution. Catalysis Science and Technology. 2020;**10**:549-559

[24] Lan Z-A, Zhang G, Wang X. A facile synthesis of Br-modified g-C3N4 semiconductors for photoredox water splitting. Applied Catalysis B Environment. 2016;**192**:116-125

Chapter 6

A Triphasic Superwetting Catalyst for Photocatalytic Wastewater Treatment

Lagnamayee Mohapatra and Jun Ha Park

Abstract

The increasing organic contamination is mainly produced by the widespread industrial, agricultural, and household applications and has become a serious worldwide issue. Therefore, we need to develop sustainable and environmentally friendly technologies to reduce waste detrimental to the environment. A promising approach is known as heterogeneous photocatalysis, inspired by natural photosynthesis. For this purpose, the challenges raised to synthesize appropriate surface nano/micro-structured materials with long-term stability and mechanical durability for practical use. The traditional photocatalytic system is diphasic (dependent upon the solid-liquid phase), where the solid-liquid reaction interface depends upon the mass transfer. Especially, the low concentrations of oxygen in water and the slow diffusion rate limit the removal of electrons which decreases the photocatalytic reaction rates even if the presence of high light intensities. Therefore, the work aims to develop novel triphasic superwetting photocatalytic materials where the photocatalytic reaction is carried out at gas-liquid-solid joint interfaces. This triphasic contact line can allow oxygen from the air to this reaction interface and minimize electron-hole recombination even at high light intensities. Herein, we intend to discuss the importance of a novel superwetting triphasic nanoarrays catalyst that will be developed and implemented.

Keywords: heterogeneous photocatalysis, triphasic system, gas-liquid-solid joint interfaces

1. Introduction

Photocatalysis has wide applications in environmental, fuel production, and chemical synthesis [1, 2]. Photocatalytic materials that can convert photon energy to chemical energy are employed to split water to produce hydrogen and also produce highly reactive intermediates for chemical synthesis and reactive oxygen radicals for the degradation of organic pollutants [3]. Light absorption generates holes and electrons within the valence and conduction bands in heterogeneous photocatalysis. Those charges may migrate within the semiconductor particle and be trapped at surface sites. They'll also participate in the interfacial electron transfer processes involving the molecules of electron acceptor (A) and donor (D). within the photocatalytic organic degradation process, oxygen act as A and water as D, creating anion

($O_2^{\bullet -}$) and chemical group ($OH^{\bullet}$) radicals. Those reactive radicals are liable for most of the oxidation of organic substances. Oxidation under these conditions is sometimes complete, giving H_2O and CO_2 because of the final products.

The reactions can therefore be used in water, air, or surface purification. But, the recombination of photogenerated charge carriers competes with this photocatalytic degradation process, which is a critical factor in limiting the kinetics of photocatalysts and reaction rate. To overcome this limitation, numerous photocatalysts have been developed to enhance efficiency. The nanomaterials with large surface areas, abundant surface states, and specific morphologies have emerged as pioneering photocatalysts for the dye degradation process. The optimal structure-properties relationships are essential for efficient organic pollutant degradation [4]. Significantly, the hierarchical hetero-nanostructured materials called nanoarrays give rise to separating the photogenerated electron-hole pairs to improve the photocatalytic activity further [5]. To reduce the recombination of charge carriers, the photocatalyst surface need suitable and sufficient acceptors. Since these processes involve some complicated steps, but improvement of separation and transportation of photogenerated charge carriers are the main challenges to designing highly effective photocatalysts for practical applications. Another critical issue that induces the photocatalytic activity of a catalyst is the nature of its surface/interface chemistry. The surface energy and chemisorption properties are vital in transferring electrons and energy between substances at the interface. This process allows the overpotential of redox reactions on the photocatalyst surface, which reduces photo-corrosion. Previous studies have mainly focused on the reactivity of the catalyst. Moreover, the less diffusion rate of oxygen in water, limits the photocatalytic reactions even under high light intensity. In contrast to conventional double-phase photocatalytic systems, which consist of catalysts immersed in a bulk liquid phase, triple-phase catalytic systems by supporting catalysts at the gas-liquid boundaries have been developed and exhibited outstanding performance [6].

The wettability modification of a catalyst surface includes a vital role in improving the charge transfer ability. Superwetting behavior could be a unique wetting phenomenon that always depends upon the phases. Superwettable surfaces, like super-hydrophilic and superhydrophobic surfaces, exhibit unique transport dynamics and providing exceptional prospects for reinforcing chemical process efficiency. Therefore, these materials are dramatically different from traditional materials. Superwetting catalysts have enhanced catalytic activity when introducing an air layer between the catalyst and liquid. The charge carriers from reacting interface are very fast to radicals by oxygen, and enough oxygen within the air layer can effectively capture electrons and minimize the electron-hole recombination. Superwetting materials are commonly designed by controlling surface energy, chemical compositions, and geometric structures of solid surfaces. This chapter discussed the photocatalytic organic pollutants supported superwetting materials. The discussion mainly contains widely investigated photocatalytic reactions involving gas and water molecules as reactants and products for organic pollutant degradation.

2. Theoretical basis of wettability

Contact angle (CA) assigned as θ, which can give the quantitative measurement of wetting of a solid by a liquid. Therefore, it can be defined as the angle measured by a liquid where the liquid-gas interface meets at solid surface as shown in **Figure 1**.

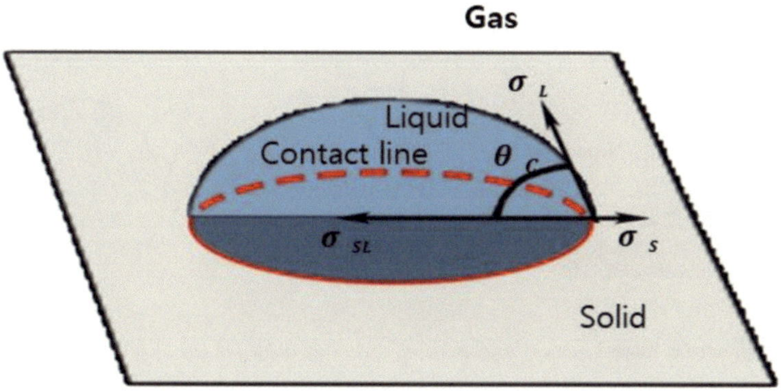

Figure 1.
Schematic illustration of the triple-phase contact line.

2.1 Theoretical models

Wettability is an essential property of solid materials. When liquid droplet contacts a solid surface in air, a three-phase contact line is formed at the three-phase junction. The contact line expands outward up to the droplet reaches a static state. In a steady-state system, a three-phase contact line is contact because of an equilibrium tangential forces created by the interfacial and surface tensions. The wetting properties of the liquid on solid surface is measured by the CA such as when $\theta < 90°$, the surface considered to be hydrophilic and when $\theta > 90°$, the surface is hydrophobic. However, when the $\theta > 150^{\circ}$, the surfaces are highly hydrophobic and called superhydrophobic. Similarly, superhydrophilic surfaces have a $\theta < 10^{\circ}$.

The Wetting properties of the solid surfaces is governed by the Young, Wenzel, and Cassie-Baxter equation. The young are derived by balancing the interfacial forces at a three-phase contact line Eq. (1).

$$cos\theta = \frac{\gamma_{sv} - \gamma_{sl}}{\gamma_{lv}} \tag{1}$$

Here θ is the CA of the liquid droplet, γ_{sv} (solid-gas) γ_{lv} (liquid-gas), and γ_{sl} (solid-liquid) interfacial force per unit length of the contact line which is surface tension. Moreover, the balance of angle formed by the liquid at the three-phase boundary is defined as young's contact angle (**Figure 2a**). According to the value of contact angle, the surface should be hydrophilic ($\theta < 90°$, $\gamma S > \gamma SL$) and hydrophobic ($\theta > 90°$, $\gamma S < \gamma SL$) (**Figure 2b** and **c**).

For the solid surface which has rough morphology, then Wenzel introduced the factor of surface roughness, r, into Young's equation after considering the influence of rough surface structure on wettability. Therefore, the definition is the actual surface area to the projected surface area is "r". In this case, the liquid fills the microstructure of the solid surface (**Figure 3a**). For a liquid droplet in the Wenzel state, the measured/apparent CA (Θw) can be expressed by the Wenzel equation Eq. (2) which can be expressed the surface roughness.

$$cos\,\Theta w = r.cos\theta \tag{2}$$

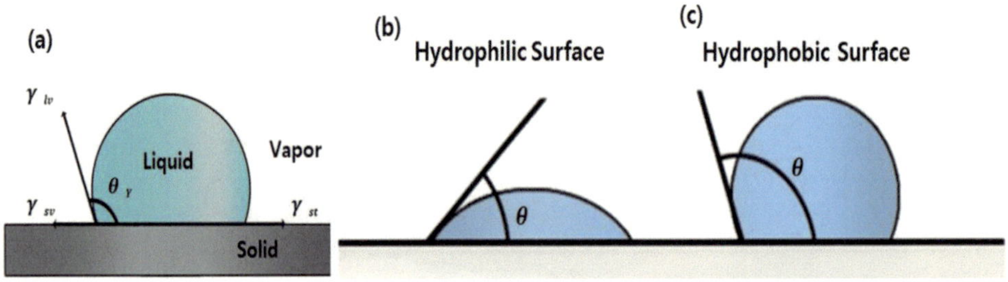

Figure 2.
(a) The wetting regime for Young's model, measurement of contact angle for (b) hydrophilic and (c) hydrophobic surfaces.

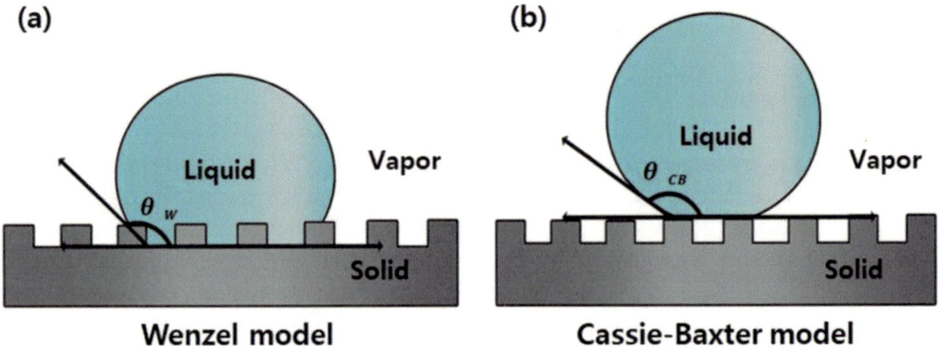

Figure 3.
Various states of droplets on a solid surface (a) Wenzel model, and (b) Cassie–Baxter model.

For hydrophobic surfaces, the roughness is very high and large r value. Sometimes, the liquid does not penetrate in to the roughness, so that the layer of trapped air between the solid surface and the liquid. In this case the contact interface between liquid and the solid surface consists of liquid-solid contact and liquid-air contact (**Figure 3b**) and expressed by Cassie equation Eq. (3)

$$cos\,\theta_c = f_1 \cdot cos\,\theta_1 + f_2 \cdot cos\,\theta_2 \tag{3}$$

Where θc is the CA, θ_1, θ_2 are the CA of the liquid droplet on the solid and air phases and f_1, f_2 are the area ratio of liquid-solid contact and the liquid-air contact, respectively. Again, for solid microstructure surface, if the liquid-solid contact area fraction is f, then liquid-air contact will be $(1 - f)$. So that, the Cassie equation will be Eq. (4):

$$cos\theta c = fcos\theta_1 + (1 - f)cos180^\circ \tag{4}$$

According to wettability, superwetting materials are generally categorized into four types: superhydrophobic, superhydrophilic, superoleophilic, and superoleophobic states. Specially, the first two kinds of superwetting materials which can be used for oil/water separation. These superwetting materials are believed to be promising materials for removing pollutants from water due to their superwetting property towards oils and water [7, 8].

2.2 Characterization of wettability

2.2.1 Classification of wettability

The most common measurements involving the static CA and the dynamic CA are utilized to know the surface property. The materials with specific wettability attain excessive attention owing to their outstanding performances for practical applications [9–11].

2.2.2 Static contact angle and dynamic contact angle

During the measurement, the contact area between liquid and solid is not changed from outside, but the dynamic contact angle can produce during wetting (advancing angle) or de-wetting (receding angle). Besides water droplets, other organic liquids can be also used for testing the surface wettability. When oil on a solid substrate, it can be observed four fundamental states: oleophilic, oleophobic, superoleophilic, and superoleophobic.

2.2.3 Contact angle hysteresis

Contact angle hysteresis (CAH) is an important physical phenomenon. Contact angle hysteresis reflects the activation energy required to move a droplet from one metastable state to another on a surface. CAH refers to the difference value between the advanced CA (θAdv) and the receding CA (θRec). An analysis software usually measures θAdv and θRec after a microscopy system reordered real-time images.

3. Importance of super wetting materials for photocatalysis

3.1 Triphase interfaces

In the liquid phase, air pockets are stuck with the rough topological surface of the superhydrophobic substrate, forming a triphasic solid–liquid-air interface. For photocatalytic reactions, this triphasic contact line usually serves as an active area for interfacial reactions and provides a vital clue to surface behavior. The oxygen supply from the air reacts with photogenerated electrons from the surface of the photocatalyst, producing oxidative reactive oxygen species (ROS) such as superoxide radicals ($O_2^{\bullet-}$) and hydroxide radicals ($OH^{\bullet}$) and resulting in the degradation of organic pollutants (**Figure 4**). Especially, during organic pollutant degradation, the carbon bonds breaks the on the superhydrophobic surface and can show long-term stability. Practically, a superhydrophobic with long-term stability catalyst is required.

3.2 Photocatalysts based on different wettability

3.2.1 Superhydrophilic photocatalysts

Inspired by the natural world's self-cleaning as well as the water-repellent properties of the lotus leaf, superwetting materials with unique wettability are believed to

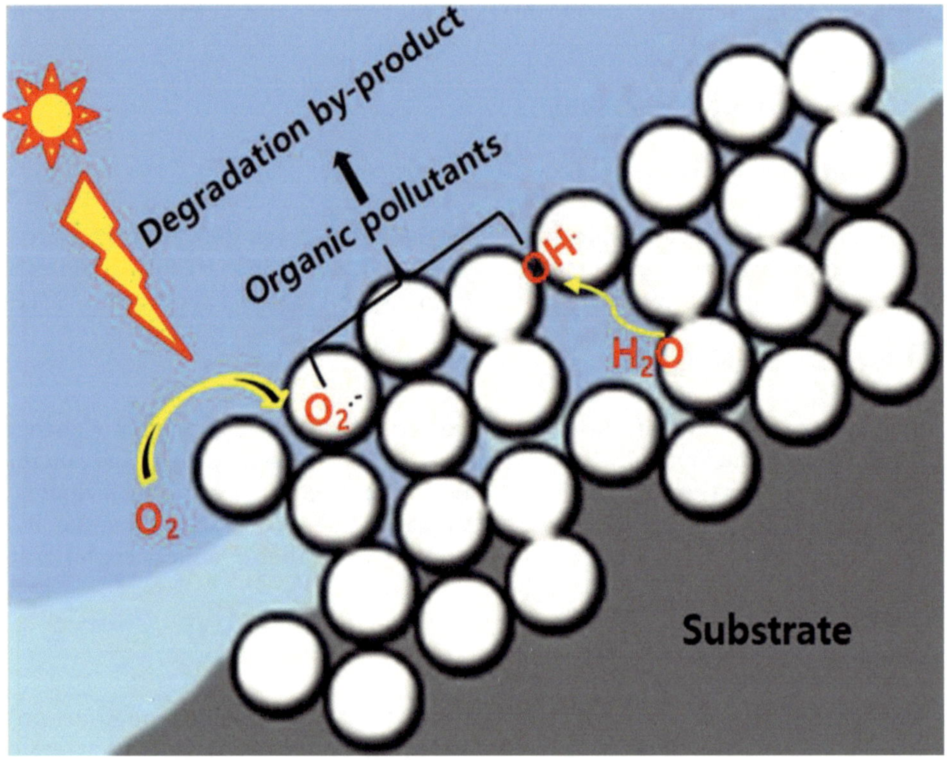

Figure 4.
Schematic representation of a triphasic nanoarray photocatalyst and the photocatalytic water purification process.

be promising materials for removing organic pollutants from water. Historically, the study of the superhydrophilicity of titanium dioxide (TiO_2) films traces back to 1997. Before illumination by UV light, the contact angle of TiO_2 surface was 72°. But after the UV illumination on the particular duration, the droplets completely spread on. This is due to creating numerous high-energy domains with hydrophilic/oleophilic properties on TiO_2 surfaces. Moreover, the wetting properties of single TiO_2 surfaces could be exchange between hydrophobicity and superhydrophilicity under the interchange of long-term dark storage and UV light irradiation.

After discovering its confirmed that, the TiO_2 surface with superamphiphilic ability has unique wetting transition under UV light [12, 13]. Furthermore, Wang and co-workers reported a hydrophilic TiO_2-coated glass with effective photogeneration, displaying antifogging and self-cleaning induced by UV illumination [14]. Then, Fujishima et al. confirmed the nanostructure TiO_2/SiO_2 films shows superwettability under UV irradiation. In the case, the upper part of TiO_2 layer and the bottom part of porous SiO_2 layer with a low refractive index providing platforms for self-cleaning and antifogging/reflection [15]. Also, Shang et al. fabricated visible active N-F doped TiO_2 Nanotube and palladium oxide is decorated on the surface of the nano array [16]. Due to their superior photocatalytic property and particular nanoarray alignment, it gives promising self-cleaning applications. Since then, the superwetting approach has been frequently used for antifogging and self-cleaning applications. Jiang's group fabricated translucent and stable Ag@AgCl/g-C_3N_4/TiO_2 ceramic films that showed superhydrophilicity and excellent photocatalytic activities for Rhodamine B degradation under visible and complete spectral irradiations. In this case, the water molecules from air can occupy the oxygen vacancies of TiO_2 of the composite and produce hydroxyl groups, which makes the TiO_2 more hydrophilic. So, by mixing P25 with g-C_3N_4 in a colloidal silica system, nano TiO_2 particles can be dispersed and attached to the g-C_3N_4 particles, leading to increased surface roughness and hydrophilicity of the film systems. In this case, the hydroxyl groups of P25 may interconnect with that of silica particles which helps to increase the bonding strength of graphitic carbon nitride composite film and silicate glass. Besides adding P25 into the film system, the catalytic efficiency is improved [17]. Then, the Zhang group studied the Polymer-based nanocomposites functionalization by organic moieties to make superhydrophilicity. Afterwards, TiO_2 nanoparticles coated with hydroxyethyl acrylate (HEA) without any solvent formed high durable superhydrophilic catalyst [18]. Compared with the bare TiO_2 films, the TiO_2 nanotube array film has excellent photocatalytic efficiency in terms of methyl orange (MO) degradation is reported [19].

Chen and co-workers reported hydrophilic interface engineering of the hydrophilic CoO_x modified hydrophobic Ta3N5, which improves its water oxidation efficiency under visible light irradiation. Compared to the pristine Ta_3N_5 surface, CoO_x deposited onto the MgO–Ta3N5 surface showed a 23-fold improvement [20]. Similarly, core-shell NaYF4:Yb, Tm@TiO2 NPS is fabricated for photocatalytic activities. Here, the hydrophilic layers of TiO2 were coated onto hydrophobic NaYF4:Yb materials and the Tm nanoplates are partially exchanging with oleic acid ligands which shows hydrophobic in nature into cetyltrimethylammonium bromide (CTAB) surfactants which is amphiphilic character. The combination of NaYF4:Yb, Tm (up conversion materials) with TiO2 (wide bandgap) with broad spectrum absorption changes the wettability of a solid surface to achieve high-quality interfaces in photocatalysts for smooth carrier migration [21].

3.2.2 Superhydrophobic photocatalysts

Superhydrophobic metal oxide like ZnO [22], exhibits advanced photocatalytic activity; however, during prolonged UV irradiation, superhydrophobicity changes into superhydrophilicity, resulting from the easy decompositions of low-surface-energy compositions under the stimulus of light. Consequently, a photocatalyst showing long-term superhydrophobicity was once considered not to exist. More generally, it is undoubtedly necessary to modify the surface of the catalyst with stable hydrophobic organics, which are chemically and directly bonded. Therefore, to achieve photocatalytically active hydrophobic materials, researchers have combined metal-oxide particles with hydrophobic polymers like Polydimethylsiloxane (PDMS) and polytetrafluoroethylene (PTFE) as composite mixtures [23–25]. Recently, metal oxides (TiO2) and nonwetting organic polymers, namely epoxy resin, followed by grafting 1H,1H,2H,2H-perfluorooctyltriethoxysilan (PFOS), we prepared an inorganic-organic superhydrophobic paint (IOS-PA) used for photocatalytic removal of three organic dyes, Nile red, methyl blue, and methyl orange [26]. PDMS, PTFE, and silicone nanofilaments have also been used to conduct long-term superhydrophobicity and photocatalysis on one surface [26–29]. Sheng and co-workers established a novel triphase photocatalytic system by creating a unique photocatalyst in which TiO2 nanoparticles (NPs) were immobilized on carbon fiber (CF) substrate treated by poly(tetrafluoroethylene) (PTFE) for water pollution remediation [30]. After immobilizing TiO_2, the surface of the materials are changed from superhydrophobic to hydrophilic. In the time of photocatalytic reactions, the TiO2 will be hydrophilic part while the substrate will be the superhydrophobic state which was connected with the atmosphere. Hence, the total system should have an abundant triphasic contact area, which allowed a sufficient oxygen transport and the rapid generation of reactive oxygen species for organic pollutants degradation. Recently, a superhydrophobic (SHB) TiO2 nanoarrays catalyst with low surface energy and rough surface microstructure was reported as a model photocatalyst. The soft surface energy and rough surface microstructures of the SHB nanoarrays give the photocatalytic system long-range hydrophobic in nature and helps to introduce the triphasic reaction interface [31]. Superhydrophobicity is an integral part of self-cleaning on a photocatalyst which showed the synergistic effect of strong water repellency and photocatalytic activity [28], where the rolling drops remove macroscopic particles and the photocatalytic degradation ensure by UV or solar light. In a superhydrophobic system's air-water-solid triphase joint interface [27, 32, 33], a continuous and steady gas channel is recognized, providing abundant gaseous reactants and the resulting quick gas transportations. This system overcomes the drawbacks of weak dissolved gas transfer and low solubility in liquid-solid diphase reaction systems. Thus, photocatalytic activity efficiency and selectivity are sharply increased. Jinxiu groups reported about the oil–water mixture separation and photocatalytic degradation of quinoline blue, rhodamine B, methyl orange and methylene blue by using [Ni(DMG)$_2$] hollow microtubes. The prepared [Ni(DMG)2] films is act as superhydrophobicity and superoleophilicity and ascribed to the Cassie–Baxter model. Similarly, Ag/TiO2@PDMS coated cotton fabric which is low-cost effective, and recyclable separation material used for water purification to degrade methylene blue (MB) [34]. The effect of a grafted PDMS layer on wetting properties of TiO_2 for photocatalytical application is studied by Butt group [35]. The most effective dual-purpose ceria nanoparticle membrane is fabricated by facile spray-deposition method on stainless steel membrane for oil-water separation and photocatalytic degradation. The prepared membrane has superwetting properties which is efficient for oil/water separation. In this case the oil is passing through the stainless-steel membrane, whereas, high column of water is blocked. Furthermore, the CeO_2 coated membrane is utilized for the efficient degradation of a dye [36].

4. Conclusions

The present chapter fully addresses the main objectives of water purification by using a triple-phase catalyst. Under irradiation, charge carriers are formed on the surface of the photocatalyst, and the success of pollutant molecule degradation critically depends on the interaction between the surface and the target molecules. Therefore, the organic pollutant degradation efficiency strongly depends on the fabrication method as it drives the shape and size of the photocatalyst and its hydrophobic or hydrophilic characteristics. Compared with state-of-the-art diphasic photocatalytic systems, for which the limited concentration and diffusion rate of oxygen reduces the degradation efficiency, the novel triphasic photocatalytic system with superhydrophobic triphasic interface architecture will allow the rapid delivery of oxygen directly from the air to the reaction interface, thus minimizing electron-hole recombination and resulting in remarkably high efficiency. Recently, a solid surface's superwettability (especially underwater superoleophobicity) has attracted much attention owing to its importance for photocatalytic. Although this is new research, it is rapidly growing and promising in future research, which enormously extends the research field of superior wettability to the triphasic system. Therefore, the surface wettability of a photocatalyst film in the liquid-liquid-solid system should also be an exciting research focus shortly.

Author details

Lagnamayee Mohapatra* and Jun Ha Park
Department of Quantum System Engineering, Jeonbuk National University, Republic of Korea

*Address all correspondence to: mlagnamayee@gmail.com

References

[1] Mohapatra L, Parida K. A review on the recent progress, challenges and perspective of layered double hydroxides as promising photocatalysts. Journal of Materials Chemistry A. 2016;**4**:10744-10766

[2] Mohapatra L, Patra D. Multifunctional hybrid materials based on layered double hydroxide towards photocatalysis. John Wiley & Sons Inc.; 2019. pp. 215-241

[3] Tao X, Zhao Y, Wang S, Li C, Li R. Recent advances and perspectives for solar-driven water splitting using particulate photocatalysts. Chemical Society Reviews. 2022;**51**:3561-3608

[4] Pasternak S, Paz Y. On the similarity and dissimilarity between photocatalytic water splitting and photocatalytic degradation of pollutants. ChemPhysChem. 2013;**14**:2059-2070

[5] Zhu L, Li H, Xia P, Liu Z, Xiong D. Hierarchical ZnO decorated with CeO2 nanoparticles as the direct Z-scheme heterojunction for enhanced photocatalytic activity. ACS Applied Materials & Interfaces. 2018;**10**:39679-39687

[6] Wen L, Tian Y, Jiang L. Bioinspired super-wettability from fundamental research to practical applications. Angewandte Chemie (International Edition in English). 2015;**54**:3387-3399

[7] Zhu H, Tu Y, Luo C, Dai L, Lou X, Huang Y, et al. Temperature-triggered switchable superwettability on a robust paint for controllable photocatalysis. Cell Reports Physical Science. 2021;**2**:100669

[8] He H, Zhang TC, Ouyang L, Yuan S. Superwetting and photocatalytic Ag2O/TiO2@CuC2O4 nanocomposite-coated mesh membranes for oil/water separation and soluble dye removal. Materials Today Chemistry. 2022;**23**:100717

[9] Zhang T, Lin P, Wei N, Wang D. Enhanced photoelectrochemical water-splitting property on TiO2 nanotubes by surface chemical modification and wettability control. ACS Applied Materials & Interfaces. 2020;**12**:20110-20118

[10] Wu Y, Feng J, Gao H, Feng X, Jiang L. Superwettability-based interfacial chemical reactions. Advanced Materials. 2019;**31**:1800718

[11] Chen L, Sheng X, Wang D, Liu J, Sun R, Jiang L, et al. High-performance Triphase bio-photoelectrochemical assay system based on superhydrophobic substrate-supported TiO2 nanowire arrays. Advanced Functional Materials. 2018;**28**:1801483

[12] Wang R, Hashimoto K, Fujishima A, Chikuni M, Kojima E, Kitamura A, et al. Photogeneration of highly amphiphilic TiO2 surfaces. Advanced Materials. 1998;**10**:135-138

[13] Fujishima A, Zhang X, Tryk DA. TiO2 photocatalysis and related surface phenomena. Surface Science Reports. 2008;**63**:515-582

[14] Wang R, Hashimoto K, Fujishima A, Chikuni M, Kojima E, Kitamura A, et al. Light-induced amphiphilic surfaces. Nature. 1997;**388**:431-432

[15] Zhang X, Fujishima A, Jin M, Emeline AV, Murakami T. Double-layered TiO2–SiO2 nanostructured films with self-cleaning and antireflective properties. The Journal of Physical Chemistry B. 2006;**110**:25142-25148

[16] Li Q, Shang JK. Composite photocatalyst of nitrogen and fluorine Codoped titanium oxide nanotube arrays with dispersed palladium oxide nanoparticles for enhanced visible light photocatalytic performance. Environmental Science & Technology. 2010;**44**:3493-3499

[17] Lv X, Wang T, Jiang W. Preparation of Ag@AgCl/g-C3N4/TiO2 porous ceramic films with enhanced photocatalysis performance and self-cleaning effect. Ceramics International. 2018;**44**:9326-9337

[18] Zhang Y, Zhang S, Wu S. Room-temperature fabrication of TiO2-PHEA nanocomposite coating with high transmittance and durable superhydrophilicity. Chemical Engineering Journal. 2019;**371**:609-617

[19] Zhuang H-F, Lin C-J, Lai Y-K, Sun L, Li J. Some critical structure factors of titanium oxide nanotube Array in its photocatalytic activity. Environmental Science & Technology. 2007;**41**:4735-4740

[20] Chen S, Shen S, Liu G, Qi Y, Zhang F, Li C. Interface engineering of a CoOx/Ta3N5 photocatalyst for unprecedented water oxidation performance under visible-light-irradiation. Angewandte Chemie International Edition. 2015;**54**:3047-3051

[21] Su W, Zheng M, Li L, Wang K, Qiao R, Zhong Y, et al. Directly coat TiO2 on hydrophobic NaYF4:Yb, Tm nanoplates and regulate their photocatalytic activities with the core size. Journal of Materials Chemistry A. 2014;**2**:13486-13491

[22] Kenanakis G, Vernardou D, Katsarakis N. Light-induced self-cleaning properties of ZnO nanowires grown at low temperatures. Applied Catalysis A: General. 2012;**411-412**:7-14

[23] Lamberti A. Microfluidic photocatalytic device exploiting PDMS/TiO2 nanocomposite. Applied Surface Science. 2015;**335**:50-54

[24] Kamegawa T, Shimizu Y, Yamashita H. Superhydrophobic surfaces with photocatalytic self-cleaning properties by nanocomposite coating of TiO2 and polytetrafluoroethylene. Advanced Materials. 2012;**24**:3697-3700

[25] Wooh S, Encinas N, Vollmer D, Butt H-J. Stable hydrophobic metal-oxide photocatalysts via grafting polydimethylsiloxane brush. Advanced Materials. 2017;**29**:1604637

[26] Zhu H, Huang Y, Zhang S, Jin S, Lou X, Xia F. A universal, multifunctional, high-practicability superhydrophobic paint for waterproofing grass houses. NPG Asia Materials. 2021;**13**:47

[27] Zhu H, Wu L, Meng X, Wang Y, Huang Y, Lin M, et al. An anti-UV superhydrophobic material with photocatalysis, self-cleaning, self-healing and oil/water separation functions. Nanoscale. 2020;**12**:11455-11459

[28] Liu J, Ye L, Sun Y, Hu M, Chen F, Wegner S, et al. Elastic superhydrophobic and photocatalytic active films used as blood repellent dressing. Advanced Materials. 2020;**32**:1908008

[29] Zhang X, Liu S, Salim A, Seeger S. Hierarchical structured multifunctional self-cleaning material with durable Superhydrophobicity and photocatalytic functionalities. Small. 2019;**15**:1901822

[30] Sheng X, Liu Z, Zeng R, Chen L, Feng X, Jiang L. Enhanced photocatalytic reaction at air–liquid–solid joint interfaces. Journal of the American Chemical Society. 2017;**139**:12402-12405

[31] Zhou H, Sheng X, Xiao J, Ding Z, Wang D, Zhang X, et al. Increasing the efficiency of photocatalytic reactions via surface microenvironment engineering. Journal of the American Chemical Society. 2020;**142**:2738-2743

[32] Tan Y, Liu M, Wei D, Tang H, Feng X, Shen S. A simple green approach to synthesis of sub-100 nm carbon spheres as template for TiO2 hollow nanospheres with enhanced photocatalytic activities. Science China Materials. 2018;**61**:869-877

[33] Preethi LK, Mathews T. Electrochemical tuning of heterojunctions in TiO2 nanotubes for efficient solar water splitting, catalysis. Science & Technology. 2019;**9**:5425-5432

[34] Zhang M, Jiang S, Han F, Chen H, Wang N, Liu L, et al. Recyclable, superhydrophobic and effective Ag/TiO2@PDMS coated cotton fabric with visible-light photocatalyst for efficient water purification. Cellulose. 2022;**29**:3529-3544

[35] Liu J, Ye L, Wooh S, Kappl M, Steffen W, Butt H-J. Optimizing hydrophobicity and photocatalytic activity of PDMS-coated titanium dioxide. ACS Applied Materials & Interfaces. 2019;**11**:27422-27425

[36] Baig U, Matin A, Gondal MA, Zubair SM. Facile fabrication of superhydrophobic, superoleophilic photocatalytic membrane for efficient oil-water separation and removal of hazardous organic pollutants. Journal of Cleaner Production. 2019;**208**:904-915

Chapter 7

Hybrid Magnetic-Semiconductor Oxides Nanomaterial: Green Synthesis and Environmental Catalytic

Thi Huong Nguyen, Minh Thanh Vu and Ngoc Son Nguyen

Abstract

Semiconductor oxide nanoparticles with various properties are used in applications such as photocatalysis, lithium-ion batteries, antimicrobial materials, magnetic and antibacterial materials, sensors, thermally conductive and anti-ferromagnetic films and photocatalysis. Coprecipitation, sol-gel, electrospray synthesis, laser ablation, hydrothermal and green synthesis methods have been developed for the synthesis of semiconductor oxide nanoparticles. The photocatalytic treatment method is environmentally friendly, thorough, and has high efficiency. Hybrid magnetic-Semiconductor oxide nanoparticles materials are aggregated. This results in the combination of the superparameters of the iron oxide nanosystem and the ability to rapidly adsorb pollutants and separate them from environmental water, solid. This is due to characteristic properties such as low superparameter, low toxicity, low electrical conductivity, and large specific surface area. In addition, biosynthesis has attracted attention because it takes advantage of nonhazardous, environmentally friendly biological systems like bacteria, fungi, leaves, vitamins, and yeast to synthesize metal oxide nanopartilces and combined with sonosumbers to increase the dispersion of the system, reduce the size of the catalytic particles, and reduce the reaction time.

Keywords: hybrid magnetic-semiconductor nanomaterials, green synthesis, environmental catalysis

1. Introduction

Pollution treating technologies have been studied and applied for a long time. Up to now, they still attract a lot of attention from scientists to aim for a new material, technology with high efficiency, thorough and convenient processing. In particular, the appearance of nanomaterials has a positive impact on almost every aspect of engineering, technology and life. Environmental treatment technology is no exception to that influence. Nanotechnology involves the fabrication of nanoscale materials, along with the means and techniques to control them, in order to utilize the superior properties of this material. Nanotechnology was discovered in the late 1960s and has been applied in many different fields from science, engineering, technology, medicine, bringing great efficiency [1]. In the field of environmental remediation, nanotechnology also promises to bring new advances. With nanomaterials having the advantage of large specific surface area, chemical activity of the surface of the material as well as the ability to transform and functionalize the surface, it opens countless research directions to bring a high performance on pollutant treatment and increase the ability to handle some stubborn pollutants. There have been many studies on this research direction, different water pollution treatment technologies have been used. In the past, the removal of pollutants often relied on adsorption technology. Accordingly, nanomaterials with adsorption affinity for pollutants are fabricated, they adsorb pollutants on the surface to help clean water flow. Of course, their limitation is that usually all materials have a certain adsorption capacity. Post-treatment material collection and an additional step of handling this material is required. A recent new direction uses materials with photocatalytic activity to combine the adsorption and treatment of organic pollutants right on the surface of the material. This type of material is usually semiconducting metal oxides, which have a wide band gap suitable for excitations by sunlight. However, nanomaterials also have disadvantages in the treatment of pollutants, especially in aqueous solution. It is because they are nano-sized that recovery will be very difficult, which easily leads to secondary contamination. To overcome this situation, scientists have thought of magnetic nanomaterials. Thanks to its magnetism, this material is easily recovered after processing.

2. Green synthesis of hybrid magnetic-semiconductor oxides nanomaterials

The synthesis method plays a vital role in guiding the application of MNPs. Because this stage determines the basic properties such as particle size, size distribution, morphology, stability, and even surface properties of the MNPs. There are several different methods for synthesizing MNPs, which have been discussed in detail in many papers [2–4]. It can be observed that the common techniques used for synthesizing MNPs include coprecipitation, thermal decomposition, sonochemical, microemulsion, chemical vapor deposition, etc. In addition to these methods, a new and compatible approach to materials synthesis is using agents of natural origin such as plant extracts, bacteria, and fungi instead of chemical agents. The critical role of natural agents is as stabilizers by complexing with metal ions. Intriguingly, the natural agent-metal ion complex system reacts with the hydroxyl group in solution, forming a metal oxide instead of a metal hydroxide. This hypothesis seems plausible since metal oxides are formed at low temperatures, not by high-temperature metal hydroxide decomposition. In addition, natural agents act as capping agents, avoiding the agglomeration of nanoparticles. This approach aims to take advantage of available, renewable, natural agents and limit dependence on chemical agents to move towards a "greener" and more environmentally friendly world. Accordingly, the process of synthesizing MNPs is also done in the simplest way. It makes even more sense in environmental applications to use a "green" approach to clean the environment instead of potentially hazardous methods.

Plant extracts were the most common among the natural agents used to synthesize MNPs.

Mangosteen peel is a natural product widely used in the synthesis of nanoparticles. They are rich in phenolic compounds, flavonoids, and terpenoids. Phenolic compounds are all potent antioxidants, while flavonoids, benzophenones, and anthocyanins are thought to be closely involved in reducing metal cations to nanoparticles [5, 6]. Yusefi et al. [7] synthesized Fe_3O_4 from mangosteen peel extract, with extract concentrations varying from 0.1, 2, 5, and 10 wt.%. The results show that the obtained ferromagnetic iron oxide nanoparticles have an average size of 13.42 nm. The saturation magnetization values varied from 49.80 emu/g to 69.42 emu/g as the extract concentration increased from 0 to 10 wt.%. Nanoparticles have high stability and a long storage time. In another publication by this group of scientists, they demonstrated that the extract of *Garcinia Mangostana* fruit peel could be used as a bio-stabilizer and capping agent to enhance physicochemical properties and stability of Fe_3O_4 nanofluid [8].

The peel of pomegranate (*Punica granatum L.*) is generally of no value in consumption but has been found to be highly medicinal. Several studies indicated that they contain many valuable bioactive compounds, such as polyphenols, flavonoids, proanthocyanidins, and hydrolyzable tannins [9–11]. In addition to the anti-cancer, anti-fungal, and inflammatory activities, the antioxidant activity is quite intense, which is attributed to polyphenols compounds such as punicalagin and ellagic acid. Thanks to that, it attracted many scientists' attention to synthesizing nanomaterials [12–14]. Yusefi et al. [15] synthesized magnetic iron oxide nano using Punica Granatum Fruit Peel Extract. The biologically active substances in the extract act as stabilizers, forming complexes with metal ions (Fe^{2+}, Fe^{3+}). This was further confirmed when the results confirmed that a significant number of organic compounds were present in the ferromagnetic oxide mixture. The organic matter content increased gradually with the concentration of extract used. In contrast, magnetism (magnetic saturation) shows in the opposite direction. Another work has also been carried out with similar results [16].

Lathyrus sativus is a well-known plant that contains alkaloids, carotenoids, flavonoids, starch, carbohydrates, essential oil, leguminvicilin, legumelin, vitamin C, oleoresin, gum resin, tannins, terpenes, phenols, riboflavin, beta-carotene, proteins, and amino acids, which act as capping, reducing and stabilizing agents [17, 18]. Thanks to this unique feature, many research groups have used L. sativus shell extract to synthesize magnetite nanoparticles [19, 20].

Potatoes are mainly composed of carbohydrates, mostly starch. These macromolecules with hydroxyl groups are expected to facilitate metal complexation. Sharma et al. [21] used potato extract to synthesize ferromagnetic nanoparticles with the assistance of ultrasonication which facilitated quick and easy complexation.

3. Application of hybrid magnetic-semiconductor

3.1 Removing heavy metals

Heavy metals are naturally formed in the earth's crust, they are named for their high density. Some heavy metals are common ingredients on earth such as tin, copper, gold, silver.... They are widely used in manufacturing and agriculture. Some of them are essential components of the human body, but in large doses they can be toxic, especially to children and unborn babies.

Some heavy metals are on the list of chemicals harmful to public health published by WHO including lead, cadmium, mercury, arsenic, manganese, chromium. There are different types of magnetic nanoparticles (MNPs) based on the magnetic metal element. In heavy metal remediation field, ferromagnetic oxide attracts more attention thanks to its universality, high magnetism, low toxicity, easy synthesis, and modification.

3.1.1 Arsenic

Arsenic (As) is a relatively common element. They are notable for their toxicity and carcinogenic potential. Long-term use of As-contaminated drinking water will lead to cancers of the liver, lung, kidney, bladder and a number of other non-cancerous diseases related to heart, brain, diabetes [22].

In order to limit the harmful effects of As on public health, the World Health Organization (WHO) has recommended that the concentration of As in drinking water be no more than 10 μg L^{-1}. In natural water sources, Arsenic exists in inorganic form with two main oxidation forms, arsenate AsO_4^{3-} (As(V)) and arsenite AsO_3^{3-} (As(III)). In particular, As(V) is more commonly found in surface water rich in dissolved oxygen, while As(III) is more present in groundwater. There are differences between these two existences of As. As(III) is more toxic, soluble, and mobile than As(V). However, there is a conversion process from As(III) to As(V), especially in the condition of water rich in dissolved oxygen. This process is also thermodynamically favorable, but this conversion time can also take days, weeks, or months, depending on the specific conditions [23, 24]. Various techniques can be applied to remove Arsenic such as precipitation, co-precipitation, ion exchange, adsorption, ultrafiltration, or reverse osmosis. Among these, adsorption is one of the most promising technologies because of its simple operation, low cost, and ease of research and improvement with new adsorbent materials. However, this technology is almost only effective with As(V). It is very inefficient to remove As(III), so a pretreatment process is often required to convert from As(III) to As(V) before adsorption to remove Arsenic. This process is possible using oxidizing agents or oxidizing systems. Among them, manganese dioxide emerges as a potential candidate for arsenic treatment [25, 26]. This is explained by the relatively low oxidation potential of MnO_2, which is consistent with the oxidation state of As(III) [26]. Taking advantage of this, studies on making nanomaterials from the binary metal oxide in which MnO_2 and magnetic materials are combined in arsenic treatment have shown effective results. In addition to the magnetism that facilitates material recovery, iron oxides have also shown high adsorption features for As(V) [27, 28]. Zhang et al. [29] synthesized Fe-Mn binary oxide (MFM) adsorbent for arsenic treatment. The adsorbent obtained has an average particle size of 26 μm, a specific surface area of 265 $m^2\ g^{-1}$, and the maximum adsorption capacity for As(V) and As(III) is 0.93 mmol g^{-1} and 1.77 mmol g^{-1}. Also, Kong et al. synthesized adsorbent materials on Fe-Mn binary oxides-loaded zeolite carriers [30]. The MFM-loaded zeolite material has good magnetism before and after arsenic adsorption. They are easily recovered by an external magnetic field. Some other characteristics such as specific surface area 340 $m^2\ g^{-1}$, higher than most other adsorbents used in arsenic removal, particle size distribution in the range of 20–100 nm, ratio The Mn/Fe atom is 2:9. The zeolite substrate content was varied from 10%, 20%, and 30%. The magnetic properties of the material depend on the zeolite content. Specifically, the magnetic saturation is 50.104, 31.779 and 16.165 emu g^{-1}, respectively. Their coercivity forces (Hc) are 21,307, 24,823 and 28,338 Oe, and the magnetic remanences (MR) are 2.3628, 1.8266 and 1.2903 emu g^{-1}, respectively. These show that the magnetic hysteresis of the material is negligible, i.e., the magnetic field is almost zero after removing the external magnetic field. Regarding the ability to adsorb arsenic, the test shows that almost As(III) is adsorbed on the MFM surface, gradually oxidizing to As(V). After 30 minutes of treatment, As(III) and As(V) concentrations decreased from 2 mg L^{-1} to 3.8 µg L^{-1} and 6.3 µg L^{-1}, respectively. The authors have proven that the oxidation of As (III) into As (V) is due to MnO_2, instead of Fe (II). These results show the potential application of MFM materials in arsenic treatment in water. In another work, Kumar et al. [31] synthesized magnetic nanohybrids from monolayer graphene oxide (GO) and manganese ferrite MNPs (GO-$MnFe_2O_4$). The obtained adsorbent is highly magnetic, and the adsorption capacity of As(III), As(V) as well as Pb(II) are very high and significantly increased thanks to denaturation by GO.

Recently, the trend of using synthetic methods or green materials in synthesizing materials is increasingly attractive. New adsorbent materials are not out of that trend, intending to prepare inexpensive and more environmentally materials successfully. Some raw materials such as natural cellulose, biochar, and plant extracts are studied and modified with MNPs to make heavy metal adsorbents. Cellulose is a renewable biopolymer with a wide range of applications. The presence of hydroxyl groups in the main chain makes them easily modified with other materials [32]. Hokkanen et al. [33] synthesized MNPs adsorbents from modified iron oxide nanoparticles with microfibrillated cellulose. Some characteristics of synthesized adsorbent materials such as improved adsorption capacity with As(V), best adsorption conditions in a low pH environment, and experimental data show that the adsorption process follows tissue pattern. Langmuir model, the kinetics is consistent with the pseudo quadratic model, regenerate the adsorbent with NaOH solution, after three cycles of use, the adsorption efficiency still reaches over 98%. With the same adsorbent system as cellulose iron oxide nanocomposite, Yu et al. [34] proposed a one-step synthesis method, using NaOH-thiourea-urea solution to dissolve cellulose. This method provides a "green" manufacturing process. The obtained adsorbent had good magnetic sensitivity. Its adsorption capacity for arsenite and arsenate are 23.16 and 32.11 mg g^{-1}, respectively. Lunge et al. [35] synthesized magnetic iron oxide nanoparticles from tea waste (MION-Tea) for arsenic removal by a straightforward method. The synthesized MNPs have a very small size, only about 5–25 nm, with a magnetization saturation value from 6.9 emu g^{-1}. The FTIR spectroscopy results indicate that traces of organic fractions of tea waste are still present on the iron oxide surface. The arsenic adsorption test gave an impressive adsorption capacity with 188.69 mg g^{-1} for As(III) and 153.8 mg g^{-1} for As(V). With a simple synthesis method, using an inexpensive tea waste agent and especially with a very high adsorption capacity of As, MION-Tea shows excellent application potential in removing As from water sources. In another interesting study by Zeng et al. [36], the iron source was obtained from iron-rich sludge water treatment. They were treated and synthesized into MNPs and re-applied to remove arsenic domestic. At pH 6.6, more than 90% of As(V) solution with a concentration of 400 g L^{-1} could be easily removed by the synthesized adsorbent (0.2 gL^{-1}) in 60 min. Although the maximum adsorption capacity is not ideal with about 12–13 mg g^{-1}, this is still considered a promising direction to take advantage of the wastewater filter residue to treat As in water compared to chemical agents. A similar approach is to attach ferromagnetic nanoparticles to plant-based adsorbents to add magnetism to facilitate adsorbent separation. M. Zang and his research team [37] synthesized porous biochar from woodcotton and loaded ferromagnetic nanoparticles. The material was introduced magnetic with saturation magnetization of 69.2 emu g^{-1}, which was used as an arsenic adsorbent with a reasonably good adsorption capacity, reaching 3.147 mg kg^{-1} for As(V). Similar work was done by Nham et al.

who modified biochar which was synthesized from slow pyrolysis of rice straw with $FeCl_3$ to form a biochar material system carrying ferromagnetic nanoparticles. The results show that the magnetic addition is favorable for the separation process, and the modified biochar material also has a more significant As(V) adsorption capacity [38]. Other studies were also carried out with similar purposes but using different raw materials such as pinewood and natural hematite [39], eucalyptus extract [40], red mud [41], and agricultural biomass [42]. This result promises to provide an inexpensive, effective, and environmental solution for making arsenic adsorbents in water purification.

The arsenic adsorption mechanism on MNPs has been studied by the research group of Liu et al. [43]. They used spectroscopic techniques, including X-ray absorption near edge structure (XANES), EXAFS, and X-ray photoelectron spectroscopy (XPS), along with batch sorption experiments and thermodynamic calculations. The results show that the adsorption of As(V) and As(III) takes place very quickly at the beginning, then reaches equilibrium after about 2 hours, which is consistent with the pseudo-second-order kinetic model. The experimental data also show that As adsorption on the MNP surface is monolayer and endothermic. The results of this study are consistent with the kinetics of As adsorption on the surface of MNPs published in the above studies. The study also demonstrated that no oxidation-reduction reaction occurs on the surface of MNPs when As is adsorbed on it. Instead, oxidation-reduction reactions can slowly occur when As is exposed to the atmosphere. The size of the nanocrystalline magnetic also dramatically affects the adsorption and desorption characteristics of As(III) and As(V). Mayo et al. [44], particle size has a profound influence on the arsenic removal process. When the particle size decreased from 300 nm to 12 nm, As(III) and As(V) adsorption capacity increased nearly 200 times. It is worth mentioning that this increase is higher than the corresponding increase in specific surfaces with such a change in grain size. This result is similar to that observed in the study of Tuutijarvi et al. [45]. In their conclusion, Tuutijarvi attributed this to the fact that for particles with a size of 12 nm, their dispersion in solution is better, while with larger particles (20 nm and 300 nm, respectively) nm) they are more easily aggregated. This explanation seems unsatisfactory. Meanwhile, J. T. Mayo's view is that the adsorption of arsenic on the surface of ferromagnetic nanoparticles is not simply adsorption on the surface of the particles but also through other means. The desorption results also support this point. Accordingly, the delay of the desorption path is more significant in the case of smaller particle sizes. The authors attributed this phenomenon to the greater affinity of arsenic for Fe_3O_4 nanoparticles. This is also in agreement with previously published research results [46].

3.1.2 Chromium

Chromium is a trace element that is very important for human health. However, in large doses can lead to serious health problems. Prolonged exposure to this metal can lead to higher accumulation levels in human and animal tissues, causing toxicity and impacting human metabolism, reducing crop yields [47]. Ingesting Cr-contaminated foods can lead to liver damage, lung congestion, and skin irritation. Chromium and its compounds are widely used in many industrial applications such as plating, tanning, metal finishing, photography [48]. Wastewater from these industries can contain Cr with concentrations ranging from tens to hundreds of mg of L^{-1}. Chromium exists in water mainly in two trivalent forms Cr(III) and hexavalent Cr(VI). Cr(III) is less mobile, non-toxic, and even a trace element for humans and animals to exist in this form. In contrast, Cr(VI) is present in anionic forms as chromates (CrO_4^{2-}), dichromates ($Cr_2O_7^{2-}$), and bichromates ($HCrO_4^-$). They are more mobile, highly soluble, and toxic to living organisms. In recent years, Cr pollution, especially Cr(VI) form, in both water and soil environments has increased due to human production activities. Therefore, the use of chromium in industrial production is being restricted gradually in some areas of the world. The treatment of Cr(VI) pollution in soil or water is an urgent issue, attracting much attention from scientists. Many different methods can be used to treat Cr-contaminated water. Among them, the common method is chemical redox combined with immobility. The disadvantages of this method are the prohibitive cost of the system, the consumption of chemicals, the generation of much sludge, the potential risk of re-pollution due to the leakage of sludge into the soil, and especially the recovery of metals to reuse after treatment is hardly feasible [49]. From many research results, the adsorption method has great potential for application in chromium removal, which overcomes the disadvantages of the traditional precipitation method. Shen et al. [50] synthesized Fe_3O_4 of different sizes using the co-precipitation technique (8 nm) and the polgol method using propylene glycol (35 nm). The Cr(VI) adsorption efficiency was tested, and the results showed that the adsorption capacity of Cr(VI) reached 35.46 and 7.45 mg g^{-1}, respectively. The effect of pH on Cr(VI) removal efficiency was investigated. As a result, a low pH value (2–3) is optimal for adsorption. This is explained by the fact that under acidic conditions, H^+ ions adsorb to the adsorbent surface, making the surface positively charged, from which Cr(VI) exists in the form of oxyanions, which are readily adsorbed by interactions electrostatic action. This explanation agrees with the Cr(VI) adsorption mechanism previously proposed by Chen et al. [51]. A similar study has been done by Rajput et al. [52]. The ferromagnetic oxide nanoparticles are synthesized by the co-precipitation method. The obtained MNPs are spherical in shape, 15–30 nm in size, and the specific surface area is about 12.7 m^2 g^{-1}, the point of zero charges (pH_{PZC})) 7.4. Adsorbent materials were regenerated in an alkaline environment. Unfortunately, their adsorption capacity decreases quite quickly after each regeneration. The results of these studies show that Fe_3O_4 can be used to adsorb Cr(VI), but the efficiency is not high, and the reusability is relatively poor. This can be explained because this adsorption process is physisorption, lacking the specific affinity of Cr(VI) for Fe_3O_4. In addition, the medium favors the adsorption of Cr(VI) under low pH, which easily leads to the dissolution of Fe_3O_4. Therefore, the study and enhancement of the adsorption capacity of Fe_3O_4 for Cr(VI) by denaturing, increasing the specific surface area, or introducing organic species with a particular affinity for this metal are being studied.

The research group of Shi et al. [53] attached $Fe_3O_4@SiO_2$-NH_2 to carboxylated biochar to form magnetic biochar to remove Cr(VI) and Cr(III) metals in the solution acid. The results show that $Fe_3O_4@SiO_2$-NH_2 not only enhances the adsorption capacity of Cr(VI) anions but also immobilizes Cr(III) cations. The proposed process mechanism consists of 3 steps: (1) adsorption of Cr(VI) anions on the surface by protonated functional groups; (2) reduction of the Cr(VI) anion to the Cr(III) cation by electron donor groups; (3) complexation and immobilization of Cr(III) by amine and carboxyl groups on magnetic biochar.

Polypyrrole (PPy) is an organic polymer synthesized by oxidative polymerization of pyrrole. PPy possesses many interesting properties such as high electrical conductivity, environmental stability, non-toxicity, and ease of preparation. These advantages make PPy very popular in many different applications [54, 55]. PPy is noted to be positively charged at the N atoms on the main chain, which is highly preferred in synthesizing adsorbents [56]. The PPy/Fe_3O_4 nanocomposite adsorbent was first synthesized by Bhaumik et al. to remove Cr(VI) [57]. The ferromagnetic nanoparticles were encapsulated by PPy through the in-situ polymerization of the pyrrole monomer. The removal efficiency of Cr(VI) was very high, reaching 100% with a 200 mg/L Cr(VI) solution at pH 2. The team also proposed that the main mechanism in Cr(VI) adsorption is the reduction and ion exchange on the PPy/ Fe_3O_4 nanocomposite surface. The adsorbent also showed the ability to reuse after two cycles of adsorption-desorption with almost no reduction in the adsorption capacity. The research group of Wang et al. [58] have synthesized a tertiary magnetic nanocomposite consisting of reduced graphene oxide (rGO), polypyrrole (PPy), and Fe_3O_4 nanoparticles (PPy-Fe_3O_4/rGO) for Cr(VI) removal application. The material's magnetism has been evaluated, showing a significant reduction in saturation magnetization value compared with Fe_3O_4/rGO. The Cr(VI) removal efficiency increased significantly after the Fe_3O_4/rGO nanocomposite was modified with PPy by an in situ polymerization. The results of surveying the influence of foreign ions showed that cations such as Na^+, K^+, Ca^{2+}, and anions such as Cl^-, and NO^{3-} hardly affect the removal efficiency of Cr(VI). However, the presence of SO_4^{2-} anion inhibited Cr(VI) adsorption.

3.1.3 Simultaneous removal of multiple other heavy metals

In addition to chromium and arsenic, magnetic nanomaterials have been studied for the simultaneous removal of many other heavy metals. This aspect of research is interesting and has significant practical implications. Because, in practice, wastewater cannot exist as a single metal cation, they are always a complex mixture. The studies also tested the effects of impurities, including other metal ions commonly found in water sources such as Na^+, Ca^{2+}, Mg^{2+} cations, or anions such as Cl^-, SO_4^{2-}, CO_3^{2-}. Several studies have also conducted surveys to evaluate the effect of organic species on the removal efficiency of heavy metals.

Liu et al. [59] synthesized humic acid (HA) coated Fe_3O_4 nanoparticles by co-precipitation method to remove some heavy metal ions such as Hg(II), Cd(II), Cu (II). The results show that the Fe_3O_4/HA material system is nano-sized with a Fe_3O_4 core of approximately 10 nm. In solution, they form particles with hydrodynamic sizes up to 140 nm. Their magnetic saturation is relatively high with 79.6 emu g^{-1}, which makes them easily recovered by an external magnetic field in a short time. The tests also show that Fe_3O_4/HA has high stability in tap water, natural water, and acidic environments from 0.1 M HCl to alkaline 2 M NaOH with low leaching (Fe $\leq$ 3.7%; HA $\leq$5.3%). The removal efficiency of heavy metals of Fe_3O_4/HA material system compared to Fe_3O_4 in other publications has been significantly enhanced. Specifically, the removal efficiency is up to over 99% for Hg(II) and Pb(II) and over 95% for Cu(II) and Cd(II) in tap water at optimum pH. The desorption of these metals to an aqueous medium is not significant. This shows the ideal potential of the Fe_3O_4/HA material system in treating polluted heavy metals in water sources. In another study, Ge et al. [60] synthesized Fe_3O_4 nanoparticles by co-precipitation method that followed functionalizing surface with a 3-aminopropyltriethoxysilane (APS) agent and then attaching a copolymer of acrylic acid and crotonic acid (AA-co-CA)tail. The material has a uniform size of 15–20 nm, saturation from 52 emu g^{-1-}, and slightly lower with Fe_3O_4 of 79.67 emu g^{-1}. Fe_3O_4@APS@AA-co-CA is used to remove heavy metal ions (Cd^{2+}, Zn^{2+}, Pb^{2+}, and Cu^{2+}) from an aqueous solution. Experimental results show that the synthesized adsorbent is highly effective with Pb^{2+} and Cu^{2+} ions. Their adsorption capacity reaches 166.1 mg g^{-1} and 126.9 mg g^{-1}, respectively. For Cd^{2+} and

Zn^{2+}, the adsorption capacity was lower, 29.6 mg g^{-1} and 43.4 mg g^{-1}, respectively. The limitation of this study is that the method and conditions for the adsorbent regeneration have not been specified. However, the authors have stated that a pH lower than two causes this material to be inactive.

The iron oxide nanomaterial facilitates quick and easy recovery of the adsorbent. However, they have the disadvantage that the chemical activity is quite sensitive. Iron oxides are easily dissolved in an acidic environment, leading to ineffective pollutant removal. Therefore, when using ferromagnetic as a pollutant treatment agent, it is necessary to take accompanying measures to enhance the durability and stability of this material. Especially in the heavy metal elution situation, regeneration of the adsorbent is usually carried out in an acidic medium. A core-shell structure magnetic nanomaterial Fe_3O_4@SiO_2-NH_2 was synthesized by Wang et al. [61]. The layer of SiO_2 with acid resistance is used as a protective shell for Fe_3O_4 ferromagnetic cores. In addition, SiO_2 rich in hydroxyl groups on the surface also helps to facilitate the functionalization of organic agents. The obtained Fe_3O_4@SiO_2-NH_2 had a specific surface area 216.2 m^2 g^{-1}, average size 18.4 nm. This material has high stability in an acidic environment. The solubility of Fe in 1 M HCl solution after 24 h is only 1.57%, much lower than 90.7% of bare Fe_3O_4. The adsorption capacity, affinity for heavy metal cations including Cu(II), Pb(II), and Cd(II), as well as the influence of pH, and foreign electrolytes, were evaluated. The results show that this material has good strength, high selectivity, and the ability to regenerate by acid agent (1 mol/L HCl acid solution) are all very effective. In particular, experiments show that after adsorption of energetic metals, Fe_3O_4@SiO_2-NH_2 is easily recovered quickly by an external magnetic field.

Some organic species that have the advantage of complexing with heavy metal ions are also used to modify the surface of nanoparticles with the expectation of high efficiency in heavy metal removal. Among them attracting much attention from researchers is ethylene diamine tetraacetic acid (EDTA). Liu et al. [62] synthesized core-shell magnetic nanomaterials based on Fe_3O_4, and the surface of the SiO_2 shell was directly modified with EDTA with the ratio of Fe_3O_4:SiO_2:EDTA components instead. in which the ratio of these components 2:5:1 gives the best adsorption capacity. Some characteristics of this material include a specific surface area of 24.07 m^2 g^{-1}, an average pore size of 15.40 nm, a total pore volume of 0.09 cm^3 g^{-1}, and saturation magnetization from 34.49 emu g^{-1}. Evaluating the effects of other metal cations such as K^+, Na^+, Mg^{2+}, and natural organic matter (NOM) such as humic acid and sodium alginate had proved that the Fe_3O_4@SiO_2-EDTA material had high selectivity for heavy metals such as Pb(II) and Cu(II). In another study, Ren et al. used EDTA as an affinity enhancer for heavy metal ions [63]. Accordingly, the magnetic nanostructure with core-shell structure Fe_3O_4-SiO_2 was surface modified with chitosan before being added with EDTA tail. The descriptive results show that the magnetic adsorbent has a size from 200 to 400 nm, and specific surface area, pore diameter, and pore volume are 1.04 m^2 g^{-1}, 8.28 nm, and 2.2×10^{-3} cm^3 g^{-1}, respectively. The acid stability of the synthesized adsorbent was significantly enhanced after surface modification with chitosan and EDTA. Specifically, after soaking for 12 hours in 1 M HCl solution, the solubility of Fe_3O_4 of SiO_2/Fe_3O_4, functionalized chitosan SiO_2/Fe_3O_4 (CMS), and EDTA-modified CMS (EDCMS) were 1.28%, 1.02%, and 0.77%, respectively. The saturation magnetization of the materials after the modification steps of Fe_3O_4 microspheres, SiO_2/Fe_3O_4 microspheres, CMS and EDCMS are 69.0, 56.3, 20.7, and 18.2 emu g^{-1}, respectively. There is a decrease in magnetic saturation after each step of material modification. This phenomenon is also explained similarly to the previous studies, as the mass content of magnetic Fe_3O_4 decreases with each addition of other agents. Although the magnetic saturation is reduced, they are still large enough to facilitate the separation of the nano adsorbents from the aqueous medium quickly and easily by the external magnetic field. The results of the heavy metal adsorption test showed that the nano adsorbents particles whose surfaces were modified by EDTA gave the adsorption capacity for Cu(II), Pb(II), and Cd(II) 0.699, 0.596, 0.563 mmol g^{-1}, respectively. While for undenatured chitosan SiO_2-Fe_3O_4 (CMS) nanoparticles, these values are only 0.495, 0.045, 0.040 mmol g^{-1}. This enhancement is attributed to the presence of EDTA, which provides ease of complexation of metal ions. Zhang et al. [64] synthesized magnetic nanomaterials for mercury adsorption in the aqueous medium. Accordingly, the Fe_3O_4 magnetic cores are surrounded by SiO_2 shells and attached to thiol (−SH) bridges on their surface. This design is built based on Pearson's acid-base theory [65], in which mercury belongs to the group of soft acids, meaning that they quickly form strong bonds with soft Lewis base groups such as −CN, −RS, −SH. The results show that the Fe_3O_4@SiO_2-SH material is nanoscale, with an average diameter of about 10 nm despite having a large size dispersion. Magnetic redundancy and reluctance are almost zero. Magnetic saturation is relatively high, 55.05, 25.45, and 20.47 emu g^{-1} correspond to Fe_3O_4, Fe_3O_4@SiO_2, and Fe_3O_4@SiO_2-SH. The study on mercury adsorption capacity showed that this material has a large adsorption capacity with mercury, even under changing pH of the solution. Especially under low pH conditions, with competitive adsorption of H+ and other cations such as K^+, Na^+, and Ca^{2+}, the mercury adsorption capacity remained at a high level, 110 mg g^{-1}.

3.2 Degradation of organic pollutants

Organic pollutants are toxic organic compounds that cause human health problems and diseases when residues exceed permissible limits. Industrial products such as detergents, organic solvents, dyes, pesticides, or some organic pollutants, which are biochemical products of bacteria, fungi, and mainly algae secreted, etc., are toxic and carcinogenic. They can exist in different forms in environments where typical physicochemical or biological techniques have become ineffective in removing them [66, 67].

Magnetic adsorbents are usually designed in a core-shell structure. Magnetite nanoparticles will have a core of magnetic material and a shell that adsorbs and treats pollutants outside. Zhang and Kong [68] have synthesized Fe_3O_4/C magnetic adsorbent in which the carbon (C) adsorbent layer is covered on the surface of the magnetic nanoparticle. The synthesized nanoparticles, which have an average diameter of about 250 nm, are good dispersion in aqueous media and are quickly and easily separated by external magnetic fields. The adsorption efficiency was tested with MB and CR pollutants. The adsorption capacity of Fe_3O_4/C for these pollutants was 44.38 mg g^{-1} and 11.22 mg g^{-1}, respectively.

Microcystins (MCs) are a class of toxins produced by certain species of freshwater cyanobacteria, commonly known as blue-green algae. There are more than 50 different compounds belonging to this group. They are disocyclic heptapeptides with a molecular weight of about 1000 Da, such as MC-RR, MC-YR, and MC-LR. Among them, MC-LR is the most popular. Microcystin belongs to a group of toxins that are very dangerous to humans, livestock, and pets if ingested [69]. In order to remove them from water, methods such as coagulation and mechanical filtration can be used, but only the insoluble particles are removed. For dissolved poisons, activated carbon can be used, but effective removal of MC requires a large amount of activated carbon adsorbent. With chemicals for chlorination or ozonation, there are similar limitations. In addition, using these methods carries the risk of creating secondary toxic products [70]. Deng et al. [71] synthesized mesoporous microspheres, sandwich structures of approximately 500 nm in size, to treat this type of organic poison. In this study, the core-shell structure Fe_3O_4@SiO_2 was fabricated by the sol-gel method. The porous structure of the shell is formed by a composite layer of cetyltrimethylammonium bromide (CTAB) and silica after the removal of CTAB by acetone extraction. As a result, this porous shell has a uniform diameter, about 70 nm thick. The authors also proved that these porous capillaries have a direction perpendicular to the material's surface. The pore size is about 2.3 nm, the specific surface area is 365 m^2 g^{-1}, and the total pore volume is 0.29 cm^3. g^{-1}, magnetic saturation is 53.3 emu g^{-1}. MC treatment results show that with a relatively low dose, about 0.05 mg μg^{-1}, the adsorbent can remove MCs in solution, and the removal efficiency is more than 95%. After extraction with acetonitrile/water mixture regenerates, the adsorbent can be reused with MC removal efficiency above 90% after eight cycles of use. This result shows that the magnetic material synthesized by the research team has very high efficiency in removing toxic MCs, convenient recovery and regeneration, can be reused many times, and is effective in economic and technical aspects.

Recently, photocatalysis technology has received significant attention for treating polluted organic compounds. Photocatalysis uses the excitation energy from sunlight to decomposing organic compounds into harmless products of CO_2 and water or other particles along the water flow. Their advantages are that they do not use additional treatment chemicals, do not generate secondary pollution products, and the toxic organic compounds are always decomposed on the catalyst surface without having to be recovered for treatment at another step [72, 73]. However, like in heavy metal treatment, the catalyst material must also be recovered to avoid environmental emissions. Therefore, synthesizing materials with both photocatalytic activity and magnetism to facilitate recovery and treatment has attracted significant attention from researchers.

In a publication by Chi et al. [74], the research team synthesized magnetic nanomaterials dopped by active silver metal centers, which are catalysts for 4-nitrophenol (4-AP) treatment. With the help of polyvinylpyrrolidone (PVP) as a reducing agent and stabilizing agent, the synthesized Fe_3O_4@SiO_2-Ag composite material has a spherical shape, a core size of about 200 nm, and a SiO_2 shell. The average thickness is about 35 nm, and the size of Ag NPs is controlled at about 3.65 nm and is evenly distributed on the Fe_3O_4@SiO_2 background. The magnetism of the materials is not significantly reduced compared to the magnetic core, whereby the magnetic saturation of Fe_3O_4, Fe_3O_4@SiO_2, and Fe_3O_4@SiO_2-Ag is 78.5, 66.4, and 63.8 emu g^{-1}, respectively. The treatment efficiency of 4-AP is excellent. The rate constant of this compound degradation is higher than that of some Ag catalysts carried on other substrates published previously. The catalytic activity of the material was also maintained stably after eight treatment cycles. The 4-AP conversion still reached over 99%. After that, a decrease in activity was initiated, but this was negligible.

Titanium oxide is considered one of the materials that was attracting much attention in its use as photocatalysts in treating organic pollutants. It has high chemical and biological inertness and solid oxidizing force. The main disadvantage of these metal oxides is that they have a reasonably wide band gap, about 3.2 eV, which means that their excitation energy must be ultraviolet radiation, not radiation in the visible region. In addition, poor recovery efficiency is also one of the limitations of this material. Many studies used magnetic metals/metal oxides integrated with TiO_2 to facilitate the recovery of this material after treatment. Mortazavi-Derazkola et al. [75] synthesized a core-shell magnetic photocatalyst Fe_3O_4@SiO_2@TiO_2@Ho to treat rhodamine B and methyl orange dyes under UV radiation conditions. This catalyst material is synthesized from the magnetic core Fe_3O_4 covered by the SiO_2 shell. This SiO_2 shell, which protects the Fe_3O_4 core from acid attacks, is also used to add silane coupling agents to facilitate the deposition of TiO_2 on the surface. Holmium (Ho) is doped onto the surface of the adsorbent, which acts as electron traps to separate the electron-hole by creating local electric fields. The descriptive results show that the particle size of Fe_3O_4@SiO_2@TiO_2@Ho is about 52 nm, the outermost Ho shell is about 3 nm thick, and the TiO_2 NPs layer is 3.5 nm thick. Their saturation magnetism decreased quite deeply after coating many outer shells of the Fe_3O_4 magnetic core. Specifically, the saturation magnetization decreased from 57.42 to 24.5 emu g^{-1} when coated with SiO_2. This value decreased to 15.9 and 6.2 emu g^{-1} when TiO_2 and Ho were coated, respectively. Although the saturation magnetization is relatively low, they are still sufficient to separate the adsorbent from the aqueous solution by an external magnetic field. Testing the ability to handle pollutant compounds rhodamine B and methyl orange showed that under dark conditions, without UV irradiation, this material only adsorbs up to 4% of organic pollutants in the water solution. Meanwhile, with an irradiation time of more than 2 h, the decomposition efficiency of Rhodamine B and Methyl Orange was 92.1% and 78.4%. In particular, this type of catalyst is readily regenerated by washing with clean water and ethanol and used again after being recovered by an external magnetic field. After each cycle of use, the catalytic activity and the degree of conversion decreased very little. This stable shows a great potential application of this material in treating organic dyes that pollute water sources. Zinc oxide (ZnO) is a semiconductor with unique properties such as a stable hexagonal wurtzite structure and a wide band gap of 3.37 eV, which is the large binding energy of 60 meV at room temperature. Hence, ZnO possesses some unique abilities, such as bactericidal properties and photocatalytic activity [76, 77]. Studies on the ability in water treatment, photocatalytic activity, the influence of factors such as dopped element, synthesis method, particle size, etc. were published [78–81]. However, in order to lead to a possible practical application, magnetic nanomaterials based on ZnO have been thought of as a means of increasing the ability to recover and reuse this potential material. The Fe-Zn binary oxide material was synthesized by Kumar et al. [82]. The optical, magnetic, and photocatalytic properties and the influence of Fe-doped content in methylene blue (MB) dye treatment were evaluated. The survey results show that Fe-Zn binary oxide has enhanced photocatalytic performance compared to simple ZnO under UV radiation and sunlight. The magnetism of this material was found to be dependent on the Fe dose. Similar material was studied by Falak et al. but with a specific and more Fe content [83]. A magnetic ZnO-$ZnFe_2O_4$ binary composite was created by the research team. The catalytic activity of this composite was also studied on MB. The results show more than 40% of their catalytic activity compared to ZnO nanoparticles. In addition, the magnetization saturation value of ZnO-ZnF_2O_4 was about 5.8 emu/g, which is high enough that they can be collected by applying an external magnetic field. In another study, Boutra et al. [84] synthesized a nanocomposite photocatalyst from ZnO, manganese ferrite ($MnFe_2O_4$), and tannic acid (TA) by hydrothermal method. The photocatalytic activity was evaluated through its ability to decompose Cong Red (CR) under visible light irradiation. The results showed that CR decomposition efficiency reached 84.2%, higher than simple ZnO. The catalyst is easily separated and reused without even washing, drying, or any other technique to remove CR. The performance after the fifth reuse remains high, up to 77.5%. Dlugosz et al. [85] synthesized Fe_3O_4/ZnO magnetite nanoparticles and tested their photocatalytic activity on series of organic dyes, including MB, MO, Quinoline Yellow, Eriochromic Black T(EBT), and Trypani Blue (TB). The synthetic magnetic nanomaterial is 30% Fe_3O_4 by mass, with saturation magnetization of about 9.5 emu/g. The recovery of Fe_3O_4/ZnO reached 83.91%, slightly lower than Fe_3O_4 (94.80%). Notably, the catalytic activity of this material was found to increase with the molecular weight of the dye. Specifically, the photodegradation efficiency of Fe_3O_4/ZnO reached 76.90% for TB (872.9 g/mol), 63.02% for EBT (461.4 g/mol) and 13.23% for MB (319.9 g/mol). We have studies the synthesis of (polyethylene glycol)–Fe_3O_4/ZnO material [86]. The main objective of this study is the preparation of a PEG (polyethylene glycol)–Fe_3O_4/ZnO magnetic nanocomposite using a green sonochemical synthesis method with rambutan peel extract as a stabilizing agent for photocatalytic methylene blue degradation. The result showed the size of nanocomposite was 20–30 nm and had the band gap ennergy of 2.58 eV. Measurements of the degradation efficiency of the photocatalyst showed that the photocatalytic degradation of methylene blue follows pseudo-first order kinetics with good correlation and linear regression coefficient. This study found that the maximum degradation of the methylene blue dye was approximately 96%, with pH = 4.0, a PEG–Fe_3O_4/ZnO concentration of 1.0 g L^{-1}, a methylene blue concentration of 200 mg L^{-1}, and a time of 90 min. In the dark, the Langmuir adsorption constant and the maximum adsorbable methylene blue quantity were calculated as K_L = 0.0451 L mg^{-1} (and $K_L = 11.275K_{LH}$) and Q_{max} = 21.05 mg g^{-1}. This study concludes that for the Fe_3O_4/ZnO magnetic nanocomposite, the adsorption process supports catalytic methylene blue degradation, reducing the decomposition time, increasing the efficiency of the catalytic process, and increasing the sample recovery due to the magnetic properties of the material.

From these research results, integrating ferromagnetic oxide nanoparticles into ZnO will yield a new magnetic material. The degree of magnetism will usually depend on the Fe_3O_4 content in the material. Therefore, in practice, if the material must achieve certain levels of magnetism in order to facilitate a particular separation requirement, consideration should be given to adjusting the mass ratio of this component in the material. In addition, in the presence of Fe_3O_4, the photocatalytic activity of the material increases quite significantly.

4. Conclusion and outlook

It can be seen from the summarized statements that using “green” agents to fabricate materials is a possible route. However, many studies have shown that the efficiency in both the reaction yield and the properties of the obtained MNPs is very impressive, the plant source is immense, and scientists still need to do more research to improve the effectiveness and activity of MNPs products.

The results show that integrating a sufficiently large amount of Fe_3O_4 will make the composites magnetic enough to separate them from the aqueous solution by an external magnetic field. This characteristic is significant, overcoming the inherent disadvantage of nanomaterials which disperse too well to be challenging to recover. Since then, this defect has been resolved smoothly. Recent studies have also considered the applicable conditions closer to reality when the effects of cations, anions, and natural organic matter (NOM) are thoroughly investigated and evaluated. In addition, with many contaminants such as arsenic, and several organic pollutants, ferromagnetic oxide not only acts as a recovery aid but also enhances the treatment efficiency thanks to its unique properties. The summarized research results also show an excellent combination between a magnetic metal oxide and a semiconductor metal oxide, which plays a crucial catalytic role in the photodegradation of pollutant compounds, especially organic pollutants. Some studies have shown the mechanism of the entire treatment process, including the adsorption of pollutants on the material's surface and the photodegradation reaction taking place at the catalytic sites under the catalysis activation of UV radiation or visible light.

5. Conclusion

In this chapter, a small portion of the space is devoted to reviewing green methods using plant-based agents to prepare MNPs.

The rest of the chapter is devoted to generalizing the results of research on synthesizing magnetic nanomaterials, mainly ferromagnetic oxides, to remove pollutants such as heavy metals, toxic substances, etc. organic pollution.

Author details

Thi Huong Nguyen*, Minh Thanh Vu and Ngoc Son Nguyen
Institute of Chemistry and Materials, Hanoi, Vietnam

*Address all correspondence to: nguyenhuong0916@gmail.com

References

[1] Bayda S, Adeel M, Tuccinardi T, Cordani M, Rizzolio F. The history of nanoscience and nanotechnology: From chemical-physical applications to nanomedicine. Molecules. 2019;**25**(1): 112. DOI: 10.3390/molecules25010112

[2] Kudr J, Haddad Y, Richtera L, Heger Z, Cernak M, Adam V, et al. Magnetic nanoparticles: From design and synthesis to real world applications. Nanomaterials (Basel). 2017;**7**(9):243. DOI: 10.3390/nano7090243

[3] Ali A, Shah T, Ullah R, Zhou P, Guo M, Ovais M, et al. Review on recent progress in magnetic nanoparticles: Synthesis, characterization, and diverse applications. Frontiers in Chemistry. 2021;**9**. DOI: 10.3389/fchem.2021.629054

[4] Alonso J, Barandiarán JM, Fernández Barquín L, García-Arribas A. Magnetic nanoparticles, synthesis, properties, and applications. In: El-Gendy AA, Barandiarán JM, Hadimani RL, editors. Magnetic Nanostructured Materials. London: Elsevier; 2018. pp. 1-40

[5] Chung I-M, Park I, Seung-Hyun K, Thiruvengadam M, Rajakumar G. Plant-mediated synthesis of silver nanoparticles: Their characteristic properties and therapeutic applications. Nanoscale Research Letters. 2016;**11**(1):1-14. DOI: 10.1186/s11671-016-1257-4

[6] Jung H-A, Su B-N, Keller WJ, Mehta RG, Kinghorn AD. Chemistry f. antioxidant xanthones from the pericarp of Garcinia mangostana (Mangosteen). Journal of Agricultural. 2006;**54**(6):2077-2082. DOI: 10.1021/jf052649z

[7] Yusefi M, Shameli K, Su Yee O, Teow S-Y, Hedayatnasab Z, Jahangirian H, et al. Green synthesis of Fe_3O_4 nanoparticles stabilized by a Garcinia mangostana fruit peel extract for hyperthermia and anticancer activities. International Journal of Nanomedicine. 2021;**16**:2515-2532. DOI: 10.2147/IJN.S284134

[8] Yusefi M, Su Yee O, Shameli K. Bio-mediated production and characterisation of magnetic nanoparticles using fruit peel extract. Journal of Research in Nanoscience and Nanotechnology. 2021;**1**(1):53-61. DOI: 10.37934/jrnn.1.1.5361

[9] Gullon B, Pintado ME, Pérez-Álvarez JA, Viuda-Martos M. Assessment of polyphenolic profile and antibacterial activity of pomegranate peel (*Punica granatum*) flour obtained from co-product of juice extraction. Food Control. 2016;**59**:94-98. DOI: 10.1016/j.foodcont.2015.05.025

[10] Jalili S, Naini A, Ashrafi M, Aminlari M. Technology. Antioxidant activity of pericarp extract from different varieties of pomegranate fruit. Journal of Agricultural Science. 2020;**22**:95-107

[11] Das AK, Nanda PK, Chowdhury NR, Dandapat P, Gagaoua M, Chauhan P, et al. Application of pomegranate by-products in muscle foods: Oxidative indices, colour stability, shelf life and health benefits. Molecules. 2021;**26**(2):467. DOI: 10.3390/molecules26020467

[12] Bawazeer S, Rauf A, Nawaz T, Khalil AA, Javed MS, Muhammad N, et al. Punica granatum peel extracts mediated the green synthesis of gold nanoparticles and their detailed in vivo biological activities %J green processing and synthesis. Green Processing and Synthesis. 2021;**10**(1):882-892. DOI: 10.1515/gps-2021-0080

[13] Chau TP, Veeraragavan GR, Narayanan M, Chinnathambi A, Alharbi SA, Subramani B, et al. Green synthesis of zirconium nanoparticles using *Punica granatum* (pomegranate) peel extract and their antimicrobial and antioxidant potency. Environmental Research. 2022;**209**:112771. DOI: 10.1016/j.envres.2022.112771

[14] Hassan H, Alibraheemi F, Abd A, Madhloom A, Yosif Z, Abdulla N. Biosynthesis of silver nanoparticles by *Punica granatum* peel extract and their biological activity on different pathogenic bacteria. NeuroQuantology. 2021;**19**:38-45. DOI: 10.14704/nq.2021.19.9.NQ21135

[15] Yusefi M, Shameli K, Ali RR, Pang S-W, Teow S-Y. Evaluating anticancer activity of plant-mediated synthesized iron oxide nanoparticles using *Punica granatum* fruit peel extract. Journal of Molecular Structure. 2020;**1204**:127539. DOI: 10.1016/j.molstruc.2019.127539

[16] Bouafia A, Laouini SE, Tedjani ML, Ali GAM, Barhoum A. Green biosynthesis and physicochemical characterization of Fe_3O_4 nanoparticles using *Punica granatum* L. fruit peel extract for optoelectronic applications. Textile Research Journal. 2021;**92**(15-16): 2685-2696. DOI: 10.1177/004051752 11006671

[17] Rybiński W, Karamać M, Sulewska K, Börner A, Amarowicz R. Antioxidant potential of grass pea seeds from European countries. Foods (Basel, Switzerland). 2018;**7**(9):142. DOI: 10.3390/foods7090142

[18] Bhattacharjee S, Waqar A, Barua K, Das A, Bhowmik S, Debi SR. Phytochemical and pharmacological evaluation of methanolic extract of *Lathyrus sativus* L. seeds. Clinical Phytoscience. 2018;**4**(1):20. DOI: 10.1186/s40816-018-0081-z

[19] Dhar PK, Saha P, Hasan MK, Amin MK, Haque MR. Green synthesis of magnetite nanoparticles using *Lathyrus sativus* peel extract and evaluation of their catalytic activity. Cleaner Engineering and Technology. 2021;**3**:100117. DOI: 10.1016/j.clet.2021.100117

[20] Selvaraj R, Pai S, Murugesan G, Pandey S, Bhole R, Gonsalves D, et al. Green synthesis of magnetic α–Fe2O3 nanospheres using Bridelia retusa leaf extract for Fenton-like degradation of crystal violet dye. Applied Nanoscience. 2021;**11**(8):2227-2234. DOI: 10.1007/s13204-021-01952-y

[21] Sharma RK, Yadav S, Gupta R, Arora G. Synthesis of magnetic nanoparticles using potato extract for dye degradation: A green chemistry experiment. Journal of Chemical Education. 2019;**96**(12):3038-3044. DOI: 10.1021/acs.jchemed.9b00384

[22] States JC. Exposure Sources, Health Risks, and Mechanisms of Toxicity. Hoboken, NJ: Wiley; 2015

[23] McNeill LS, Edwards M. Soluble arsenic removal at water treatment plants. Journal-American Water Works Association. 1995;**87**(4):105-113. DOI: 10.1002/j.1551-8833.1995.tb06346.x

[24] Mudhoo A, Sharma SK, Garg VK, Tseng C-H. Arsenic: An overview of applications, health, and environmental concerns and removal processes. Critical Reviews in Environmental Science and Technology. 2011;**41**(5):435-519. DOI: 10.1080/10643380902945771

[25] Oscarson DW, Huang PM, Defosse C, Herbillon A. Oxidative power of Mn(IV) and Fe(III) oxides with respect to As(III) in terrestrial and aquatic environments. Nature.

1981;**291**(5810):50-51. DOI: 10.1038/291050a0

[26] Scott MJ, Morgan JJ. Reactions at oxide surfaces. 1. Oxidation of As(III) by synthetic Birnessite. Environmental Science & Technology. 1995;**29**(8):1898-1905. DOI: 10.1021/es00008a006

[27] Pierce ML, Moore CB. Adsorption of arsenite and arsenate on amorphous iron hydroxide. Water Research. 1982;**16**(7):1247-1253. DOI: 10.1016/0043-1354(82)90143-9

[28] Raven KP, Jain A, Loeppert RH. Arsenite and arsenate adsorption on Ferrihydrite: Kinetics, equilibrium, and adsorption envelopes. Environmental Science & Technology. 1998;**32**(3):344-349. DOI: 10.1021/es970421p

[29] Zhang G, Qu J, Liu H, Liu R, Wu R. Preparation and evaluation of a novel Fe–Mn binary oxide adsorbent for effective arsenite removal. Water Research. 2007;**41**(9):1921-1928. DOI: 10.1016/j.watres.2007.02.009

[30] Kong S, Wang Y, Hu Q, Olusegun AK. Magnetic nanoscale Fe–Mn binary oxides loaded zeolite for arsenic removal from synthetic groundwater. Colloids and Surfaces A: Physicochemical and Engineering Aspects. 2014;**457**:220-227. DOI: 10.1016/j.colsurfa.2014.05.066

[31] Kumar S, Nair RR, Pillai PB, Gupta SN, Iyengar MAR, Sood AK. Graphene oxide–$MnFe_2O_4$ magnetic nanohybrids for efficient removal of lead and arsenic from water. ACS Applied Materials & Interfaces. 2014;**6**(20):17426-17436. DOI: 10.1021/am504826q

[32] Habibi Y, Lucia LA, Rojas OJ. Cellulose nanocrystals: Chemistry, self-assembly, and applications. Chemical Reviews. 2010;**110**(6):3479-3500. DOI: 10.1021/cr900339w

[33] Hokkanen S, Repo E, Lou S, Sillanpää M. Removal of arsenic(V) by magnetic nanoparticle activated microfibrillated cellulose. Chemical Engineering Journal. 2015;**260**:886-894. DOI: 10.1016/j.cej.2014.08.093

[34] Yu X, Tong S, Ge M, Zuo J, Cao C, Song W. One-step synthesis of magnetic composites of cellulose@iron oxide nanoparticles for arsenic removal. Journal of Materials Chemistry A. 2013;**1**(3):959-965. DOI: 10.1039/C2TA00315E

[35] Lunge S, Singh S, Sinha A. Magnetic iron oxide (Fe_3O_4) nanoparticles from tea waste for arsenic removal. Journal of Magnetism and Magnetic Materials. 2014;**356**:21-31. DOI: 10.1016/j.jmmm.2013.12.008

[36] Zeng H, Zhai L, Qiao T, Yu Y, Zhang J, Li D. Efficient removal of As(V) from aqueous media by magnetic nanoparticles prepared with iron-containing water treatment residuals. Scientific Reports. 2020;**10**(1):9335. DOI: 10.1038/s41598-020-65840-1

[37] Zhang M, Gao B, Varnoosfaderani S, Hebard A, Yao Y, Inyang M. Preparation and characterization of a novel magnetic biochar for arsenic removal. Bioresource Technology. 2013;**130**:457-462. DOI: 10.1016/j.biortech.2012.11.132

[38] Nham NT, Tahtamouni TMA, Nguyen TD, Huong PT, Jitae K, Viet NM, et al. Synthesis of iron modified rice straw biochar toward arsenic from groundwater. Materials Research Express. 2019;**6**(11):115528. DOI: 10.1088/2053-1591/ab4b98

[39] Wang S, Gao B, Zimmerman AR, Li Y, Ma L, Harris WG, et al. Removal of

arsenic by magnetic biochar prepared from pinewood and natural hematite. Bioresource Technology. 2015;**175**:391-395. DOI: 10.1016/j.biortech.2014.10.104

[40] Martínez-Cabanas M, López-García M, Barriada JL, Herrero R, Sastre de Vicente ME. Green synthesis of iron oxide nanoparticles. Development of magnetic hybrid materials for efficient As(V) removal. Chemical Engineering Journal. 2016;**301**:83-91. DOI: 10.1016/j.cej.2016.04.149

[41] Akin I, Arslan G, Tor A, Ersoz M, Cengeloglu Y. Arsenic(V) removal from underground water by magnetic nanoparticles synthesized from waste red mud. Journal of Hazardous Materials. 2012;**235-236**:62-68. DOI: 10.1016/j.jhazmat.2012.06.024

[42] Noor NM, Othman R, Mubarak NM, Abdullah EC. Agricultural biomass-derived magnetic adsorbents: Preparation and application for heavy metals removal. Journal of the Taiwan Institute of Chemical Engineers. 2017;**78**:168-177. DOI: 10.1016/j.jtice.2017.05.023

[43] Liu C-H, Chuang Y-H, Chen T-Y, Tian Y, Li H, Wang M-K, et al. Mechanism of arsenic adsorption on magnetite nanoparticles from water: Thermodynamic and spectroscopic studies. Environmental Science & Technology. 2015;**49**(13):7726-7734. DOI: 10.1021/acs.est.5b00381

[44] Mayo JT, Yavuz C, Yean S, Cong L, Shiple H, Yu W, et al. The effect of nanocrystalline magnetite size on arsenic removal. Science and Technology of Advanced Materials. 2007;**8**(1-2):71-75. DOI: 10.1016/j.stam.2006.10.005

[45] Tuutijärvi T, Lu J, Sillanpää M, Chen G. As(V) adsorption on maghemite nanoparticles. Journal of Hazardous Materials. 2009;**166**(2):1415-1420. DOI: 10.1016/j.jhazmat.2008.12.069

[46] Yean S, Cong L, Yavuz CT, Mayo JT, Yu WW, Kan AT, et al. Effect of magnetite particle size on adsorption and desorption of arsenite and arsenate. Journal of Materials Research. 2005;**20**(12):3255-3264. DOI: 10.1557/jmr.2005.0403

[47] Prasad S, Yadav KK, Kumar S, Gupta N, Cabral-Pinto MMS, Rezania S, et al. Chromium contamination and effect on environmental health and its remediation: A sustainable approaches. Journal of Environmental Management. 2021;**285**:112174. DOI: 10.1016/j.jenvman.2021.112174

[48] Vercoutere K, Cornelis R. Chromium speciation in environmental and biological samples. In: Quevauviller P, Maier EA, Griepink B, editors. Techniques and Instrumentation in Analytical Chemistry. Vol. 17. Netherlands: Elsevier; 1995. p. 195-213

[49] Petruzzelli D, Passino R, Tiravanti G. Ion exchange process for chromium removal and recovery from tannery wastes. Industrial & Engineering Chemistry Research. 1995;**34**(8):2612-2617. DOI: 10.1021/ie00047a009

[50] Shen YF, Tang J, Nie ZH, Wang YD, Ren Y, Zuo L. Preparation and application of magnetic Fe_3O_4 nanoparticles for wastewater purification. Separation and Purification Technology. 2009;**68**(3):312-319. DOI: 10.1016/j.seppur.2009.05.020

[51] Hu J, Lo CG. Fast removal and recovery of Cr(VI) using surface-modified jacobsite ($MnFe_2O_4$) nanoparticles. Langmuir. 2005;**21**(24):11173-11179. DOI: 10.1021/la051076h

[52] Rajput S, Pittman CU, Mohan D. Magnetic magnetite (Fe_3O_4) nanoparticle synthesis and applications for lead (Pb2+) and chromium (Cr6+) removal from water. Journal of Colloid and Interface Science. 2016;**468**:334-346. DOI: 10.1016/j.jcis.2015.12.008

[53] Shi S, Yang J, Liang S, Li M, Gan Q, Xiao K, et al. Enhanced Cr(VI) removal from acidic solutions using biochar modified by Fe3O4@SiO2-NH2 particles. Science of the Total Environment. 2018;**628-629**:499-508. DOI: 10.1016/j.scitotenv.2018.02.091

[54] TyV V, Efimov ON. Polypyrrole: A conducting polymer; its synthesis, properties and applications. Russian Chemical Reviews. 1997;**66**(5):443-457. DOI: 10.1070/rc1997v066n05abeh000261

[55] Rapi S, Bocchi V, Gardini GP. Conducting polypyrrole by chemical synthesis in water. Synthetic Metals. 1988;**24**(3):217-221. DOI: 10.1016/0379-6779(88)90259-7

[56] Zhang X, Bai. Surface electric properties of Polypyrrole in aqueous solutions. Langmuir. 2003;**19**(26):10703-10709. DOI: 10.1021/la034893v

[57] Bhaumik M, Maity A, Srinivasu VV, Onyango MS. Enhanced removal of Cr(VI) from aqueous solution using polypyrrole/Fe_3O_4 magnetic nanocomposite. Journal of Hazardous Materials. 2011;**190**(1):381-390. DOI: 10.1016/j.jhazmat.2011.03.062

[58] Wang H, Yuan X, Wu Y, Chen X, Leng L, Wang H, et al. Facile synthesis of polypyrrole decorated reduced graphene oxide–Fe_3O_4 magnetic composites and its application for the Cr(VI) removal. Chemical Engineering Journal. 2015;**262**:597-606. DOI: 10.1016/j.cej.2014.10.020

[59] Liu J-f, Zhao Z-s, Jiang G-b. Coating Fe_3O_4 magnetic nanoparticles with humic acid for high efficient removal of heavy metals in water. Environmental Science & Technology. 2008;**42**(18):6949-6954. DOI: 10.1021/es800924c

[60] Ge F, Li M-M, Ye H, Zhao B-X. Effective removal of heavy metal ions Cd2+, Zn2+, Pb2+, Cu2+ from aqueous solution by polymer-modified magnetic nanoparticles. Journal of Hazardous Materials. 2012;**211-212**:366-372. DOI: 10.1016/j.jhazmat.2011.12.013

[61] Wang J, Zheng S, Shao Y, Liu J, Xu Z, Zhu D. Amino-functionalized Fe_3O_4@ SiO_2 core–shell magnetic nanomaterial as a novel adsorbent for aqueous heavy metals removal. Journal of Colloid and Interface Science. 2010;**349**(1):293-299. DOI: 10.1016/j.jcis.2010.05.010

[62] Liu Y, Fu R, Sun Y, Zhou X, Baig SA, Xu X. Multifunctional nanocomposites Fe_3O_4@SiO_2-EDTA for Pb(II) and Cu(II) removal from aqueous solutions. Applied Surface Science. 2016;**369**:267-276. DOI: 10.1016/j.apsusc.2016.02.043

[63] Ren Y, Abbood HA, He F, Peng H, Huang K. Magnetic EDTA-modified chitosan/SiO_2/Fe_3O_4 adsorbent: Preparation, characterization, and application in heavy metal adsorption. Chemical Engineering Journal. 2013;**226**:300-311. DOI: 10.1016/j.cej.2013.04.059

[64] Zhang S, Zhang Y, Liu J, Xu Q, Xiao H, Wang X, et al. Thiol modified Fe_3O_4@SiO_2 as a robust, high effective, and recycling magnetic sorbent for mercury removal. Chemical Engineering Journal. 2013;**226**:30-38. DOI: 10.1016/j.cej.2013.04.060

[65] Pearson RG. Hard and soft acids and bases. Journal of the American Chemical

Society. 1963;**85**(22):3533-3539. DOI: 10.1021/ja00905a001

[66] Carmen Z, Daniela S. Textile organic dyes—Characteristics, polluting effects and separation/elimination procedures from industrial effluents – A critical overview. In: Organic Pollutants Ten Years After the Stockholm Convention—Environmental and Analytical Update. Croatia: InTech; 2012

[67] Wani KA, Jangid NK, Bhat AR. Impact of textile dyes on public health and the environment. IGI Global. 2019. pp. 176-7.c

[68] Zhang Z, Kong J. Novel magnetic Fe_3O_4@C nanoparticles as adsorbents for removal of organic dyes from aqueous solution. Journal of Hazardous Materials. 2011;**193**:325-329. DOI: 10.1016/j.jhazmat.2011.07.033

[69] Chen L, Chen J, Zhang X, Xie P. A review of reproductive toxicity of microcystins. Journal of Hazardous Materials. 2015;**301**:10.1016/j.jhazmat.2015.08.041

[70] Cheng R, Zhu H, Shutes B, Yan B. Treatment of microcystin (MC-LR) and nutrients in eutrophic water by constructed wetlands: Performance and microbial community. Chemosphere. 2021;**263**:128139. DOI: 10.1016/j.chemosphere.2020.128139

[71] Deng Y, Qi D, Deng C, Zhang X, Zhao D. Superparamagnetic high-magnetization microspheres with an Fe_3O_4@SiO_2 core and perpendicularly aligned mesoporous SiO_2 shell for removal of microcystins. Journal of the American Chemical Society. 2008;**130**(1):28-29. DOI: 10.1021/ja0777584

[72] Amos SGE, Garreau M, Buzzetti L, Waser J. Photocatalysis with organic dyes: Facile access to reactive intermediates for synthesis. Beilstein Journal of Organic Chemistry. 2020;**16**:1163-1187. DOI: 10.3762/bjoc.16.103

[73] Nicewicz DA, Nguyen TM. Recent applications of organic dyes as photoredox catalysts in organic synthesis. ACS Catalysis. 2014;**4**(1):355-360. DOI: 10.1021/cs400956a

[74] Chi Y, Yuan Q, Li Y, Tu J, Zhao L, Li N, et al. Synthesis of Fe_3O_4@SiO_2-Ag magnetic nanocomposite based on small-sized and highly dispersed silver nanoparticles for catalytic reduction of 4-nitrophenol. Journal of Colloid and Interface Science. 2012;**383**(1):96-102

[75] Mortazavi-Derazkola S, Salavati-Niasari M, Amiri O, Abbasi A. Fabrication and characterization of Fe_3O_4@SiO_2@TiO_2@Ho nanostructures as a novel and highly efficient photocatalyst for degradation of organic pollution. Journal of Energy Chemistry. 2017;**26**(1):17-23. DOI: 10.1016/j.jechem.2016.10.015

[76] Bhatia S, Verma N. Photocatalytic activity of ZnO nanoparticles with optimization of defects. Materials Research Bulletin. 2017;**95**:468-476. DOI: 10.1016/j.materresbull.2017.08.019

[77] Elmi F, Alinezhad H, Moulana Z, Salehian F, Mohseni Tavakkoli S, Asgharpour F, et al. The use of antibacterial activity of ZnO nanoparticles in the treatment of municipal wastewater. Water Science and Technology. 2014;**70**(5):763-770. DOI: 10.2166/wst.2014.232

[78] Uribe-López MC, Hidalgo-López MC, López-González R, Frías-Márquez DM, Núñez-NogueiraG, Hernández-CastilloD, et al. Photocatalytic activity of ZnO nanoparticles and the role of the synthesis method on their physical

and chemical properties. Journal of Photochemistry and Photobiology A: Chemistry. 2021;**404**:112866. DOI: 10.1016/j.jphotochem.2020.112866

[79] Nguyen NT, Nguyen VA. Synthesis, characterization, and photocatalytic activity of ZnO nanomaterials prepared by a green, nonchemical route. Journal of Nanomaterials. 2020;**2020**:1768371. DOI: 10.1155/2020/1768371

[80] Al-Rawashdeh NAF, Allabadi O, Aljarrah MT. Photocatalytic activity of graphene oxide/zinc oxide nanocomposites with embedded metal nanoparticles for the degradation of organic dyes. ACS Omega. 2020;**5**(43):28046-28055. DOI: 10.1021/acsomega.0c03608

[81] Pham VV, Nguyen TD, Ha La PP, Thi CM. A comparison study of the photocatalytic activity of ZnO nanoparticles for organic contaminants degradation under low-power UV-A lamp. Advances in Natural Sciences: Nanoscience and Nanotechnology. 2020;**11**(1):015005. DOI: 10.1088/2043-6254/ab6163

[82] Kumar K, Chitkara M, Sandhu IS, Mehta D, Kumar S. Photocatalytic, optical and magnetic properties of Fe-doped ZnO nanoparticles prepared by chemical route. Journal of Alloys and Compounds. 2014;**588**:681-689. DOI: 10.1016/j.jallcom.2013.11.127

[83] Falak P, Hassanzadeh-Tabrizi SA, Saffar-Teluri A. Synthesis, characterization, and magnetic properties of ZnO-$ZnFe_2O_4$ nanoparticles with high photocatalytic activity. Journal of Magnetism and Magnetic Materials. 2017;**441**:98-104. DOI: 10.1016/j.jmmm.2017.05.044

[84] Boutra B, Güy N, Özacar M, Trari M. Magnetically separable $MnFe_2O_4$/TA/ZnO nanocomposites for photocatalytic degradation of Congo red under visible light. Journal of Magnetism and Magnetic Materials. 2020;**497**:165994. DOI: 10.1016/j.jmmm.2019.165994

[85] Długosz O, Szostak K, Krupiński M, Banach M. Synthesis of Fe_3O_4/ZnO nanoparticles and their application for the photodegradation of anionic and cationic dyes. International Journal of Environmental Science and Technology. 2021;**18**(3):561-574. DOI: 10.1007/s13762-020-02852-4

[86] Pham TMH, Vu MT, Cong TD, Nguyen NS, Doan TA, Trung TD, et al. Green sonochemical process for preparation of polythylenen glycol-Fe_3O_4/ZnO removal of methylene blue in solution. Bulletin of Materials Science. 2022;**45**(1):13. DOI: 10.1007/s12034-021-02584-2

Chapter 8

Nanomaterials of Carbon and Metal Sulfides in Photocatalysis

Ana Cristina Estrada, Joana Lúcia Lopes and Tito Trindade

Abstract

Heterogeneous semiconductor photocatalysis has received much interest because of its applications in important global energy and environmental challenges in a cost-effective sustainable way. The photocatalytic efficiency of semiconductor photocatalysts under solar irradiation has been pointed out by difficulties associated with low visible-light absorption range, fast recombination of photogenerated carriers, and low chemical stability in operational conditions. Graphitic materials have attracted great interest due to properties, such as high surface area, mechanical strength, and photochemical stability. Thus, their combination with metal sulfides, has been explored as promising strategies to produce new photocatalysts. These nanocomposites show great potential in photodegradation of contaminants of emerging concern (CEC), which might be detected in water sources, such as traces of Pharmaceutics and pesticides. Here, we briefly review fundamental principles photocatalysis in general, with the focus on the use of carbon-nanomaterials of distinct structural dimensionalities associated with nano-crystalline metal sulfides, envisaging their application as heterogeneous photocatalysts for water remediation. Key aspects concerning the photocatalyst properties, such as light absorption, charge separation and transfer, and stability, are also approached. Graphene and graphene derivatives have demonstrated great potential for increasing photogenerated charge-carrier separation and migration efficiency, as well as in extending the light absorption range and adsorption capacity.

Keywords: metal sulfide, carbon nanomaterials, photocatalysis, water treatment

1. Introduction

Photocatalyst is a term that combines two words—*photo*, which is related to light, and *catalyst*, which is a compound that does not change the thermodynamics of the reaction but changes its kinetics, by establishing new reaction routes with lower activation energy, without being consumed during the process. Hence, semiconducting photocatalysis involves chemical reactions that occur at the surfaces of certain semiconductor compounds when irradiated with light of a selected wavelength range. Typically, these reactions occur in a liquid medium using the photocatalyst in the solid state, thus the chemical process is generally termed heterogeneous photocatalysis. In this work, the semiconductor photocatalyst is considered as part of a colloid or suspension, though this has not been always the case. For instance, thin films have been also applied namely for air purification. Examples of heterogeneous photocatalytic processes using semiconductor particles include photooxidation reactions, which have been exploited for the degradation of organic pollutants present in water [1–5]. Compared to more conventional water treatment methods, such as those based on adsorption and flocculation, which might require a subsequent step for the chemical degradation of the pollutant, in photocatalysis the pollutant is eliminated by aerobic photooxidation. Ideally, this oxidative process should generate carbon dioxide and water as the final products, that is, the complete mineralization of the organic pollutant, though this has been rarely achieved. As such, different remediation technologies can coexist in the same water treatment plant and, in several situations, their complementary role brings more efficient approaches. For example, adsorption and photocatalytic technologies can be implemented in different stages in the same water treatment plant. Even though, advanced oxidation processes based on the use of efficient photocatalysts have been regarded as a way to minimize the impact of CEC in water sources, which even in trace levels are harmful and for which conventional water treatments are ineffective.

Two main optical processes have been proposed considering the role of the semiconductor during a photocatalytic reaction, as illustrated in **Figure 1** for TiO_2 photocatalysts. In direct photocatalysis, a photon with energy higher than the band gap energy of the semiconductor ($h\nu_1$) is absorbed and an electron (e^-) is excited to the conduction band (CB), leaving a hole (h^+) in the valence band (VB). The band gap energy of the semiconductor is defined as the difference between the CB (bottom energy level) and the VB (top energy level). The photogenerated electron–hole pair ($e^-/-h^+$) is responsible for reduction and oxidation reactions that take place at the surface of the photocatalyst particle in contact with the aqueous medium. The electron in the CB migrates to the surface of the semiconductor and participates in reduction reactions, and the hole in the VB diffuses to the photocatalyst surface and is involved in oxidation reactions. In addition, the dissolved O_2 can accept photogenerated electrons to yield superoxide radicals ($O_2^{\cdot -}$) and photogenerated holes can oxidize H_2O to form strong oxidant hydroxyl radicals ($HO^\bullet$) (Eqs. (1) and (2)) [6, 7].

$$e^-_{CB} + O_2 \rightarrow O_2^{\bullet -} \tag{1}$$

$$h^+_{VB} + H_2O \rightarrow HO^\bullet + H^+ \tag{2}$$

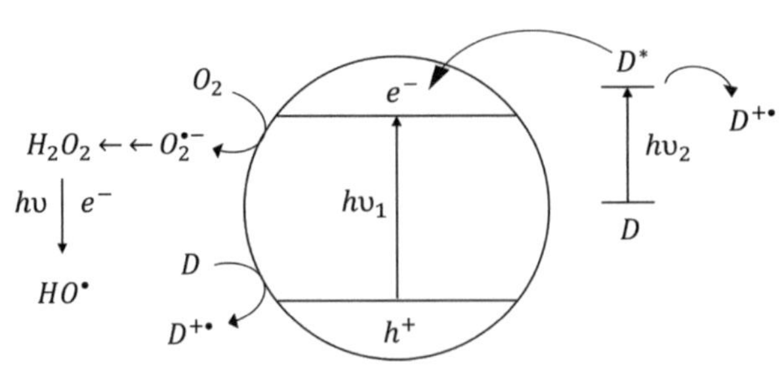

Figure 1.
The schematic representation of the direct ($h\upsilon_1$) and indirect ($h\upsilon_2$) photochemical processes occurring in light-irradiated TiO_2 nanoparticles, commonly used as photocatalysts in the form of aqueous colloids. Adapted from [6].

On the other hand, in indirect photocatalysis, also known as photosensitized photocatalysis, the mechanism involves the photoexcitation ($h\nu_2$) of a second species (P) to an excited state from which an electron is injected into the CB of the semiconductor. This process has been observed in the degradation of contaminant organic dyes, which can also act as photosensitizers for cases in which the reduction potential of the excited state is negative enough for electron injection into the CB of the semiconductor [7]. In indirect photocatalysis, there is no generation of a VB hole and the semiconductor functions as an electron relay, thereby preventing undesired back reactions [7]. Nevertheless, this process is usually less efficient than direct photocatalysis due to the lower efficiency of the electron injection. Both direct and indirect photocatalysis convert the initially generated superoxide radicals into other reactive oxygen species with high oxidative power (Eqs. (3)-(7)), for example, with reduction potentials of 0.94 V ($O_2^{\bullet-}/H_2O_2$), 1.29 V (H_2O_2/H_2O) and 1.90 V ($HO^{\bullet}/HO^{-}$) [6, 8]. Although such radicals are nonselective, they are effective in oxidizing organic contaminants, such as dye molecules [9–15], antibiotics [16–20], or pesticides [21–25], as well as for other sanitation applications, such as the elimination of pathogens [26–32].

$$O_2^{\bullet-} + H^+ \rightarrow HO_2^{\bullet} \tag{3}$$

$$HO_2^{\bullet} + HO_2^{\bullet} \rightarrow H_2O_2 + O_2 \tag{4}$$

$$O_2^{\bullet-} + HO_2^{\bullet} \rightarrow O_2 + HO_2^{-} \tag{5}$$

$$HO_2^{-} + H^+ \rightarrow H_2O_2 \tag{6}$$

$$H_2O_2 + O_2^{\bullet-} \rightarrow HO^{\bullet} + HO^{-} + O_2 \tag{7}$$

In semiconductor photocatalysis, several fundamental aspects should be considered to develop the photocatalyst based on functional and operational criteria. Hence, light absorption (absorption coefficient and wavelength range), photoinduced charge separation, charge trapping, and charge transfer are among the key parameters for designing efficient photocatalytic systems [33, 34]. For instance, photogenerated electrons are unstable species in an excited state, which tend naturally to return to the ground state either via adsorbed hydroxyl radicals or by recombination with unreacted holes or structural traps on semiconductors [35–38]. Since these species are determinants in the efficiency of a photocatalyst, several research groups have explored strategies to increase the photoinduced charge separation to avoid charge recombination and consequently increase the lifetime of photogenerated electron/hole pairs. These strategies include (i) coupling of semiconductor photocatalysts with metal nanoparticles [39–41]; (ii) sensitization of the photocatalyst surface through physical or chemical adsorption of molecules that absorb visible light and are excited either to the singlet or triplet excited state [42] and; (iii) coupling of at least two semiconductor photocatalysts with different bandgap values [43, 44]. The presence of charge trapping sites in a semiconductor photocatalyst allows also the extension of the lifetime of the charge carriers from microseconds to milliseconds since in these sites there is greater charge-carrier stability. Although such trap sites are mostly located at the surface of a semiconductor photocatalyst, they may be present also on grain boundaries or in the bulk lattice, or even present as electron scavengers, such as O_2. On the other hand, deeply stabilized trapped charges lose redox potential and increase the potential barrier for charge transfer at the semiconductor or water interface [45].

Thus, electron transfer reaction depends largely on structural parameters ascribed to the semiconductor photocatalyst, such as crystal facet structure, lattice surface, size, and morphology. Trapping mechanisms might be favorable if they allow photon activity to generate charge carriers, and permit charge carriers to reach the electron transfer regions. Otherwise, it could be disadvantageous for the overall photocatalytic process.

Several strategies have been proposed to adjust the physical and chemical properties of semiconductor photocatalysts to improve light absorption and charge transfer efficiency, reduce the recombination rate of photogenerated charge carriers, and accelerate surface reactions [46]. Examples of such strategies include metal-ion doping of the semiconductor [39–41], combination with distinct semiconductors that result in heterostructures [43, 44], and surface chemical functionalization using selected photosensitizers [42]. Noteworthy, the combination of inorganic semiconductors with carbonaceous materials, such as graphene and their structural derivatives, has also received great attention in the design of a new class of nanocomposite photocatalysts [47, 48]. The use of carbon nanostructures for supported semiconductor photocatalysts offers great advantages. Hence, depending on the carbon material, high electrically conductive nanostructures can act as scavengers of photogenerated electrons. Also, water-compatible nanomaterials promote the aqueous dispersion of the photocatalyst, which by achieving a high specific surface area enhances the adsorption capacity of the system [48]. Furthermore, surface functionalization of the carbon lattice confers functional chemical groups that might favor the subsequent attachment of semiconductor nanophases. A paradigmatic example of this situation is the application of graphene oxide as a nanoplatform for semiconductor photocatalysts, and notwithstanding limitations that can also arise such as photoreduction of the carbon substrate or the limited absorption by the photocatalyst [49–51].

2. Metal-sulfide photocatalysts

In general terms, a good photocatalyst should have the following characteristics: effective charge-carrier separation, fast charge transfer, strong optical absorption, photochemical stability, low-cost production, and nontoxicity [52]. Among the several types of photocatalysts available, inorganic semiconductors have been intensely investigated in water remediation processes because they might fulfill, at least in selected cases, the above requirements. Inorganic materials considered as semiconductors exhibit bandgap energies in the range of 0.3–3.8 eV. In particular, TiO_2 and TiO_2-based heterogeneous photocatalysts have been the most explored semiconductor materials for photocatalytic applications because of the high free energy of photogenerated charge carriers, low-cost, and high chemical stability [53, 54]. However, both TiO_2 polymorphs (anatase/rutile) show a wide bandgap (anatase 3.2 eV; rutile 3.0 eV), which limits photocatalytic applications of pure TiO_2 to UV irradiated systems. Other semiconductor photocatalysts exhibiting narrower bandgaps have been investigated, which can replace TiO_2 in certain conditions or that might act as a complementary phase in extending light absorption to the visible composite systems. Among these semiconductors, this chapter focus on the use of binary metal-sulfide compounds, with emphasis on their nanocrystalline forms. **Table 1** shows examples of metal sulfides investigated as photocatalysts and selected properties for the pure phases.

Metal sulfide	Bandgap energy (eV)*	Structure
ZnS	3.6	Cubic, Hexagonal
CdS	2.4	Cubic; Hexagonal
CuS	2.4	Hexagonal
Ag_2S	1.0	Monoclinic
Bi_2S_3	1.4	Orthorhombic

Table 1.
Characteristics of macrocrystalline metal sulfides as photocatalysts in an aqueous medium [44].

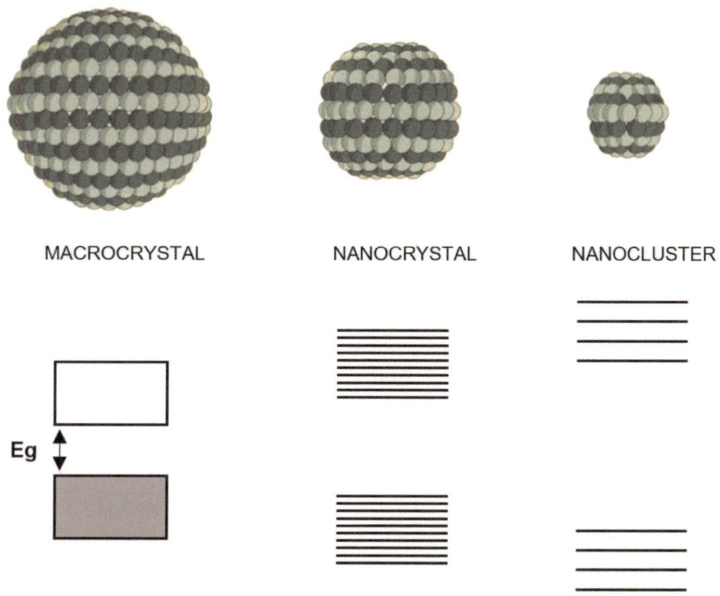

Figure 2.
The scheme illustrates the widening of the bandgap energy of a certain semiconductor material, as particle size decreases from its macrocrystalline form (left) to the nanocluster regime (right). Quantum dots are nanocrystalline semiconductors (middle) that show quantum-size effects, corresponding to the intermediate situation between macrocrystalline materials and nanoclusters.

A macrocrystalline metal sulfide (MS) semiconductor comprises a three-dimensional network of ordered atoms (metal and S atoms) characterized by a band gap energy at a certain temperature. As particle size decreases, and below a certain threshold, the electronic band structure of the semiconductor changes with the widening of the bandgap energy [55, 56]. For semiconductor nanoclusters, that is molecular-like nanocrystals depicted on the right in **Figure 2**, an analogous interpretation applies, although the energy gap is usually understood as the energy separation between the frontier molecular orbitals HOMO (highest occupied molecular orbital) and LUMO (lowest unoccupied molecular orbital). Thus, as in the case of conventional photosemiconductors, the incidence of a photon with energy greater than this energetic separation originates in semiconductor nanocrystals (and nanoclusters) the formation of an electron–hole pair, often called exciton, which in the macrocrystalline material is dimensionally characterized by the Bohr exciton radius of that semiconductor. The charge carriers in nanosized semiconductors migrate fast and participate in several photoprocesses, which include trapping and recombination [57, 58].

Metal sulfides can be explored in the macrocrystalline form as photocatalysts, for example, in aqueous suspensions, membranes, and thin films [59–61]. However, in the past decades, there has been intense research on their use as nanocrystalline materials, namely due to the possibility to explore quantum-size effects, as mentioned above. MS semiconductor nanocrystals (quantum dots) are small crystalline particles that exhibit quantum size-dependent optical and electronic properties [62, 63]. With typical dimensions in the range of 1–100 nm, these nanocrystals bridge the gap between those of molecules and micrometric crystals, displaying distinct optical behavior in relation to their bulk counterparts [64]. If the size of nanocrystals is smaller than the bulk exciton Bohr radius, the charge carriers become spatially confined, showing size-dependent absorption and fluorescence spectra with discrete electronic transitions at room temperature (**Figure 2**).

For instance, the optical spectra of colloids of nanocrystalline semiconductors show blue shifts in their absorption edges (or excitonic peaks) with decreasing particle diameters. Metal-sulfide nanocrystals that exhibit quantum size effects, that is, quantum dots, can be used as size-tuned light-absorption photosensitizers, namely in visible photocatalytic applications [44, 65–67]. Quantum size effects occurring in MS nanocrystals dispersed in aqueous suspensions, also affect the CB and VB redox levels, thus influencing redox reactions that involve the migration of photogenerated charge carriers to the particles' surfaces. Nanosized semiconductors have dimensions considerably superior to conventional molecular photosensitizers, which in comparison to the latter, present a broader absorption wavelength range, large density of states, and high optical extinction coefficients [62], hence favoring photon harvesting in photocatalytic applications.

Colloidal synthesis offers a wide range of chemical methods to obtain MS nanocrystals with controlled particle size distributions and particle shapes, thus with tailored bandgaps for diverse semiconductors and their solid solutions [68–71]. Furthermore, such colloids can be selected as nanodispersed systems showing strong visible-light absorption and size-tuned bandgap. However, these systems also show limitations, which deserve further research aiming their application as more efficient photocatalysts. Although certain MS is used as visible-light photocatalysts, the photogenerated electron–hole pairs are also susceptible to recombination. The occurrence of charge-carrier recombination limits their mobility from the bulk lattice to the particles' surface, thus decreasing the efficiency of the photocatalyst. Moreover, surface-sulfide anions (S^{2-}) in aqueous MS colloids are prone to oxidation, a process that gains more relevance due to the oxidative role of photogenerated holes at the surface [72, 73]. In fact, under light irradiation, sulfide anions can oxidize forming sulfate (SO_4^{2-}) or elemental sulfur (S^0), causing the deactivation of the photocatalyst.

The inhibition of metal-sulfide photocorrosion is an important requirement for photocatalytic reactions, namely because the long lifetime of photogenerated electron–hole pairs and the chemical stability are essential for producing efficient photocatalysts. Several strategies have been reported that tackle this problem, such as modifying the crystal structure, size, and morphology of semiconductors [74, 75], combining with transition metal ions or cocatalysts [76, 77], producing heterojunctions, [78–80] and by adjusting the reaction parameters [81–83]. For instance, Bo *et al.* have reported that the interfacial interaction between both semiconductors in the MoS_2/CdS heterostructures restrains the photocorrosion of MoS_2.

The authors have shown that electrons photogenerated on the CB of CdS are transferred to the CB of MoS_2 to participate in the H_2 evolution reaction, while the holes on the VB of MoS_2 migrate to the VB of CdS [79]. Huang *et al.* have shown that the growth of a larger bandgap semiconductor, such as ZnO, on a core with a smaller band gap as CdS improves the stability of the hybrid nanostructure and inhibits the photocorrosion of CdS particles [84]. In turn, Yi and Wang have found that the photocorrosion of CdS is significantly inhibited when cobalt ions or molybdate are injected into the CdS-lactic acid system. The photogenerated holes in the CdS are fastly captured by the transition metal ions, reducing the oxidation of S^{2-} on the CdS surface [85, 86]. The coupling of metal-sulfide semiconductor photocatalysts with inorganic substrates might bring other advantages and several approaches have been reported [87, 88]. In this context, carbon nanomaterials have also been investigated as functional platforms that bring new potential to the application of these materials, including photocorrosion inhibition of the supported metal sulfides.

3. Carbon-based nanostructures

The development of heterogeneous photocatalysts by combining metal sulfides and different carbon nanomaterials has been explored as an effective strategy to obtain high-performance photocatalysts. Owing to delocalized electrons from the conjugative π-system, graphitic carbon nanostructures are good at accepting and shuttling the photogenerated electrons from semiconductor photocatalysts; hence, effectively separating the electron–hole pairs [89–93]. For instance, Wan *et al.* have shown that the synergistic influence of charge-carrier migration, advanced excited states, and suitable Fermi levels between CdS phases and graphene leads to enhanced photoactivity and stability [94]. Also, Lv *et al.* have shown that graphene attached to semiconductors can efficiently accommodate and transport electrons from the excited semiconductor, which not only hindered charge recombination but also improved charge transfer, giving rise to high photocatalytic efficiency [89]. These works confirmed the relevant role of graphene, among the carbon-based nanomaterials, in aqueous colloidal chemistry processes, such as heterogeneous photocatalysis. Thus, in this chapter, graphene and its derived nanostructures are used as illustrative examples in the fabrication of carbon-supported metal-sulfides photocatalysts.

Graphene is a 2D material formed by a one-atom-thick planar layer of sp^2-hybridized carbon atoms that resemble a chicken-wire-shaped lattice, presenting outstanding electronic, thermal, and mechanical properties [95]. Graphene is the basic structural material of graphite, which result from the overstacking of graphene monolayers *via* van der Waals forces, resulting in interspaced neighboring layers that are 0.34 nm far apart [96, 97]. The carbon atoms in each graphene sheet establish covalent bonds due to the overlapping of trigonal planar sp^2 hybrid orbitals. The overlapping of the perpendicular unhybridized p_z orbitals accounts for the formation of the VB and the CB, respectively composed of filled π orbitals and empty π^* orbitals [98].

The mechanical exfoliation of graphite creates free-standing graphene sheets, as shown by Novoselov and Geim, who used sequential micromechanical cleavage of graphite using the "scotch-tape method." The authors were honored with the Nobel Prize in Physics in 2004, 6 years later to such an important finding [98, 99]. The direct exfoliation of bulk graphite produces layers of graphene with good quality and crystallinity, low defect densities, and high conductivity, but frequently, at a low yield

[100]. As such, graphene layers can be obtained by the chemical exfoliation of a low-cost raw material bulk graphite, which applied together with selected chemicals produce graphene and graphene derivatives, such as GO and reduced graphene oxide (rGO) [100–102]. Although water is a first-choice medium for the production of graphene-based materials, the hydrophobic nature of pristine graphene sheets tends to promote their restacking, which makes exfoliation challenging. The use of surfactants during the exfoliation processes has been considered to overcome this limitation because they allow exfoliated layers to remain suspended and avoid overstacking [101, 103]. The success of the exfoliation processes is overcoming the van der Walls forces by increasing the distance between the layers *via* chemical intercalation. Ideally, to obtain good dispersion of graphene layers, the solvents should have surface tensions of 40 mJ/m^2 [97, 101, 104]. Therefore, graphene can be exfoliated by the sonication of graphite in dimethylformamide (DMF), N-methyl-2-pyrrolidone (NMP), pyridine, and perfluorinated compounds [98, 101, 104, 105]. For instance, Hernandez *et al*. have used sonication-based exfoliation of graphite in NMP to obtain a final material containing graphene monolayers (28%) and nanosheets less than six atomic layers thick, almost in quantitative yield [106]. Commonly used sonication exfoliation processes involve shear forces and cavitation mechanisms, which involve the growth and collapse of micrometer-sized bubbles, acting on the bulk material precursor and causing their exfoliation [97].

GO is composed of sp^2 graphene layers with a high content of oxygen-containing functional groups, such as hydroxyl, epoxy, carboxylic, and carbonyl groups [107]. The UV–visible absorption spectra of GO suspensions show an absorption peak ascribed to π-π^* electronic transitions of aromatic C-C bonds and n-π^* transitions of the oxygen-containing groups, at around 230 nm and 315 nm, respectively [108]. The aqueous suspensions of GO are normally stable due to the hydrophilic character of oxygen-containing groups present in the sheets' surfaces, namely at the edges. Colloidal stability is favored by the electrostatic repulsion that arises due to anionic groups that form due to extensive proton dissociation in such functional groups, over a certain pH range. On the other hand, the presence of out-of-planar C-O covalent bonds increases the interlayer distance from 0.34 to 0.65 nm, therefore decreasing the energy needed to separate the graphene layers [96, 98, 107]. The hydrophilic nature of oxidized graphite facilitates water to be adsorbed into its lamellar structure, showing a further increase in the interlayer distance to 1.15 nm [109]. For instance, the use of polar solvents (e.g., ethanol, acetonitrile, and dimethyl sulfoxide) allows the preparation of stable colloids but either flocculation or aggregation occur when nonpolar organic solvents are used as the dispersing medium [107].

Carbon nanotubes (CNT) are 1D materials formed by graphene sheets rolled around a common axis, with diameters reaching between 0.5 and 100 nm, and lengths extending several micrometers or even millimeters [110]. CNT can be single-(SWCNT) or multi-walled (MWCNT) according to the number of graphene sheets rolled-up, that is, a single sheet or more than one, respectively. SWCNTs have diameters in the range of 1–2 nm and MWCNT show typical diameters in the range of 10–100 nm range [111]. Pristine CNT has hydrophobic nature, and their high aspect ratio favors interparticle van de Waals forces mediated by the outer walls, which results in a tendency for CNT aggregation [112]. Thus, non-functionalized CNT dispersed in a liquid medium exists as large bundles, which limit handling and, consequently, their use in many applications. Usually, mechanical disentanglement of CNT bundles is achieved by ultrasonication of the respective dispersions in which shear forces promote the separation of CNT but can also cut such nanostructures.

Nevertheless, the debundling process depends on the modification of the CNT surface by using chemical agents that enhance the compatibility of the CNT with the dispersing medium. Hence, surface modifiers, such as surfactants, homopolymers, and block copolymers, have been used to promote the dispersion of CNT in aqueous environments. In addition, surface oxidation treatments that result in the presence of carboxylic, hydroxyl, and carbonyl functional groups at the end of the tubes and on their sidewalls, also allow better dispersions of CNT in water [113].

Powder X-ray diffraction (XRD) has been used to check the crystalline structure of graphitic materials. Bulk graphite shows a strong Bragg diffraction peak at 26.6° corresponding to the reflection of (002) planes and associated with an interlayer distance of 0.34 nm. The oxidation and exfoliation of graphite increase the interlayer distance changing the peak position of the basal (002) reflection from 26.6 to 11.2°, which corresponds to an interplanar distance of 0.79 nm, as observed for GO materials [98].

Raman spectroscopy has been a key instrumental technique to study graphene materials, such as the surface chemistry of GO and the existence of structural defects. The Raman spectra of graphitic materials are typically characterized by three distinct vibrational bands: the G-, D-, and 2D-bands. The G-band is observed around 1580 cm^{-1} and is ascribed to the in-plane bending mode of the sp^2 hybridized carbon atoms in graphene. In high-quality graphene, this band is very sharp, suggesting its high crystallinity and non-defect structure. The D-band at around 1350 cm^{-1} has been associated with the amount and type of defects in the carbon lattice, for example, the existence of sp^3 hybridization or due to vacancies [114]. The extension of such defects in the carbon sheet, either at the edges or topological defects, have been monitored by Raman measurements using such diagnosis band, namely by computing the intensity ratio between the G- and D-bands [98, 104, 114]. In the Raman spectrum of high-quality pristine graphene, the D-band is not observed or is very weak, but it is observed in GO samples due to the presence of different oxygen functional groups in the carbon sheets. Hence, the D-to-G Raman band intensity ratio provides useful information on the nature and extension of structural defects that characterize the GO samples [94]. The 2D band is an overtone of the D-band, resulting from a two-photon lattice vibrational process. For true single-layer graphene, such a band occurs as a symmetric feature below 2700 cm^{-1} [104, 114]. Overstacking of successive layers results in structures of less symmetry with a Raman shift to higher wavenumbers [98]. For example, in graphite and graphite oxide materials, it is observed a broad band at about 2800 cm^{-1}. The features of the G and 2D bands are particularly useful in exfoliation and surface modification laboratorial tasks because are the first indication for distinguishing between monolayer (or few-layer) graphene and graphite-based materials. Furthermore, it has been shown that Raman methods applied to GO modified with metal sulfides are an alternative strategy to probe the surface of nanocomposite photocatalysts [115].

4. Application of carbon-based semiconductor nanostructures in photocatalysis

Metal sulfides, such as the binary compounds CdS, Ag_2S, Bi_2S_3, and CuS, have been referenced in photocatalysis literature as efficient photon harvesters of visible-light radiation [116]. When supported on graphitic materials, these semiconductors improve the conductivity for electron capture and transport [51, 117–123]. There are several methods of synthesis of metal sulfides coupled to rGO and GO substrates, which comprise solid-state, sonochemical, microwave irradiation, solvothermal, and hydrothermal methods [113, 120, 121, 124–138]. Our research group has developed a single-source method to prepare GO-based composites having supported metal sulfides. The type of metal sulfide generated *in situ* is determined by the metal dialkyldithiocarbamate complex employed as a single-molecule precursor, thus GO-based nanocomposites of Ag_2S, CuS, Bi_2S_3, and ZnS, are examples of such materials (**Figure 3**) [131]. In fact, this method is an extension of the sonochemical method first developed by Estrada *et al.* for decorating MWCNT, GO, and graphite with CdS obtained from the precursor cadmium(II) diethyldithiocarbamate [113].

Although CdS presents serious drawbacks for practical applications due to its well-known toxicity, research on CdS-based nanomaterials provide helpful insights concerning the visible-light response and underlying mechanisms in semiconductor photocatalysis [139]. There are a number of studies reporting visible-light active heterostructures of CdS/rGO and CdS/GO, which were investigated as photocatalysts for the degradation of organic dyes [124, 126, 128]. These heterostructures showed higher photocatalytic efficiency than bare CdS and could be used for up to four cycles, without loss of activity. For instance, Zhang *et al.* developed visible-light irradiated CdS/graphene nanophotocatalysts for the photooxidation of alcohols and reduction of Cr(VI) ions in water [140]. Multicomponent photocatalysts of TiO_2/CdS/rGO have shown higher photocatalytic activity than TiO_2/rGO, for the photodegradation of RhB, MB, and *p*-chlorophenol, under visible-light irradiation [141, 142]. Wang *et al.* showed that nanocomposites based on heterojunctions of CdS and TiO_2 nanoparticles were efficiently supported on rGO [141]. Such heterostructures prevented CdS photocorrosion due to the synergy that results from supporting such coupled semiconductor nanostructures on rGO (**Figure 4**). Similarly, the coupling of semiconducting phases, such as TiO_2 and CdS or Ag_2S, improves photon harvesting and charge separation and prevents the oxidation of the metal sulfides [142, 143].

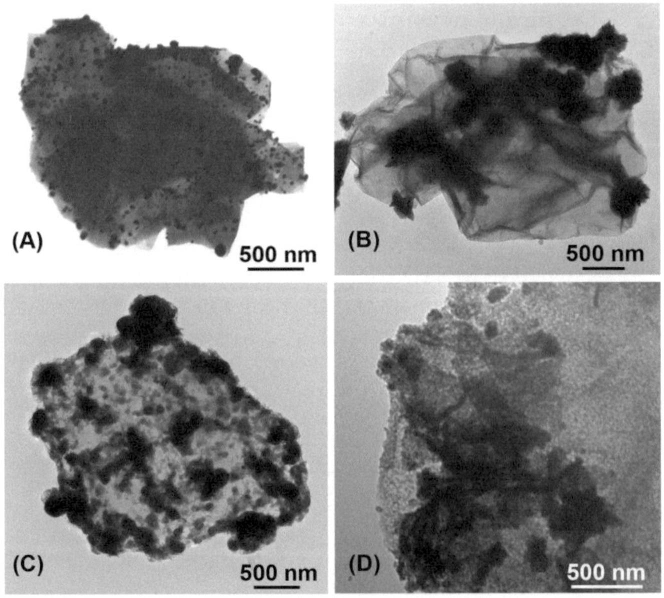

Figure 3.
TEM images of heterostructures of graphene oxide decorated with (A) Ag_2S, (B) CuS, (C) Bi_2S_3, and (D) ZnS.

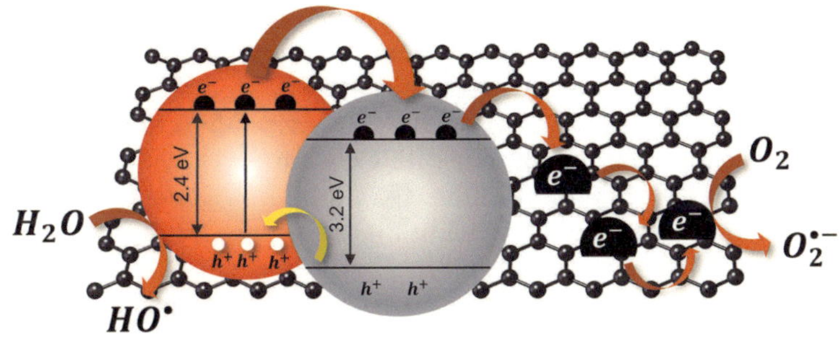

Figure 4.
The scheme illustrates visible-light photogeneration of oxygen radicals in a hybrid heterostructure composed of CdS (red)/TiO_2(gray) supported on rGO sheets dispersed in an aqueous medium.

The semiconductor Bi_2S_3 absorbs in the visible and NIR spectral range and does not pose serious toxicity concerns associated with CdS. Wang and coworkers showed that Bi_2S_3 immobilized on carbon dots have higher photocatalytic efficiency than their individual components, by investigating the degradation of MB and tetracycline under UV-, visible-, and NIR-light irradiation [144]. Khalid *et al.* synthesized nanorods of Bi_2S_3, which showed 87% efficiency in the degradation of Congo red dye, under UV-light irradiation over 90 minutes [145]. Chen *et al.* have reported improved photodegradation of 2,4-dichlorophenol irradiated with visible light in the presence of Bi_2S_3/rGO nanocomposites [137]. The authors also found that there is an optimal loading of Bi_2S_3 phases on carbon substrates, concluding that for higher contents of rGO less efficient photocatalytic systems are obtained. Similarly, for Ag_2S/graphene, it was found that the performance of the photocatalyst depended on the relative amounts of semiconductor and graphene in the nanostructure. The authors have investigated samples with distinct graphene content (wt%: 2, 4, and 6), showing that in those conditions, the photodegradation of RhB, occurred most efficiently under visible-light irradiation in the presence of the sample 4 wt% in graphene [146].

Copper sulfide is a *p*-type semiconductor with phase-dependent properties; thus, the band gap energy range between 1.2 and 2.2 eV, depending on the crystalline form present [147–151]. This is an interesting aspect for photocatalytic applications because several crystalline phases have been reported for copper sulfide, showing the metal in distinct oxidation states, such as in chalcocite (Cu_2S) and covellite (CuS). Additionaly, several nonstoichiometric phases ($Cu_{2-x}S$) have been reported showing compositions that can be easily varied depending on the experimental conditions [152]. Hybrid nanostructures composed of copper sulfide and graphene (or graphene derivatives) show high potential in photocatalysis. For instance, it has been reported that hybrid nanostructures of CuS/rGO show superior photocatalytic activity as compared to the single-phase system composed of CuS nanoparticles, for the photodegradation of organic dyes under visible-light irradiation [121, 153–155] and UV-light irradiation [156]. El-Hout *et al.* have reported that CuS/rGO photocatalysts lead to the complete mineralization of malachite green after 90 minutes, under sunlight irradiation [157]. As previously mentioned, Shi *et al.* also stated that there is an optimal loading of CuS on rGO, showing that samples with 20% of rGO have better photocatalytic activity than samples containing 30% of rGO [120]. This has been explained by the effect on the stacking of graphene sheets and metal-sulfide particle aggregation, which results from the presence of a high amount of carbon nanomaterials (rGO or graphene) employed in the composite structure [155, 158]. Wang *et al.* have found that the synergistic interaction occurring between graphene and metal-sulfide phases in CuS/graphene, with an impact on the electronic conductivity of graphene and CuS/graphene morphology, accounts for the observed stability and photoactivity of such heterostructures [159]. Matos *et al.* have synthesized hybrid composites comprising S-doped graphene decorated with CuS and Fe_3O_4 semiconductor phases, which showed higher photocatalytic ability than their individual components in the photodegradation of 4-nitrophenol, in addition, these photocatalysts could be recovered and reused in subsequent cycles [158].

Although ZnS is a non-absorbing material in the visible range due to its wide band gap energy (3.66 eV for blende structure and 3.77 eV for wurtzite structure), it has been found that ZnS coupled to carbon nanomaterials result in hybrid heterostructures with photocatalytic activity under visible-light irradiation [127, 133, 135]. Hence, Ming *et al.* have reported the degradation of ciprofloxacin, MB, and RhB under visible-light irradiation in the presence of ZnS/carbon nanostructures [160]. Also, Chen and Chakraborty have shown that under UV-light irradiation, the photodegradation of RhB and MO occurs more efficiently in the presence of the ZnS/graphene and ZnS/rGO photocatalysts, respectively [136, 161].

5. Conclusions

This chapter provided a concise overview on the use of graphene and graphene derivatives coupled to nanocrystalline semiconductors of metal sulfides in heterogeneous photocatalysis. For these applications, pure metal-sulfide nanoparticles show some limitations, which depending on the semiconductor include limited harvesting of photons in the visible region, low photocatalytic quantum yield, fast recombination of photogenerated charge carriers, and photocorrosion. Hence, several chemical strategies have been reported to improve photoefficiency and performance of these nanoparticles as photocatalysts. Metal-sulfide phases coupled with graphitic materials have been employed to prevent the photocorrosion of the chalcogenide semiconductor and to increase the photocatalytic efficiency of the resultant hybrid nanostructures. Particularly, graphene (and its derivatives) have shown great merits in improving the photogenerated charge-carriers separation and migration, also extending the light absorption range and the adsorption capacity of the photocatalysts. Furthermore, their use as supporting substrates in aqueous suspensions also inhibits the agglomeration of the particles, thus keeping exposed a high surface area to the photoactive semiconductors. The design of graphene-based materials decorated with metal sulfides as photocatalysts requires the assessment of the several parameters that might contribute to their performance in specific contexts. Hence, it has been reported that these hybrid nanostructures show optimal compositional features on the carbon nanomaterial and semiconductor, depending on the target pollutant and operational conditions. Furthermore, the surface functionalization of graphene materials also plays an important role in the development of such photocatalysts, namely the content load in the metal sulfide and their defect structure. Finally, the cost-effective production of graphene-based semiconductor nanocomposites at a large scale, envisaging their application as photostable and efficient heterogeneous photocatalysts, is still a great challenge. In the chemical design of such photocatalysts, eco-friendly up-scale strategies should be also addressed by researchers, to guarantee its future commercialization for environmental applications.

Acknowledgements

A.C. Estrada acknowledges the research position funded by national funds (OE), through FCT-Fundação para a Ciência e a Tecnologia, I.P., in the scope of the framework contract foreseen in the numbers 4, 5, and 6 of the article 23, of the Decree-Law 57/2016, of August 29, changed by Law 57/2017, of July 19. J. L. Lopes thanks FCT for the doctoral grant SFRH/BD/126241/2016. This work was developed within the scope of the project CICECO-Aveiro Institute of Materials, UIDB/50011/2020, UIDP/50011/2020 & LA/P/0006/2020, financed by national funds through the FCT/MEC (PIDDAC).

Author details

Ana Cristina Estrada*, Joana Lúcia Lopes and Tito Trindade
CICECO-Aveiro Institute of Materials, Department of Chemistry, University of Aveiro, Aveiro, Portugal

*Address all correspondence to: ana.estrada@ua.pt

References

[1] Yang X, Wang D. Photocatalysis: From fundamental principles to materials and applications. ACS Applied Energy Materials. 2018;**1**(12):6657-6693

[2] Wang H, Li X, Zhao X, Li C, Song X, Zhang P, et al. A review on heterogeneous photocatalysis for environmental remediation: From semiconductors to modification strategies. Chinese Journal of Catalysis. 2022;**43**:178-214

[3] Jabbar ZH, Esmail ES. Recent advances in nano-semiconductors photocatalysis for degrading organic contaminants and microbial disinfection in wastewater: A comprehensive review. Environmental Nanotechnology, Monitoring & Management. 2022;**17**: 100666

[4] Melchionna M, Fornasiero P. Updates on the roadmap for photocatalysis. ACS Catalysis. 2020;**10**:5493-5501

[5] Dong S, Feng J, Fan M, Pi Y, Hu L, Han X, et al. Recent developments in heterogeneous photocatalytic water treatment using visible light-responsive photocatalysts: A review. RSC Advances. 2015;**5**:14610-14630

[6] Kisch H. Semiconductor photocatalysis - Mechanistic and synthetic aspects. Angewandte Chemie International Edition. 2013;**52**:812-847

[7] Pichat P, editor. Photocatalysis and Water Purification: From Fundamentals to Recent Applications. Weinheim, Germany: Wiley-VCH; 2013

[8] Zeng X, Liu Y, Hu X, Zhang X. Photoredox catalysis over semiconductors for light-driven hydrogen peroxide production. Green Chemistry. 2021;**23**:1466-1494

[9] Rafiq A, Ikram M, Ali S, Niaz F, Khan M, Khan Q, et al. Photocatalytic degradation of dyes using semiconductor photocatalysts to clean industrial water pollution. Journal of Industrial and Engineering Chemistry. 2021;**97**:111-128

[10] Qutub N, Singh P, Sabir S, Sagadevan S, Oh WC. Enhanced photocatalytic degradation of Acid Blue dye using CdS/TiO_2 nanocomposite. Scientific Reports. 2022;**12**:5759

[11] Saeed M, Muneer M, Haq AUL, Akram N. Photocatalysis: An effective tool for photodegradation of dyes - A review. Environmental Science and Pollution Research. 2022;**29**:293-311

[12] Zhao Y, Li Y, Sun L. Recent advances in photocatalytic decomposition of water and pollutants for sustainable application. Chemosphere. 2021;**276**: 130201

[13] Rosu MC, Coros M, Pogacean F, Magerusan L, Socaci C, Turza A, et al. Azo dyes degradation using TiO_2-Pt/graphene oxide and TiO_2-Pt/reduced graphene oxide photocatalysts under UV and natural sunlight irradiation. Solid State Sciences. 2017;**70**:13-20

[14] Bai L, Pan X, Guo R, Linghu X, Shu Y, Wu Y, et al. Sunlight-driven photocatalytic degradation of organic dyes in wastewater by chemically fabricated ZnO/$Cs_4SiW_{12}O_{40}$ nanoheterojunction. Applied Surface Science. 2022;**599**:153912

[15] Yuan Y, Guo R, Hong L, Ji X, Li Z, Lin Z, et al. Recent advances and perspectives of MoS_2-based materials for photocatalytic dyes degradation: A review. Colloid Surface A. 2021;**611**: 125836

[16] Yang X, Chen Z, Zhao W, Liu C, Qian X, Zhang M, et al. Recent advances in photodegradation of antibiotic residues in water. Chemical Engineering Journal. 2021;**405**:126806

[17] Wu S, Lin Y, Hu YH. Strategies of tuning catalysts for efficient photodegradation of antibiotics in water environments: A review. Journal of Materials Chemistry A. 2021;**9**:2592-2611

[18] Roy N, Alex SA, Chandrasekaran N, Mukherjee A, Kannabiran K. A comprehensive update on antibiotics as an emerging water pollutant and their removal using nano-structured photocatalysts. Journal of Environmental Chemical Engineering. 2021;**9**:104796

[19] Zhu XD, Wang YJ, Sun RJ, Zhou DM. Photocatalytic degradation of tetracycline in aqueous solution by nanosized TiO_2. Chemosphere. 2013;**92**: 925-932

[20] Brice RP, Claire JC, Mouldi H, Vincent G, Carole CB, Gaël P. Photo-oxidation of three major pharmaceuticals in urban wastewater under artificial and solar irradiations. Journal of Photochemistry and Photobiology A: Chemistry. 2022;**425**:113673

[21] Farner Budarz J, Cooper EM, Gardner C, Hodzic E, Ferguson PL, Gunsch CK, et al. Chlorpyrifos degradation via photoreactive TiO_2 nanoparticles: Assessing the impact of a multi-component degradation scenario. Journal of Hazardous Materials. 2019; **372**:61-68

[22] Berberidou C, Kitsiou V, Kazala E, Lambropoulou DA, Kouras A, Kosma CI, et al. Study of the decomposition and detoxification of the herbicide bentazon by heterogeneous photocatalysis: Kinetics, intermediates and transformation pathways. Applied Catalysis B: Environmental. 2017;**200**: 150-163

[23] Cruz M, Gomez C, Duran-Valle CJ, Pastrana-Martínez LM, Faria JL, Silva AMT, et al. Bare TiO_2 and graphene oxide TiO_2 photocatalysts on the degradation of selected pesticides and influence of the water matrix. Applied Surface Science. 2017;**416**: 1013-1021

[24] Vaya D, Surolia PK. Semiconductor based photocatalytic degradation of pesticides: An overview. Environmental Technology and Innovation. 2020;**20**: 101128

[25] El-Saeid MH, Baqais A, Alshabanat M. Study of the photocatalytic degradation of highly abundant pesticides in agricultural soils. Molecules. 2022;**27**:634

[26] Regmi C, Joshi B, Ray SK, Gyawali G, Pandey RP. Understanding mechanism of photocatalytic microbial decontamination of environmental wastewater. Frontiers in Chemistry. 2018;**6**(33):1-6

[27] Wu H, Inaba T, Wang ZM, Endo T. Photocatalytic TiO_2@CS-embedded cellulose nanofiber mixed matrix membrane. Applied Catalysis B: Environmental. 2020;**276**:119111

[28] Hu X, Hu X, Tang C, Wen S, Wu X, Long J, et al. Mechanisms underlying degradation pathways of microcystin-LR with doped TiO_2 photocatalysis. Chemical Engineering Journal. 2017;**330**: 355-371

[29] Soleimani M, Ghasemi JB, Mohammadi Ziarani G, Karimi-Maleh H, Badiei A. Photocatalytic degradation of organic pollutants, viral and bacterial pathogens using titania nanoparticles.

Inorg Chemistry Communication. 2021; **130**:108688

[30] Jalvo B, Faraldos M, Bahamonde A, Rosal R. Antimicrobial and antibiofilm efficacy of self-cleaning surfaces functionalized by TiO2 photocatalytic nanoparticles against Staphylococcus aureus and Pseudomonas putida. Journal of Hazardous Materials. 2017; **340**:160-170

[31] Venieri D, Fraggedaki A, Kostadima M, Chatzisymeon E, Binas V, Zachopoulos A, et al. Solar light and metal-doped TiO2 to eliminate water-transmitted bacterial pathogens: Photocatalyst characterization and disinfection performance. Applied Catalysis B: Environmental. 2014;**154-155**:93-101

[32] Aftab S, Shabir T, Shah A, Nisar J, Shah I, Muhammad H, et al. Highly Efficient Visible Light Active Doped ZnO Photocatalysts for the Treatment of Wastewater Contaminated with Dyes and Pathogens of Emerging Concern. Nanomaterials. 2022;**12**:486

[33] Zhou W, Fu H. Defect-mediated electron-hole separation in semiconductor photocatalysis. Inorganic Chemistry Frontiers. 2018;**5**:1240-1254

[34] Sah C-T, Shockley W. Electron-hole recombination statistics in semiconductors through flaws with many charge conditions. Physics Review. 1958;**109**(4):1103-1115

[35] Ahmed M, Dincer I. A review on photoelectrochemical hydrogen production systems: Challenges and future directions. International Journal of Hydrogen Energy. 2019;**44**(5): 2474-2507

[36] Acar C, Dincer I, Zamfirescu C. A review on selected heterogeneous photocatalysts for hydrogen production. International Journal of Energy Research. 2014;**38**:1903-1920

[37] Bessegato GG, Guaraldo TT, de Brito JF, Brugnera MF, Zanoni MVB. Achievements and trends in photoelectrocatalysis: From environmental to energy applications. Electrocatalysis. 2015;**6**:415-441

[38] Garcia-Segura S, Brillas E. Applied photoelectrocatalysis on the degradation of organic pollutants in wastewaters. Journal of Photochemistry and Photobiology C: Photochemistry Reviews. 2017;**31**:1-35

[39] Devendran P, Selvakumar D, Ramadoss G, Sivaramakrishnan R, Alagesan T, Jayavel R, et al. A novel visible light active rare earth doped CdS nanoparticles decorated reduced graphene oxide sheets for the degradation of cationic dye from wastewater. Chemosphere. 2022;**287**: 132091

[40] Kočí K, Matějů K, Obalová L, Krejčíková S, Lacný Z, Plachá D, et al. Effect of silver doping on the TiO_2 for photocatalytic reduction of CO_2. Applied Catalysis B: Environmental. 2010;**96**: 239-244

[41] Choi W, Termin A, Hoffmann MR. The role of metal ion dopants in quantum-sized TiO_2: Correlation between photoreactivity and charge carrier recombination dynamics. The Journal of Physical Chemistry. 1984;**98**: 13669-13679

[42] Youssef Z, Colombeau L, Yesmurzayeva N, Baros F, Vanderesse R, Hamieh T, et al. Dye-sensitized nanoparticles for heterogeneous photocatalysis: Cases studies with TiO_2, ZnO, fullerene and graphene for water

purification. Dyes and Pigments. 2018; **159**:49-71

[43] Long R, English NJ, Prezhdo OV. Minimizing electron-hole recombination on TiO_2 sensitized with PbSe quantum dots: Time-domain Ab initio analysis. Journal of Physical Chemistry Letters. 2014;**5**:2941-2946

[44] Ayodhya D, Veerabhadram G. A review on recent advances in photodegradation of dyes using doped and heterojunction based semiconductor metal sulfide nanostructures for environmental protection. Mater Today Energy. 2018;**9**:83-113

[45] Mohamed HH, Bahnemann DW. The role of electron transfer in photocatalysis: Fact and fictions. Applied Catalysis B: Environmental. 2012;**128**: 91-104

[46] Zhang S, Ou X, Xiang Q, Carabineiro SAC, Fan J, Lv K. Research progress in metal sulfides for photocatalysis: From activity to stability. Chemosphere. 2022;**303**:135085

[47] Xiang Q, Yu J, Jaroniec M. Graphene-based semiconductor photocatalysts. Chemical Society Reviews. 2012;**41**:782-796

[48] Syed N, Huang J, Feng Y, Wang X, Cao L. Carbon-based nanomaterials via heterojunction serving as photocatalyst. Frontiers in Chemistry. 2019;**7**:1-7

[49] Thanh Doan Nguyen T, Nguyen D, Ngoc Doan H, Phong Vo P, Tan Huynh V, Ha Hoang V, et al. In-depth understanding of the photoreduction of graphene oxide to reduced-graphene oxide on TiO_2 surface: Statistical analysis of X-ray photoelectron and Raman spectroscopy data. Applied Surface Science. 2022;**581**:152325

[50] Spilarewicz-Stanek K, Jakimińska A, Kisielewska A, Dudek M, Piwoński I. Graphene oxide photochemical transformations induced by UV irradiation during photocatalytic processes. Materials Science in Semiconductor Processing. 2021;**123**: 105525

[51] Ahmad N, Sultana S, Sabir S, Khan MZ. Exploring the visible light driven photocatalysis by reduced graphene oxide supported Ppy/CdS nanocomposites for the degradation of organic pollutants. Journal of Photochemistry and Photobiology A: Chemistry. 2020;**386**:112129

[52] Kovačič Z, Likozar B, Hus M. Photocatalytic CO_2 reduction: A review of ab initio mechanism, kinetics, and multiscale modeling simulations. ACS Catalysis. 2020;**10**:14984-15007

[53] Lettmann C, Hildenbrand K, Kisch H, Macyk W, Maier WF. Visible light photodegradation of 4-chlorophenol with a coke-containing titanium dioxide photocatalyst. Applied Catalysis B: Environmental. 2001;**32**: 215-227

[54] Anucha CB, Altin I, Bacaksiz E, Stathopoulos VN. Titanium dioxide (TiO_2)-based photocatalyst materials activity enhancement for contaminants of emerging concern (CECs) degradation: In the light of modification strategies. Chemical Engineering Journal Advances. 2022;**10**:100262

[55] Weller H. Quantized semiconductor particles: A novel state of matter for materials science. Advanced Materials. 1993;**5**(2):88-95

[56] Esteves ACC. Trindade, T Synthetic studies on II/VI semiconductor quantum dots. Current Opinion in Solid

State & Materials Science. 2002;**6**(4): 347-353

[57] Gaya UI. Heterogeneous Photocatalysis Using Inorganic Semiconductor Solids. Drordrecht, New York: Springer; 2014. pp. 1-213

[58] El-Sayed MA. Small is different: Shape-, size-, and composition-dependent properties of some colloidal semiconductor nanocrystals. Accounts of Chemical Research. 2004; **37**:326-333

[59] Mgabi LP, Dladla BS, Malik MA, Garje SS, Akhtar J, Revaprasadu N. Deposition of cobalt and nickel sulfide thin films from thio- and alkylthio-urea complexes as precursors via the aerosol assisted chemical vapour deposition technique. Thin Solid Films. 2014;**564**: 51-57

[60] Gosavi SR, Nikam CP, Shelke AR, Patil AM, Ryu SW, Bhat JS, et al. Chemical synthesis of porous web-structured CdS thin films for photosensor applications. Materials Chemistry and Physics. 2015;**160**: 244-250

[61] Vakalopoulou E, Rath T, Kräuter M, Torvisco A, Fischer RC, Kunert B, et al. Metal sulfide thin films with tunable nanoporosity for photocatalytic applications. ACS Applied Nano Materials. 2022;**5**:1508-1520

[62] Trindade T, Brien PO, Pickett NL. Nanocrystalline semiconductors: Synthesis, properties, and perspectives. Chemistry of Materials. 2001;**13**: 3843-3858

[63] Smith AM, Nie S. Semiconductor nanocrystals: Structure, properties, and band gap engineering. Accounts of Chemical Research. 2010;**43**: 190-200

[64] Brus LE. Electron-electron and electron-hole interactions in small semiconductor crystallites: The size dependence of the lowest excited electronic state. The Journal of Chemical Physics. 1984;**80**:4403-4409

[65] Vogel R, Pohl K, Weller H. Sensitization of highly porous, polycrystalline TiO_2 electrodes by quantum sized CdS. Chemical Physics Letters. 1990;**174**:241-246

[66] Vogel R, Hoyer P, Weller H. Quantum-sized PbS, CdS, Ag_2S, Sb_2S_3, and Bi_2S_3 particles as sensitizers for various nanoporous wide-bandgap semiconductors. The Journal of Physical Chemistry. 1994;**98**:3183-3188

[67] Baker DR, Kamat PV. Photosensitization of TiO_2 nanostructures with CdS quantum dots: Particulate versus tubular support architectures. Advanced Functional Materials. 2009;**19**:805-811

[68] Tang A, Qu S, Li K, Hou Y, Teng F, Cao J, et al. One-pot synthesis and self-assembly of colloidal copper(I) sulfide nanocrystals. Nanotechnology. 2010;**21**: 285602

[69] Kadlag KP, Rao MJ, Nag A. Ligand-free, colloidal, and luminescent metal Sul fi de nanocrystals. Journal of Physical Chemistry Letters. 2013;**4**: 1676-1681

[70] Smagin VP, Davydov DA, Unzhakova NM, Biryukov AA. Synthesis and spectral properties of colloidal solutions of metal sulfides. Russian Journal of Inorganic Chemistry. 2015;**60**: 1588-1593

[71] Giansante C. Surface chemistry control of colloidal quantum dot band

gap. Journal of Physical Chemistry C. 2018;**122**:18110-18116

[72] Zhuang TT, Liu Y, Li Y, Zhao Y, Wu L, Jiang J, et al. Integration of semiconducting sulfides for full-spectrum solar energy absorption and efficient charge separation. Angewandte Chemie International Edition. 2016;**55**: 6396-6400

[73] Weng B, Qi MY, Han C, Tang ZR, Xu YJ. Photocorrosion inhibition of semiconductor-based photocatalysts: Basic principle, current development, and future perspective. ACS Catalysis. 2019;**9**:4642-4687

[74] Lang D, Xiang Q, Qiu G, Feng X, Liu F. Effects of crystalline phase and morphology on the visible light photocatalytic H2-production activity of CdS nanocrystals. Dalton Transactions. 2014;**43**:7245-7253

[75] Chen J, Wu XJ, Yin L, Li B, Hong X, Fan Z, et al. One-pot synthesis of CdS nanocrystals hybridized with single-layer transition-metal dichalcogenide nanosheets for efficient photocatalytic hydrogen evolution. Angewandte Chemie International Edition. 2015;**54**: 1210-1214

[76] Zhang K, Kim W, Ma M, Shi X, Park JH. Tuning the charge transfer route by p-n junction catalysts embedded with CdS nanorods for simultaneous efficient hydrogen and 7oxygen evolution. Journal of Materials Chemistry A. 2015;**3**(9): 4803-4810

[77] Huang H, Dai B, Wang W, Lu C, Kou J, Ni Y, et al. Oriented built-in electric field introduced by surface gradient diffusion doping for enhanced photocatalytic H_2 evolution in CdS nanorods. Nano Letters. 2017;**17**: 3803-3808

[78] Xie YP, Yu ZB, Liu G, Ma XL, Cheng HM. CdS-mesoporous ZnS core-shell particles for efficient and stable photocatalytic hydrogen evolution under visible light. Energy & Environmental Science. 2014;**7**:1895-1901

[79] Bo T, Wang Y, Wang J, Zhao Z, Zhang J, Zheng K, et al. Photocatalytic H_2 evolution on CdS modified with partially crystallized MoS_2 under visible light irradiation. Chemical Physics Letters. 2020;**746**:137305

[80] Li P, He T. Common-cation based Z-scheme ZnS@ZnO core-shell nanostructure for efficient solar-fuel production. Applied Catalysis B: Environmental. 2018;**238**:518-524

[81] Tian B, Yang B, Li J, Li Z, Zhen W, Wu Y, et al. Water splitting by CdS/Pt/WO_3-CeO_x photocatalysts with assisting of artificial blood perfluorodecalin. Journal of Catalysis. 2017;**350**:189-196

[82] Berr MJ, Wagner P, Fischbach S, Vaneski A, Schneider J, Susha AS, et al. Hole scavenger redox potentials determine quantum efficiency and stability of Pt-decorated CdS nanorods for photocatalytic hydrogen generation. Applied Physics Letters. 2012;**100**: 223903

[83] Davis AP, Huang CP. The photocatalytic oxidation of sulfur-containing organic compounds using cadmium sulfide and the effect on CdS photocorrosion. Water Research. 1991; **25**:1273-1278

[84] Huang L, Wang X, Yang J, Liu G, Han J, Li C. Dual cocatalysts loaded type I CdS/ZnS core/shell nanocrystals as effective and stable photocatalysts for H2 evolution. Journal of Physical Chemistry C. 2013;**117**:11584-11591

[85] Yi X, Li H, Wang P, Fan J, Yu H. Boosting antiphotocorrosion and hydrogen-production activity of cadmium sulfide by cobalt lactate complex. Applied Surface Science. 2020; **512**:144786

[86] Wang P, Li H, Sheng Y, Chen F. Inhibited photocorrosion and improved photocatalytic H^2-evolution activity of CdS photocatalyst by molybdate ions. Applied Surface Science. 2019;**463**:27-33

[87] Aral Dhas N, Zaban A, Gedanken A. Surface synthesis of zinc sulfide nanoparticles on silica microspheres: Sonochemical preparation, characterization, and optical properties. Chemistry of Materials. 1999;**11**(3): 806-813

[88] Monteiro OC, Esteves ACC, Trindade T. The synthesis of SiO2@CdS nanocomposites using single-molecule precursors. Chemistry of Materials. 2002;**14**(7):2900-2904

[89] Lv XJ, Fu WF, Chang HX, Zhang H, Cheng JS, Zhang GJ, et al. Hydrogen evolution from water using semiconductor nanoparticle/graphene composite photocatalysts without noble metals. Journal of Materials Chemistry. 2012;**22**:1539-1546

[90] Han L, Wang P, Dong S. Progress in graphene-based photoactive nanocomposites as a promising class of photocatalyst. Nanoscale. 2012;**4**: 5814-5825

[91] Radich JG, Krenselewski AL, Zhu J, Kamat PV. Is graphene a stable platform for photocatalysis? Mineralization of reduced graphene oxide with UV-irradiated TiO_2 nanoparticles. Chemistry of Materials. 2014;**26**:4662-4668

[92] Putri LK, Tan LL, Ong WJ, Chang WS, Chai SP. Graphene oxide: Exploiting its unique properties toward visible-light-driven photocatalysis. Applied Materials Today. 2016;**4**:9-16

[93] Tiwari S, Jhamb N, Kumar S, Ganguli AK. Synthesis of photocorrosion-resistant VS_4-MoS_2-rGO based nanocomposite with efficient photoelectrochemical water-splitting activity. ChemNanoMat; 2022;**8**:3

[94] Wang W, Tao Y, Fan J, Yan Z, Shang H, Phillips DL, et al. Fullerene–Graphene acceptor drives ultrafast carrier dynamics for sustainable CdS photocatalytic hydrogen evolution. Advanced Functional Materials. 2022;**32**: 2201357

[95] Georgakilas V, Perman JA, Tucek J, Zboril R. Broad family of carbon nanoallotropes: Classification, chemistry, and applications of fullerenes, carbon dots, nanotubes, graphene, nanodiamonds, and combined superstructures. Chemical Reviews. 2015;**115**:4744-4822

[96] Hontoria-Lucas C, López-Peinado AJ, López-González J, Rojas-Cervantes ML, Martín-Aranda RM. Study of oxygen-containing groups in a series of graphite oxides: Physical and chemical characterization. Carbon. 1995;**33**(11): 1585-1592

[97] Cai M, Thorpe D, Adamson DH, Schniepp HC. Methods of graphite exfoliation. Journal of Materials Chemistry. 2012;**22**(48):24992-25002

[98] Spyrou K, Rudolf P. An Introduction to Graphene. In: Georgakilas V, editor. Functionalization of Graphene. Weinheim, Germany: Wiley VCH; 2014. pp. 1-18

[99] Zhu Y, Murali S, Cai W, Li X, Suk JW, Potts JR, et al. Graphene and

graphene oxide: Synthesis, properties, and applications. Advanced Materials. 2010;**22**:3906-3924

[100] Boukhvalov DW, Katsnelson MI. Chemical functionalization of graphene with defects. Nano Letters. 2008;**8**(12): 4374-4379

[101] Ciesielski A, Samorì P. Graphene via sonication assisted liquid-phase exfoliation. Chemical Society Reviews. 2014;**43**(1):381-398

[102] Jilani A, Othman MHD, Ansari MO, Hussain SZ, Ismail AF, Khan IU, et al. Graphene and its derivatives: Synthesis, modifications, and applications in wastewater treatment. Environmental Chemistry Letters. 2018; **16**:1301-1323

[103] Bourlinos AB, Georgakilas V, Zboril R, Sterioti TA, Stubos AK. Liquid-phase exfoliation of graphite towards solubilized graphenes. Small. 2009; **5**(16):1841-1845

[104] Khan U, Porwal H, Óneill A, Nawaz K, May P, Coleman JN. Solvent-exfoliated graphene at extremely high concentration. Langmuir. 2011;**27**: 9077-9082

[105] Coleman JN. Liquid exfoliation of defect-free graphene. Accounts of Chemical Research. 2013;**46**:14-22

[106] Hernandez Y, Nicolosi V, Lotya M, Blighe FM, Sun Z, De S, et al. High-yield production of graphene by liquid-phase exfoliation of graphite. Nature Nanotechnology. 2008;**3**: 563-568

[107] Compton OC, Nguyen ST. Graphene oxide, highly reduced graphene oxide, and graphene: Versatile building blocks for carbon-based materials. Small. 2010;**6**:711-723

[108] Wang HX, Wang Q, Zhou KG, Zhang HL. Graphene in light: Design, synthesis and applications of photo-active graphene and graphene-like materials. Small. 2013;**8**:1266-1283

[109] Lerf A, Buchsteiner A, Pieper J, Schöttl S, Dekany I, Szabo T, et al. Hydration behavior and dynamics of water molecules in graphite oxide. Journal of Physics and Chemistry of Solids. 2006;**67**:1106-1110

[110] Saleh TA. The Role of Carbon Nanotubes in Enhancement of Photocatalysis. In: Suzuki S, editor. Syntheses and Applications of Carbon Nanotubes and Their Composites. London: IntechOpen; 2013. pp. 479–494

[111] Liu Z, Tabakman S, Welsher K, Dai H. Carbon nanotubes in biology and medicine: In vitro and in vivo detection, imaging and drug delivery. Nano Research. 2009;**2**:85-120

[112] Soulie-Ziakovic C, Nicolay R, Prevoteau A, Leibler L. Dispersible carbon nanotubes. Chemistry A European Journal. 2014;**20**(5):1210-1217

[113] Estrada AC, Mendoza E, Trindade T. Decoration of carbon nanostructures with metal sulfides by sonolysis of single-molecule precursors. European Journal of Inorganic Chemistry. 2015;**2014**:3184-3190

[114] Allen MJ, Tun VC, Kaner RB. Honeycomb carbon: A review of graphene. Chemical Reviews. 2010;**110**: 132-145

[115] Lopes JL, Fateixa S, Estrada AC, Gouveia JD, Gomes JRB, Trindade T. Surface-enhanced Raman scattering due to a synergistic effect on ZnS and graphene oxide. Journal of Physical Chemistry C. 2020;**124**(23):12742-12751

[116] Hao H, Lang X. Metal sulfide photocatalysis: Visible-Light-induced organic transformations. ChemCatChem. 2019;**11**:1378-1393

[117] Mondal A, Prabhakaran A, Gupta S, Subramanian VR. Boosting photocatalytic activity using reduced graphene oxide (RGO)/semiconductor nanocomposites: Issues and future scope. ACS Omega. 2021;**6**:8734-8743

[118] Hu Y, Zhou C, Wang H, Chen M, Zeng G, Liu Z, et al. Recent advance of graphene/semiconductor composite nanocatalysts: Synthesis, mechanism, applications and perspectives. Chemical Engineering Journal. 2021;**414**:128795

[119] Sun Y, Li G, Xu J, Lei B, Feng H, Sun Z. Impacts of graphene sheets on photoelectric and photocatalytic activities of SnS_2 nanoparticles. Materials Chemistry and Physics. 2019; **229**:92-97

[120] Chen FJ, Cao YL, Jia DZ. A room-temperature solid-state route for the synthesis of graphene oxide-metal sulfide composites with excellent photocatalytic activity. CrystEngComm. 2013;**15**(23):4747-4754

[121] Shi J, Zhou X, Liu Y, Su Q, Zhang J, Du G. Sonochemical synthesis of CuS/reduced graphene oxide nanocomposites with enhanced absorption and photocatalytic performance. Materials Letters. 2014;**126**:220-223

[122] Zou L, Wang X, Xu X, Wang H. Reduced graphene oxide wrapped CdS composites with enhanced photocatalytic performance and high stability. Ceramics International. 2016; **42**:372-378

[123] Azimi HR, Ghoranneviss M, Elahi SM, Mahmoudian MR, Jamali-Sheini F, Yousefi R. Excellent photocatalytic performance under visible-light irradiation of ZnS/rGO nanocomposites synthesized by a green method. Frontiers of Materials Science. 2016;**10**(4):385-393

[124] Liu F, Shao X, Wang J, Yang S, Li H, Meng X, et al. Solvothermal synthesis of graphene-CdS nanocomposites for highly efficient visible-light photocatalyst. Journal of Alloys and Compounds. 2013;**551**: 327-332

[125] Mo Z, Liu P, Guo R, Deng Z, Zhao Y, Sun Y. Graphene sheets/ Ag_2S nanocomposites: Synthesis and their application in supercapacitor materials. Materials Letters. 2012;**68**: 416-418

[126] Kaur M, Umar A, Mehta SK, Kansal SK. Reduced graphene oxide-CdS heterostructure: An efficient fluorescent probe for the sensing of Ag(I) and sunset yellow and a visible-light responsive photocatalyst for the degradation of levofloxacin drug in aqueous phase. Applied Catalysis B: Environmental. 2019;**245**:143-158

[127] Sagadevan S, Chowdhury ZZ, Bin JMR, Rafique RF, Aziz FA. One pot synthesis of hybrid ZnS–Graphene nanocomposite with enhanced photocatalytic activities using hydrothermal approach. Journal of Materials Science: Materials in Electronics. [Internet. 2018;**29**(11): 9099-9107. Available from:. DOI: 10.1007/s10854-018-8937-z

[128] Wei XN, Ou CL, Guan XX, Peng ZK, Zheng XC. Facile assembly of CdS-reduced graphene oxide heterojunction with enhanced elimination performance for organic pollutants in wastewater. Applied Surface Science. 2019;**469**:666-673

[129] Yu L, Ruan H, Zheng Y, Li D. A facile solvothermal method to produce ZnS quantum dots-decorated graphene nanosheets with superior photoactivity. Nanotechnology. 2013;**24**: 375601

[130] Zhang Y, Zhang N, Tang ZR, Xu YJ. Graphene transforms wide band gap ZnS to a visible light photocatalyst. the new role of graphene as a macromolecular photosensitizer. ACS Nano. 2012;**6**(11): 9777-9789

[131] Lopes J, Estrada A, Fateixa S, Ferro M, Trindade T. A general route for growing metal sulfides onto graphene oxide and exfoliated graphite oxide. Nanomaterials. 2017;**7**:245

[132] Bai S, Shen X, Zhu G, Zhou H, Xu H, Fu G, et al. Optical properties and a simple and general route for the rapid syntheses of reduced graphene oxide-metal sulfide nanocomposites. European Journal of Inorganic Chemistry. 2013;**2**: 256-262

[133] Hu H, Wang X, Liu F, Wang J, Xu C. Rapid microwave-assisted synthesis of graphene nanosheets-zinc sulfide nanocomposites: Optical and photocatalytic properties. Synthetic Metals. 2011;**161**:404-410

[134] Liu X, Pan L, Lv T, Zhu G, Sun Z, Sun C. Microwave-assisted synthesis of CdS-reduced graphene oxide composites for photocatalytic reduction of Cr(VI). Chemical Communications. 2011;**47**: 11984-11986

[135] Thangavel S, Krishnamoorthy K, Kim SJ, Venugopal G. Designing ZnS decorated reduced graphene-oxide nanohybrid via microwave route and their application in photocatalysis. Journal of Alloys and Compounds. 2016; **683**:456-462

[136] Chen F, Cao Y, Jia D, Liu A. Solid-state synthesis of ZnS/graphene nanocomposites with enhanced photocatalytic activity. Dyes and Pigments. 2015;**120**:8-14

[137] Chen Y, Tian G, Mao G, Li R, Xiao Y, Han T. Facile synthesis of well-dispersed Bi_2S_3 nanoparticles on reduced graphene oxide and enhanced photocatalytic activity. Applied Surface Science. 2016;**378**:231-238

[138] Hull S, Keen DA, Sivia DS, Madden PA, Wilson M. The high-temperature superionic behaviour of Ag_2S. Journal of Physics Condensed Matter. 2002;**14**:L9-L17

[139] Sharma S, Dutta V, Raizada P, Hosseini-Bandegharaei A, Singh P, Nguyen VH. Tailoring cadmium sulfide-based photocatalytic nanomaterials for water decontamination: A review. Environmental Chemistry Letters. 2021; **19**:271-306

[140] Zhang N, Yang MQ, Tang ZR, Xu YJ. CdS-graphene nanocomposites as visible light photocatalyst for redox reactions in water: A green route for selective transformation and environmental remediation. Journal of Catalysis. 2013;**303**:60-69

[141] Wang L, Wen M, Wang W, Momuinou N, Wang Z, Li S. Photocatalytic degradation of organic pollutants using rGO supported TiO_2-CdS composite under visible light irradiation. Journal of Alloys and Compounds. 2016;**683**:318-328

[142] Tian Z, Yu N, Cheng Y, Wang Z, Chen Z, Zhang L. Hydrothermal synthesis of graphene/TiO_2/CdS nanocomposites as efficient visible-light-driven photocatalysts. Materials Letters. 2017;**194**:172-175

[143] Liu T, Liu B, Yang L, Ma X, Li H, Yin S, et al. RGO/Ag_2S/TiO_2 ternary heterojunctions with highly enhanced UV-NIR photocatalytic activity and stability. Applied Catalysis B: Environmental. 2017;**204**:593-601

[144] Wang J, Zhang X, Wu J, Chen H, Sun S, Bao J, et al. Preparation of Bi2S3/carbon quantum dot hybrid materials with enhanced photocatalytic properties under ultraviolet-, visible- and near infrared-irradiation. Nanoscale. 2017;**9**: 15873-15882

[145] Khalid A, Akhtar N, He K, Liu B, Ahmad M, Ambreen J, et al. Bismuth sulfide photocatalysis water treatment under visible irradiation. Research on Chemical Intermediates. 2021;**47**: 3395-3409

[146] Hu W, Zhao L, Zhang Y, Zhang X, Dong L, Wang S. Preparation and photocatalytic activity of graphene-modified Ag_2S composite. Journal of Experimental Nanoscience. 2016;**11**: 433-443

[147] Cao Q, Che R, Chen N. Scalable synthesis of Cu_2S double-superlattice nanoparticle systems with enhanced UV/visible-light-driven photocatalytic activity. Applied Catalysis B: Environmental. 2015;**162**:187-195

[148] Srinivas B, Kumar BG, Muralidharan K. Stabilizer free copper sulphide nanostructures for rapid photocatalytic decomposition of rhodamine B. Journal of Molecular Catalysis A: Chemical. 2015;**410**:8-18

[149] Ayodhya D, Venkatesham M, Santoshi kumari A, Reddy GB, Ramakrishna D, Veerabhadram G. Photocatalytic degradation of dye pollutants under solar, visible and UV lights using green synthesised CuS nanoparticles. Journal of Experimental Nanoscience. 2016;**11**(6):418-432

[150] Li S, Zhang Z, Yan L, Jiang S, Zhu N, Li J, et al. Fast synthesis of CuS and Cu9S5 microcrystal using subcritical and supercritical methanol and their application in photocatalytic degradation of dye in water. Journal of Supercritical Fluids. 2017;**123**:11-17

[151] Wu H, Li Y, Li Q. Facile synthesis of CuS nanostructured flowers and their visible light photocatalytic properties. Applied Physics A: Materials Science & Processing. 2017;**123**:196

[152] Estrada AC, Silva FM, Soares SF, Coutinho JAP, Trindade T. An ionic liquid route to prepare copper sulphide nanocrystals aiming at photocatalytic applications. RSC Advances. 2016;**6**(41): 34521-34528

[153] Zhang Y, Tian J, Li H,Wang L, Qin X, Asiri AM, et al. Biomolecule-assisted, environmentally friendly, one-pot synthesis of CuS/reduced graphene oxide nanocomposites with enhanced photocatalytic performance. Langmuir. 2012;**28**:12893-12900

[154] Hu XS, Shen Y, Zhang YT, Nie JJ. Preparation of flower-like CuS/reduced graphene oxide(RGO) photocatalysts for enhanced photocatalytic activity. Journal of Physics and Chemistry of Solids. 2017; **103**:201-208

[155] Saranya M, Ramachandran R, Kollu P, Jeong SK, Grace AN. A template-free facile approach for the synthesis of CuS–rGO nanocomposites towards enhanced photocatalytic reduction of organic contaminants and textile effluents. RSC Advances. 2015;**5**: 15831-15840

[156] Siong VLE, Lee KM, Juan JC, Lai CW, Tai XH, Khe CS. Removal of

methylene blue dye by solvothermally reduced graphene oxide: A metal-free adsorption and photodegradation method. RSC Advances. 2019;**9**: 37686-37695

[157] El-Hout SI, El-Sheikh SM, Gaber A, Shawky A, Ahmed AI. Highly efficient sunlight-driven photocatalytic degradation of malachite green dye over reduced graphene oxide-supported CuS nanoparticles. Journal of Alloys and Compounds. 2020;**849**:156573

[158] Matos R, Nunes MS, Kuźniarska-Biernacka I, Rocha M, Guedes A, Estrada AC, et al. Graphene@metal sulfide/oxide nanocomposites as novel photo-fenton-like catalysts for 4-nitrophenol degradation. European Journal of Inorganic Chemistry. 2021;**47**: 4915-4928

[159] Wang Y, Zhang L, Jiu H, Li N, Sun Y. Depositing of CuS nanocrystals upon the graphene scaffold and their photocatalytic activities. Applied Surface Science. 2014;**303**:54-60

[160] Ming F, Hong J, Xu X, Wang Z. Dandelion-like ZnS/carbon quantum dots hybrid materials with enhanced photocatalytic activity toward organic pollutants. RSC Advances. 2016;**6**: 31551-31558

[161] Chakraborty K, Chakrabarty S, Das P, Ghosh S, Pal T. UV-assisted synthesis of reduced graphene oxide zinc sulfide composite with enhanced photocatalytic activity. Materials Science & Engineering B. 2016;**204**:8-14

Chapter 9

Photocatalysis in the Skin Related to UVA Photoaging

Satoshi Yoshimoto, Moemi Yoshida and Masamitsu Ichihashi

Abstract

Skin aging is classified into chronological aging and photoaging, involving ultraviolet radiation (UV), visible light, and others. UVA and UVA-photosensitizers (involving photocatalysis) contribute to the production of chronically induced skin damage that results in photoaging, especially wrinkles that are associated with histopathological actinic elastosis in the dermis. Hydrogen peroxide produced by the photosensitization involving photocatalysis, such as flavin, has been proposed as a risk factor for photoaging. It was also revealed that hydrogen peroxide production by UVA is amplified through the following reactions. The photosensitization of type I and type II by riboflavin as an initiator oxidizes coexisted amino acids and vitamins. The oxidized amino acids and vitamins produce reactive oxygen species (ROS), including hydrogen peroxide, through secondary UVA-photosensitization. Finally, we proposed a screening method for detecting the effects of antioxidants on UVA-photosensitization. In our previous study, histidine and other antioxidants did not inhibit UVA-photosensitized by riboflavin, even though they have been reported to scavenge singlet oxygen and superoxide. In contrast, we demonstrated that ergothioneine suppressed the production of hydrogen peroxide by UVA-photosensitization. The purpose of this report is to provide new findings for the prevention of photoaging by discussing the characteristics of UVA-photocatalysts in the skin.

Keywords: photoaging, photosensitization, riboflavin, UVA, hydrogen peroxide

1. Introduction

In living organisms, photosensitizing reactions using photosensitizers (involving photocatalysts in this review) are used in a wide variety of ways. Beneficial uses include treatment of skin disease [1], elimination of cancer cells [2], and construction of tough collagen structures [3]. On the other hand, the photosensitizing reaction by amino acids and vitamin components in the living body through exposure to ultraviolet rays can cause skin aging. In this chapter, we reviewed the photosensitizing reactions in living organisms related to photoaging.

The important role of the skin is to protect the body from various external environmental factors. In other words, the skin has the role of preventing physical, chemical, and bacteriological invasion into the body and preventing water loss due to evaporation. The skin is composed of three layers: the outermost layer of the epidermis, the dermis, and the subcutaneous adipose tissue. Keratinocytes and melanocytes are well-known cells that make up the epidermis. Keratinocytes contribute to the barrier function of the skin by differentiating, and melanocytes produce melanin pigment to protect the epidermis and dermis from ultraviolet rays. In the dermis, the extracellular components, produced by dermal fibroblasts, have collagen fibers, elastic fibers, hyaluronic acid, and proteoglycan as the main constituents. These extracellular components have a high water retention effect and contribute to the maintenance of the hydrophilic environment of the dermis. Subcutaneous adipose tissue is rich in mature adipocytes and has the role of reducing external pressure. In addition to those essential capabilities, the skin has also a role in thermoregulation, immune response, and social communication [4–8].

Like many other organs, the skin undergoes adverse changes over time in response to changes in lifestyle and hormonal balance. However, unlike most other organs, the skin receives major changes due to exposure to the environment, especially UV rays from the sun. Chronic exposure to UV rays causes an early aging phenotype (photoaging) that resembles the aging caused by the passage of time (chronological aging) [9]. As a result, areas of the body that are routinely exposed to the sun, such as the face, neck, and forearms, show the visible manifestation of aging (senile lentigo, wrinkles, sagging, etc.) faster than other areas of the body [10].

The effects of chronological aging and photoaging induce serious alteration in the dermis with detrimental changes to the extracellular matrix [11]. Collagen accounts for the majority of the dermal matrix. However, with age, normal collagen content decreases, and the ratio of collagen degenerated by oxidation, carbonyl modification, and glycation increases [12, 13]. In addition, the ability of fibroblasts to generate collagen is diminished by environmental factors in addition to chronological aging. In particular, the upper layers of the dermis on the face, neck, and back of the hands, which have been exposed to the sun's rays, are characterized by the accumulation of glycated elastic fibers (solar elastosis) [14]. Furthermore, photoaging is mainly induced by long-term UV exposure. UVA, a long-wavelength UV rays, causes serious damage to the skin due to ROS produced by the reaction with photosensitizers in the skin (**Figure 1**) [15]. However, due to the wide variety of *in vivo* photosensitizers associated with ROS production, understanding the mechanisms of ROS production and effective quenching methods is very complicated. In this chapter, we focused on the hydrogen peroxide generated by the photosensitization reaction by multiple photosensitizers and UVA in the living body. We also introduced a simple screening method for discovering active ingredients that are effective against the photosensitization reaction through UVA.

2. UVA-photosensitization reaction and photoaging

Sunlight is now considered to be one of the most harmful extrinsic factors that can induce ROS production [16]. Other well-known factors include tobacco smoke [17, 18], PM2.5, and air pollutants [19, 20]. The spectrum of sunlight includes infrared energy (greater than 760 nm), visible light (400–760 nm), and ultraviolet (UV) light (less than 400 nm). UVs are further classified into UVA (400–315 nm), UVB (315–280 nm), and UVC (280–100 nm) [21]. Photobiological reactions are primarily produced by exposure to UVB and UVA radiation. UV is a major cause of DNA damage in the epidermal skin cells [22, 23]. Furthermore, UV rays contribute to change in the stem cell niche, which can lead to photoaging [1, 24]. UVA accounts for about 95% of the UV rays that reach the surface of the ground and is likely to contribute to the risk of the initiation of human skin cancer [25].

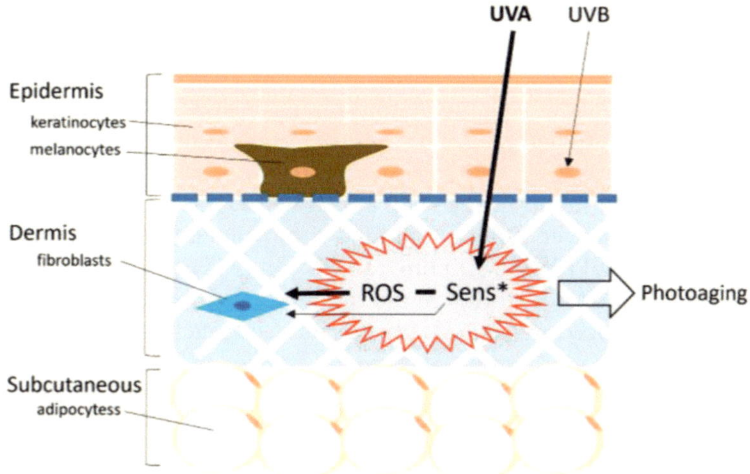

Figure 1.
Photoaging of the skin: Photoaging of the dermis is mainly induced by long-term UV exposure. UVA, a long-wavelength ultraviolet light, causes serious damage to the dermal skin due to ROS produced by the reaction with photosensitizers in the body. Sens: Activated photosensitizers.*

UVA indirectly damages DNA [26], in contrast to UVB, which is absorbed by DNA and causes direct cytotoxicity [27]. UVA-induced damage is mainly caused through interactions with the photosensitizers, which produce ROS [28]. UVA causes various changes in the dermis, which appear to be primarily involved in the initiation and progression of photoaging. These photosensitizers absorb photons/energy, resulting in a photosensitizer excited state called the singlet excited state [29, 30]. Two reactions can occur following this first reaction. One is a reaction that emits either heat or fluorescence and returns to the ground state, and the other is a triplet excited state due to intersystem crossing. This triplet excited state reacts with both DNA and molecular oxygen, resulting in DNA modification or the production of ROS, such as superoxide, hydroxyl radical, singlet oxygen, and hydrogen peroxide [31]. The term "photoaging" was coined to emphasize the importance of UV and the resulting ROS formation in the skin-aging process (**Figure 2**) [32].

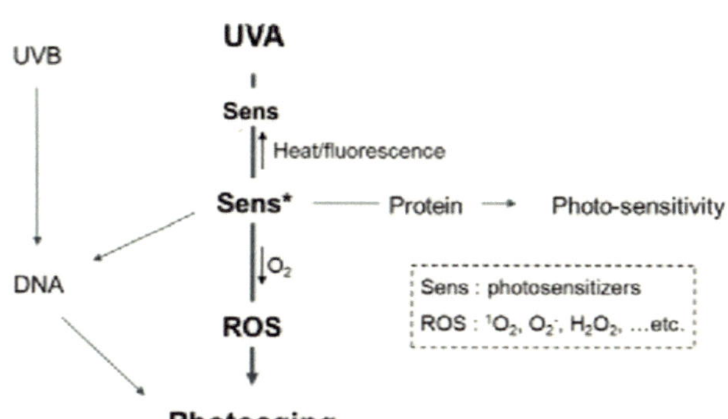

Figure 2.
Scheme of photosensitization and Photoaging: UVA that penetrates the epidermis and reaches the dermis is absorbed by the photosensitizers in the skin tissue and produces ROS under existing O_2 molecules.

3. Photosensitizing reactions and photosensitizing components that occur in the presence of oxygen molecules

Under certain circumstances, endogenous photosensitizers, such as porphyrins, melanin, urocanic acid, bilirubin, flavins, pterins, and amino acid, such as tryptophan, act as photosensitizers [31–33]. Photosensitization, such as melanin and bilirubin, out of the major pigments of the skin, are known as the major absorbers of visible regions of the spectrum over 300 to 600 nm. On the other hand, other photosensitizers, such as urocanic acid (250 to 300 nm), riboflavin (355 nm), and pterin (345 to 375 nm), show maximum absorption in the UV range and are hardly absorbed in the visible region of the spectrum [34, 35]. Photosensitized reactions involving oxygen molecules are reported as either type I or type II. Previously, the definition of type II reaction involved the formation of singlet oxygen (major reaction) and superoxide (minor reaction) [36, 37]. Currently, it was revised, and the definition of type I reaction now involves the formation of superoxide because we define type I on the basis of the formation of radicals. Type II is now established as the sensitized formation of singlet oxygen. This review followed the definition of guidelines in Baptista et al. [38].

Dermal fibroblasts are often used as a research target for skin aging [39]. The senescence of dermal fibroblasts is thought to have a significant effect on dermal matrix metabolism and degenerate dermal structure [40]. It is well-known that matrix degradation by activation of matrix metalloproteinases (MMPs) contributes to the formation of wrinkles and sagging skin [41]. In addition, it has been reported that aging fibroblasts, which showed an increase of SA-β-galactosidase [42], intercellular ROS, p16 expression [43], DNA damage, and other typical cellular senescence phenotypes, are present in the dermis at the photoaging site. *In vitro*, UVA and UVA-photocatalysts are often used to induce cellular senescence (**Figure 3**). Thus, the ROS generated by photosensitization reactions are considered to be important targets for the development of anti-photoaging agents.

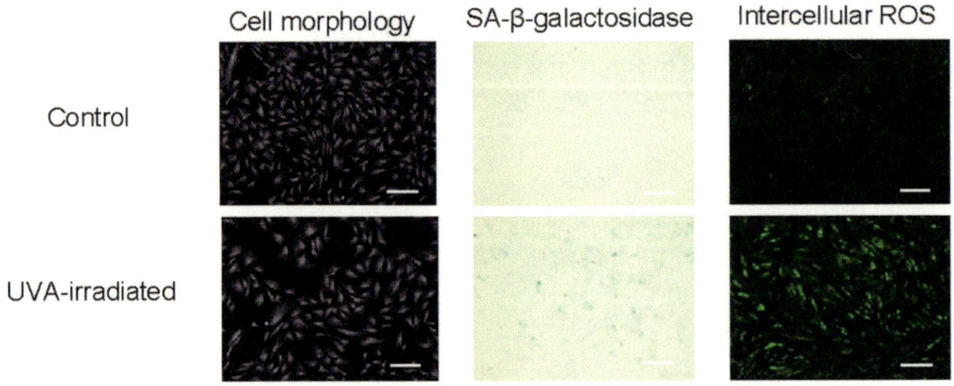

Figure 3.
Senescence phenotype of fibroblast induced by UVA irradiation in vitro: Repeated UVA irradiation of human dermal fibroblasts at a dose of 36 J/cm²/10 days in the condition of riboflavin coexisted to amino acids and vitamins induced typical phenotype of cellular senescence. Left: Flattened and larger cells have a greater diameter ratio compared to nonirradiated control cells (calcein-AM staining). Center: Increased of blue coloration cells (senescence-associated β-galactosidase staining) compared to nonirradiated control cells. Right: Increased higher level of green fluorescence intensity (dihydrorhodamine 123 staining) compared to nonirradiated control cells. Scale bar; 200 μm.

4. Study of UVA-photosensitization *in vitro*

Many researchers have investigated photosensitization reactions *in vitro* system using a single photocatalyst [44–50]. On the other hand, some research groups have shown that the coexistence of multiple components induces stronger cytotoxicity than a single photosensitizer condition [44, 51]. It has been reported that a low concentration of photocatalyst, about 0.4 μg/mL riboflavin, can induce cell damage in coexisting conditions with amino acids and vitamins [52]. On the other hand, in single riboflavin conditions, it takes about 100 μg/mL of riboflavin to induce cell toxicity *in vitro* [53]. In the condition that photocatalysts and multiple components coexist, the type I mechanism and the type II mechanism may be amplified at the same time in photosensitization. Therefore, it is necessary to consider which of the type I and type II reactions is the main reaction under the coexistence conditions of photocatalysis and multiple components.

4.1 Combination of components with type I as the main reaction (riboflavin and folic acid)

One type of photosensitization occurs in aqueous solution containing riboflavin and folic acid. Folic acid coexisting riboflavin in aqueous solution easily undergoes oxidative degradation upon UVA exposure to produce a pterin derivative [54]. Pterin derivatives are reported to accumulate in vitiligo skin [55], and to induce UV stress in melanocytes. Pterin derivatives are known to produce superoxide, hydrogen peroxide, and other types of ROS through UVA-photosensitizing [56]. In our previous study, when aqueous solution containing both folic acid and riboflavin was exposed to UVA, blue fluorescence derived from pterin derivatives appeared earlier than in aqueous solution containing folic acid alone. Those results indicate that the oxidative degradation of folic acid proceeds only very slowly in HBSS containing folic acid alone, but occurs rapidly in the presence of the photosensitizer riboflavin. Since this reaction was not inhibited by NaN_3, a singlet oxygen scavenger, it was thought that the oxidative degradation of folic acid was possibly promoted via photosensitization of type I generated by riboflavin photosensitization [52]. Therefore, it is considered that the superoxide quencher is effective for these types of reactions.

4.2 Combination of components with type II as the main reaction (riboflavin and tryptophan)

The other type of photosensitization occurs between riboflavin and tryptophan. Since tryptophan has a maximum absorption wavelength in the UVB region (especially at 280 nm), exposure to UVB is known to produce tryptophan oxides, such as FICZ and kynurenine derivatives. These tryptophan oxides have absorption wavelengths in the UVA region, and it has been reported that UVA exposure produces superoxide, H_2O_2, and other types of ROS [57–59]. In our previous study, since tryptophan does not have a UV absorption region, UV exposure to tryptophan alone did not cause the oxidative degradation of tryptophan and did not produce kynurenine. However, exposure to UVA in the presence of riboflavin decreased the 280 nm absorption by tryptophan and increased the 360 nm absorption by kynurenine. Those results indicate that the oxidative degradation of tryptophan, which does not occur in aqueous solution with tryptophan alone, may be initiated by the photosensitization of riboflavin. This phenomenon was markedly suppressed by the addition of NaN_3, which suggests that the oxidative degradation of tryptophan may be promoted by a singlet oxygen generated at an earlier time by the photosensitization of riboflavin [52]. Therefore, the singlet oxygen quencher is considered to be effective for these types of reactions.

As a point to be noted, it has been reported that HEPES and phenol red can enhance the cytotoxicity and the production of ROS by photosensitization reaction under the coexistence condition with riboflavin [60, 61]. When considering a photosensitizing reaction with multiple components, it is necessary to consider the possibility that a component other than the object to be evaluated may become noise.

5. Antioxidants and inhibition of the UVA-photosensitization reaction

The following is a summary of typical antioxidants (**Table 1**). Oxygen radicals scavenger (trolox [62], lutein [63], allicin [64], resveratrol [65], isoflavones [66], quercetin [67], catechin [68], theaflavin [69], curcumin [70], chlorogenic acid [71], and superoxide dismutase [72]), singlet oxygen quencher (astaxanthin, histidine [73], and lycopene [74]), hydrogen peroxide scavengers (catalase and glutathione peroxidase), and scavengers for all type of ROS (ascorbic acid [75], ergothioneine [76], L-cysteine, glutathione [77], and tocopherol [78]) are suggested to be effective in suppressing UVA-photosensitization.

However, these antioxidants have not been investigated for the effects of the photosensitizing reaction. Therefore, it is necessary to know if these antioxidants can promote the photosensitizing reaction when coexisting with a photocatalyst, such as riboflavin.

6. Screening method for antioxidants that suppress the photosensitizer reaction *in vitro*

We have proposed an *in vitro* assay using cytotoxicity and hydrogen peroxide as detection indicators in a screening method for compounds that suppress cell

Target	Antioxidants		
Oxygen radicals	Allicin	Resveratrol	
	Catechin	Quercetin	
	Chlorogenic acid	Superoxide dismutase	
	Curcumin	Theaflavin	Ascorbic acid
	Isoflavones	Trolox	L-cysteine
	Lutein		Ergothioneine
Singlet oxygen	Astaxanthin		Glutathione
	Histidine	Lycopene	Tocopherol
Hydrogen peroxide	Ascorbate peroxidase	Glutathione peroxidase	
	Catalase		

Table 1.
The classification of typical antioxidants.

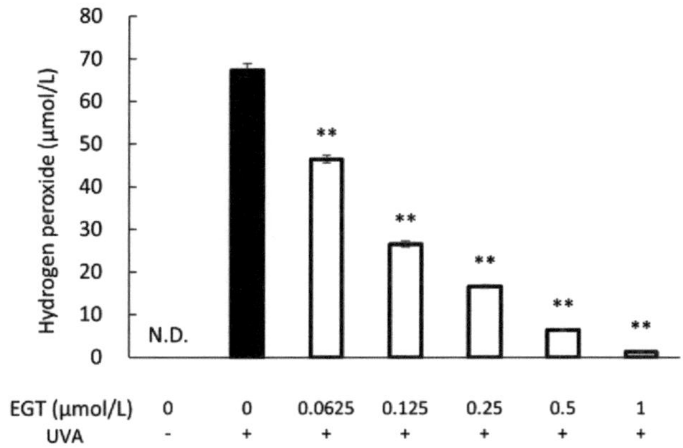

Figure 4.
*Efficacy of ergothioneine against the production of hydrogen peroxide through UVA-photosensitization: Samples were added to HBSS containing 1 μmol/L riboflavin, 30 μmol/L folic acid, and 100 μmol/L tryptophan. Hydrogen peroxide is detected in UVA irradiation control in this assay. For samples, each dose of ergothioneine (EGT) was used. A lamp (Toshiba Lighting & Technology Corporation, Yokosuka, Japan) emitting a UVA spectrum (340–410 nm) was used as the UVA source and was adjusted to an intensity of 1.0 mW/cm^2. Each sample solution was irradiated for 1 hour in an ice box for temperature control (3.6 J/cm^2). The amount of hydrogen peroxide generated in each solution after UVA irradiation with 3.6 J/cm^2 was measured by ADHP/HRP methods. Data are expressed as means ± S.D. statistical analysis was performed by Student's t-test with p-value <0.010 (**p < 0.010) considered statistically significant differences.*

damage caused by UVA-photosensitization. This is because past studies have suggested that hydrogen peroxide generated in the solvent by the UVA photosensitization reaction is the main cause of cell damage [52, 79]. Furthermore, by coexisting with the target compound during UVA exposure, it is possible to evaluate the photosensitization reaction between the target active ingredients and other components.

As an interesting example, histidine, which is used as a singlet oxygen quencher, has not been shown as an effective compound in this *in vitro* assay. This was because histidine enhances the photosensitizing response to riboflavin during UVA exposure [51]. These indicate that they may not be useful depending on the conditions of amino acids and vitamins in which existing antioxidants that are expected to have a photosensitizing effect coexist. Ergothioneine is a powerful antioxidant that has been reported to eliminate singlet oxygen, superoxide, and hydrogen peroxide. A previous study reported that ergothioneine has a protective ability against hydrogen peroxide and other ROS [76, 80]. It was indicated that ergothioneine has an anti-photosensitization efficacy because ergothioneine was treated only during UVA irradiation [81]. In our *in vitro* assay, using hydrogen peroxide as an index, the amount of hydrogen peroxide produced was suppressed in a concentration-dependent manner, without increasing hydrogen peroxide production at any concentration (**Figure 4**). These findings suggest that ergothioneine may prevents the progression of photoaging in the skin.

7. Discussions

In this paper, we discussed the possibility that ROS production through photosensitization reaction in the living body may be important factor for photoaging of the skin, especially damage to dermal fibroblasts caused by UVA.

In addition, the concentration of amino acids and vitamins in the human skin, especially in the dermis, must be clarified to discuss the role of sunlight on skin photoaging. There are some reports of the concentration of vitamins in the blood. It has been reported that riboflavin may exist at approximately 1–300 ng/mL [82, 83], folic acid may exist at approximately 13–57 ng/mL [84], and tryptophan may exist at approximately 12 μg/mL [85]. However, as far as we have investigated, there have been no reports of detailed verification of the concentrations of amino acids and vitamins in the dermis. In the future, we hope that the concentrations of amino acids and vitamins in the skin must be clarified by detailed studies on the mechanism of production of ROS by the UVA-photosensitization reaction in the skin.

Photosensitivity, unlike photoaging, is an acute response to light. Among photosensitivity research, there are also reports on the reduction of phototoxicity by studying combinations of ketoprofen with several antioxidants. The report investigates the effects of eight known radical scavengers on UV-induced photodegradation of ketoprofen and the production of ROS. Interestingly, quercetin was the only one that simultaneously suppressed the photolysis of ketoprofen and the production of ROS. Tocopherols eliminated ROS but did not suppress the photolysis of ketoprofen [86]. It can be inferred that quercetin directly quenched the photosensitizing reaction of ketoprofen.

It should be noted that it is important to look for antioxidants that suppress the reaction of the photosensitizer, which is effective not only for photoaging caused by ROS production through photosensitization but also for the prevention of photosensitivity. We concluded that understanding of the photosensitizing mechanism of environmental components, such as amino acids and vitamins in the skin, will be effective in reducing or preventing harmful skin symptoms induced by phototoxicity, which is caused by UVA.

In addition, in recent years, much attention has been paid to treatments targeting senescent cells, such as senolytics [87]. In 2018, Yoon reported that the elimination of nearby aging fibroblasts was effective in improving senile pigmented spots [88]. This strongly suggests that the phenomenon of skin aging may be caused by cell aging of fibroblasts. Therefore, the prevention of cellular senescence, especially the prevention of photoaging and photodamage caused by ultraviolet rays, is considered to be an even more important issue than before. For anti-photoaging to be effectively implemented, we need studies to elucidate the photosensitizing-reaction mechanism considering various components in the skin.

8. Conclusion

Until now, the role of photosensitizers involving photocatalysis in the skin components is poorly understood in photoaging. In this review, we focused on the hydrogen peroxide generated by the UVA-photosensitization reaction by multiple photosensitizers, such as riboflavin, amino acids and vitamins. We also introduced a simple screening method for discovering compounds that are effective for UVA-photosensitization, using ergothioneine as an example.

Acknowledgements

We are grateful to Professor Ando Hideya of the Okayama University of Science for helpful discussions and comments on the manuscript.

Author details

Satoshi Yoshimoto[1], Moemi Yoshida[2]* and Masamitsu Ichihashi[3,4,5]

1 Nikko Chemicals Co., Tokyo, Japan

2 School of Pharmacy and Pharmaceutical Sciences, Mukogawa Women's University, Nishinomiya, Japan

3 Laboratory of Stem Cell Biology, Graduate School of Pharmaceutical Sciences, Kobe Gakuin University, Kobe, Japan

4 Anti-Aging Medical Research Center, Doshisha University, Kyoto, Japan

5 BTR Arts Ginza Clinic, Tokyo, Japan

*Address all correspondence to: yoshisato@nikkolgroup.com

References

[1] Parrish JA, Fitzpatrick TB, Tanenbaum L, Pathak MA. Photochemotherapy of psoriasis with Oral Methoxsalen and longwave ultraviolet light. The New England Journal of Medicine. 1974;**291**(23):1207-1211. DOI: 10.1056/nejm197412052912301

[2] Dougherty TJ, Gomer CJ, Henderson BW, et al. Photodynamic therapy. Journal of the National Cancer Institute. 1998;**90**(12):889-905. DOI: 10.1093/JNCI/90.12.889

[3] Sorkin N, Varssano D. Corneal collagen crosslinking: A systematic review. Ophthalmologica. 2014;**232**(1):10-27. DOI: 10.1159/000357979

[4] Rittié L, Fisher GJ. Natural and sun-induced aging of human skin. Cold Spring Harbor Perspectives in Medicine. 2015;**5**(1):a015370. DOI: 10.1101/CSHPERSPECT.A015370

[5] Kligman AM. Perspectives and problems in cutaneous gerontology. The Journal of Investigative Dermatology. 1979;**73**(1):39-46. DOI: 10.1111/1523-1747.EP12532758

[6] Ongrádi J, Stercz B, Kövesdi V, Vértes L. Immunosenescence and vaccination of the elderly, I. age-related immune impairment. Acta Microbiologica et Immunologica Hungarica. 2009;**56**(3):199-210. DOI: 10.1556/AMICR.56.2009.3.1

[7] Shaw AC, Joshi S, Greenwood H, Panda A, Lord JM. Aging of the innate immune system. Current Opinion in Immunology. 2010;**22**(4):507-513. DOI: 10.1016/J.COI.2010.05.003

[8] Mahbub S, L. Brubaker A, J. Kovacs E. Aging of the innate immune system: An update. Curr. Immunological Reviews. 2011;**7**(1):104-115. DOI: 10.2174/157339511794474181

[9] Cao C, Xiao Z, Wu Y, Ge C. Diet and skin aging-from the perspective of food nutrition. Nutrients. 2020;**12**(3):810. DOI: 10.3390/NU12030870

[10] Chantalat J, Bruning E, Sun Y, Liu JC. Application of a topical biomimetic electrical signaling technology to photo-aging: A randomized, double-blind, placebo-controlled trial of a galvanic zinc-copper complex. Journal of Drugs in Dermatology. 2012;**11**(1):30-37

[11] Quan T, He T, Kang S, Voorhees JJ, Fisher GJ. Solar ultraviolet irradiation reduces collagen in photoaged human skin by blocking transforming growth factor-beta type II receptor/Smad signaling. The American Journal of Pathology. 2004;**165**(3):741-751. DOI: 10.1016/S0002-9440(10)63337-8

[12] Ogura Y, Kuwahara T, Akiyama M, et al. Dermal carbonyl modification is related to the yellowish color change of photo-aged Japanese facial skin. Journal of Dermatological Science. 2011;**64**(1):45-52. DOI: 10.1016/J.JDERMSCI.2011.06.015

[13] Fisher GJ, Wang Z, Datta SC, Varani J, Kang S, Voorhees JJ. Pathophysiology of premature skin aging induced by ultraviolet light. The New England Journal of Medicine. 1997;**337**(20):1419-1429. DOI: 10.1056/NEJM199711133372003

[14] Talwar HS, Griffiths CEM, Fisher GJ, Hamilton TA, Voorhees JJ. Reduced type I and type III procollagens in photodamaged adult human skin. The

Journal of Investigative Dermatology. 1995;**105**(2):285-290. DOI: 10.1111/1523-1747.EP12318471

[15] Prasad A, Pospísil P. Ultraweak photon emission induced by visible light and ultraviolet A radiation via photoactivated skin chromophores: in vivo charge coupled device imaging. Journal of Biomedical Optics. 2012;**17**(8):085004. DOI: 10.1117/1.JBO.17.8.085004

[16] Nishigori C, Hattori Y, Arima Y, Miyachi Y. Photoaging and oxidative stress. Experimental Dermatology. 2003;**12**(Suppl 2(2)):18-21. DOI: 10.1034/J.1600-0625.12.S2.3.X

[17] Morita A. Tobacco smoke causes premature skin aging. Journal of Dermatological Science. 2007;**48**(3):169-175. DOI: 10.1016/J.JDERMSCI.2007.06.015

[18] Morita A, Torii K, Maeda A, Yamaguchi Y. Molecular basis of tobacco smoke-induced premature skin aging. The Journal of Investigative Dermatology. Symposium Proceedings. 2009;**14**(1):53-55. DOI: 10.1038/jidsymp.2009.13

[19] Martens DS, Cox B, Janssen BG, et al. Prenatal air pollution and newborns' predisposition to Accelerated biological aging. JAMA Pediatrics. 2017;**171**(12):1160-1167. DOI: 10.1001/JAMAPEDIATRICS.2017.3024

[20] Martens DS, Nawrot TS. Air pollution stress and the aging phenotype: The telomere connection. Current Environmental Health Reports. 2016;**3**(3):258-269. DOI: 10.1007/S40572-016-0098-8

[21] Krutmann J, Schroeder P. Role of mitochondria in photoaging of human skin: The defective powerhouse model. The Journal of Investigative Dermatology. Symposium Proceedings. 2009;**14**(1):44-49. DOI: 10.1038/JIDSYMP.2009.1

[22] Panich U, Sittithumcharee G, Rathviboon N, Jirawatnotai S. Ultraviolet radiation-induced skin aging: The role of DNA damage and oxidative stress in epidermal stem cell damage mediated skin aging. Stem Cells International. 2016;**2016**:7370642. DOI: 10.1155/2016/7370642

[23] Moriwaki S, Takahashi Y. Photoaging and DNA repair. Journal of Dermatological Science. 2008;**50**(3):169-176. DOI: 10.1016/J.JDERMSCI.2007.08.011

[24] Shindo Y, Witt E, Packer L. Antioxidant defense mechanisms in murine epidermis and dermis and their responses to ultraviolet light. The Journal of Investigative Dermatology. 1993;**100**(3):260-265. DOI: 10.1111/1523-1747.EP12469048

[25] Schmitz S, Garbe C, Tebbe B, Orfanos CE. [long-wave ultraviolet radiation (UVA) and skin cancer]. Der Hautarzt. Zeitschrift fur Dermatologie, Venerologie, und verwandte Gebiete. 1994;**45**(8):517-525. DOI: 10.1007/S001050050118

[26] Brem R, Guven M, Karran P. Oxidatively-generated damage to DNA and proteins mediated by photosensitized UVA. Free Radical Biology & Medicine. 2017;**107**:101-109. DOI: 10.1016/J.FREERADBIOMED.2016.10.488

[27] Cadet J, Douki T, Ravanat JL. Oxidatively generated damage to cellular DNA by UVB and UVA radiation. Photochemistry and Photobiology. 2015;**91**(1):140-155. DOI: 10.1111/PHP.12368

[28] Chen L, Hu JY, Wang SQ. The role of antioxidants in photoprotection:

A critical review. Journal of the American Academy of Dermatology. 2012;**67**(5):1013-1024. DOI: 10.1016/J.JAAD.2012.02.009

[29] Kruft BI, Greer A. Photosensitization reactions In vitro and In vivo. Photochemistry and Photobiology. 2011;**87**(6):1204-1213. DOI: 10.1111/J.1751-1097.2011.00993.X

[30] Wondrak GT, Jacobson MK, Jacobson EL. Endogenous UVA-photosensitizers: Mediators of skin photodamage and novel targets for skin photoprotection. Photochemical & Photobiological Sciences. 2006;**5**(2):215-237. DOI: 10.1039/B504573H

[31] Wang RJ, Stoien JD, Landa F. Lethal effect of near-ultraviolet irradiation on mammalian cells in culture. Nature. 1974;**247**(5435):43-45. DOI: 10.1038/247043a0

[32] Sohal RS, Weindruch R. Oxidative stress, caloric restriction, and aging. Science. 1996;**273**(5271):59-63. DOI: 10.1126/SCIENCE.273.5271.59

[33] Mccormick JP, Fischer JR, Pachlatko JP, Eisenstark A. Characterization of a cell-lethal product from the photooxidation of tryptophan: Hydrogen peroxide. Science. 1976;**191**(4226):468-469. DOI: 10.1126/SCIENCE.1108203

[34] Baier J, Maisch T, Maier M, Engel E, Landthaler M, Bäumler W. Singlet oxygen generation by UVA light exposure of endogenous photosensitizers. Biophysical Journal. 2006;**91**(4):1452-1459. DOI: 10.1529/biophysj.106.082388

[35] Ou-Yang H, Stamatas G, Saliou C, Kollias N. A Chemiluminescence study of UVA-induced oxidative stress in human skin in vivo. The Journal of Investigative Dermatology. 2004;**122**(4):1020-1029. DOI: 10.1111/j.0022-202X.2004.22405.x

[36] Foote CS. Definition of type I and type II photosensitized oxidation. Photochemistry and Photobiology. 1991;**54**(5):659-659. DOI: 10.1111/J.1751-1097.1991.TB02071.X

[37] Kawanishi S, Hiraku Y, Oikawa S. Mechanism of guanine-specific DNA damage by oxidative stress and its role in carcinogenesis and aging. Mutation Research/Reviews in Mutation Research. 2001;**488**(1):65-76. DOI: 10.1016/S1383-5742(00)00059-4

[38] Baptista MS, Cadet J, di Mascio P, et al. Type I and type II photosensitized oxidation reactions: Guidelines and mechanistic pathways. Photochemistry and Photobiology. 2017;**93**(4):912-919. DOI: 10.1111/PHP.12716

[39] Cristofalo VJ, Lorenzini A, Allen RG, Torres C, Tresini M. Replicative senescence: a critical review. Mechanisms of Ageing and Development. 2004;**125**(10-11):827-848. DOI: 10.1016/J.MAD.2004.07.010

[40] Millis AJT, Hoyle M, McCue HM, Martini H. Differential expression of metalloproteinase and tissue inhibitor of metalloproteinase genes in aged human fibroblasts. Experimental Cell Research. 1992;**201**(2):373-379. DOI: 10.1016/0014-4827(92)90286-H

[41] Brenneisen P, Wenk J, Klotz LO, et al. Central role of ferrous/ferric iron in the ultraviolet B irradiation-mediated signaling pathway leading to increased interstitial collagenase (matrix-degrading metalloprotease (MMP)-1) and stromelysin-1 (MMP-3) mRNA levels in cultured human dermal fibroblasts. The Journal of Biological Chemistry. 1998;**273**(9):5279-5287. DOI: 10.1074/JBC.273.9.5279

[42] Dimri GP, Lee X, Basile G, et al. A biomarker that identifies senescent

human cells in culture and in aging skin in vivo. Proceedings of the National Academy of Sciences of the United States of America. 1995;**92**(20):9363-9367. DOI: 10.1073/PNAS.92.20.9363

[43] Coppé JP, Patil CK, Rodier F, et al. Senescence-associated secretory phenotypes reveal cell-nonautonomous functions of oncogenic RAS and the p53 tumor suppressor. PLoS Biology. 2008;**6**(12):2853-2868. DOI: 10.1371/JOURNAL.PBIO.0060301

[44] Graindorge D, Martineau S, Machon C, et al. Singlet oxygen-mediated oxidation during UVA radiation alters the dynamic of genomic DNA replication. PLoS One. 2015;**10**(10):e0140645. DOI: 10.1371/JOURNAL.PONE.0140645

[45] Gęgotek A, Atalay S, Domingues P, Skrzydlewska E. The differences in the proteome profile of Cannabidiol-treated skin fibroblasts following UVA or UVB irradiation in 2D and 3D cell cultures. Cell. 2019;**8**(9):995. DOI: 10.3390/CELLS8090995

[46] Zhang C, Yuchi H, Sun L, Zhou X, Lin J. Human amnion-derived mesenchymal stem cells protect against UVA irradiation-induced human dermal fibroblast senescence, in vitro. Molecular Medicine Reports. 2017;**16**(2):2016. DOI: 10.3892/MMR.2017.6795

[47] Gasparrini M, Forbes-Hernandez TY, Afrin S, et al. Strawberry-based cosmetic formulations protect human dermal fibroblasts against UVA-induced damage. Nutrients. 2017;**9**(6):605. DOI: 10.3390/NU9060605

[48] Huang CH, Li HJ, Wu NL, et al. Photoprotective effects of Cycloheterophyllin against UVA-induced damage and oxidative stress in human dermal fibroblasts. PLoS One. 2016;**11**(9):e0161767. DOI: 10.1371/JOURNAL.PONE.0161767

[49] Liu X, Zhang R, Shi H, et al. Protective effect of curcumin against ultraviolet A irradiation-induced photoaging in human dermal fibroblasts. Molecular Medicine Reports. 2018;**17**(5):7227. DOI: 10.3892/MMR.2018.8791

[50] Khan A, Bai H, Shu M, Chen M, Khan A, Bai Z. Antioxidative and antiphotoaging activities of neferine upon UV-A irradiation in human dermal fibroblasts. Bioscience Reports. 2018;**38**:20181414. DOI: 10.1042/BSR20181414

[51] Seo SW, Park SK, Oh SJ, Shin OS. TLR4-mediated activation of the ERK pathway following UVA irradiation contributes to increased cytokine and MMP expression in senescent human dermal fibroblasts. PLoS One. 2018;**13**(8):e0202323. DOI: 10.1371/JOURNAL.PONE.0202323

[52] Yoshimoto S, Kohara N, Sato N, Ando H, Ichihashi M. Riboflavin plays a pivotal role in the UVA-induced cytotoxicity of fibroblasts as a key molecule in the production of H2O2 by UVA radiation in collaboration with amino acids and vitamins. International Journal of Molecular Sciences. 2020;**21**(2):554. DOI: 10.3390/IJMS21020554

[53] Sato K, Taguchi H, Maeda T, et al. The primary cytotoxicity in ultraviolet-A-irradiated riboflavin solution is derived from hydrogen peroxide. The Journal of Investigative Dermatology. 1995;**105**(4):608-612. DOI: 10.1111/1523-1747.EP12323724

[54] Juzeniene A, Thu Tam TT, Iani V, Moan J. The action spectrum for folic acid photodegradation in aqueous

solutions. Journal of Photochemistry and Photobiology. B. 2013;**126**:11-16. DOI: 10.1016/J.JPHOTOBIOL.2013.05.011

[55] Rokos H, Beazley WD, Schallreuter KU. Oxidative stress in vitiligo: Photo-oxidation of pterins produces H(2)O(2) and pterin-6-carboxylic acid. Biochemical and Biophysical Research Communications. 2002;**292**(4):805-811. DOI: 10.1006/bbrc.2002.6727

[56] Thomas AH, Serrano MP, Rahal V, et al. Tryptophan oxidation photosensitized by pterin. Free Radical Biology & Medicine. 2013;**63**:467-475. DOI: 10.1016/J.FREERADBIOMED.2013.05.044

[57] Park SL, Justiniano R, Williams JD, Cabello CM, Qiao S, Wondrak GT. The tryptophan-derived endogenous arylhydrocarbon receptor ligand 6-formylindolo[3,2-b]carbazole (FICZ) is a nanomolar UVA-photosensitizer in epidermal keratinocytes. The Journal of Investigative Dermatology. 2015;**135**(6):1649. DOI: 10.1038/JID.2014.503

[58] Walrant P, Santus R. N-formyl-kynurenine, a tryptophan photooxidation product, as a photodynamic sensitizer. Photochemistry and Photobiology. 1974;**19**(6):411-417. DOI: 10.1111/J.1751-1097.1974.TB06533.X

[59] Plowman JE, Deb-Choudhury S, Grosvenor AJ, Dyer JM. Protein oxidation: Identification and utilisation of molecular markers to differentiate singlet oxygen and hydroxyl radical-mediated oxidative pathways. Photochemical & Photobiological Sciences. 2013;**12**(11):1960-1967. DOI: 10.1039/C3PP50182E

[60] Roberts JE, Wielgus AR, Boyes WK, Andley U, Chignell CF. Phototoxicity and cytotoxicity of Fullerol in human lens epithelial cells. Toxicology and Applied Pharmacology. 2008;**228**(1):49. DOI: 10.1016/J.TAAP.2007.12.010

[61] Mahns A, Melchheier I, Suschek C, v., Sies H, Klotz LO. Irradiation of cells with ultraviolet-A (320-400 nm) in the presence of cell culture medium elicits biological effects due to extracellular generation of hydrogen peroxide. Free Radical Research. 2003;**37**(4):391-397. DOI: 10.1080/1071576031000064702

[62] Giordano ME, Caricato R, Lionetto MG. Concentration dependence of the antioxidant and Prooxidant activity of Trolox in HeLa cells: Involvement in the induction of apoptotic volume decrease. Antioxidants (Basel). 2020;**9**(11):1-12. DOI: 10.3390/ANTIOX9111058

[63] Fuad NIN, Sekar M, Gan SH, Lum PT, Vaijanathappa J, Ravi S. Lutein: A comprehensive review on its chemical, biological activities and therapeutic potentials. Pharmacognosy Journal. 2020;**12**(6s):1769-1778. DOI: 10.5530/pj.2020.12.239

[64] Chung LY. The antioxidant properties of garlic compounds: Alyl cysteine, alliin, allicin, and allyl disulfide. Journal of Medicinal Food. 2006;**9**(2):205-213. DOI: 10.1089/JMF.2006.9.205

[65] Xia N, Daiber A, Förstermann U, Li H. Antioxidant effects of resveratrol in the cardiovascular system. British Journal of Pharmacology. 2017;**174**(12):1633-1646. DOI: 10.1111/BPH.13492

[66] He H, Li J, Xie Y, Li Z, Shi H, Lu CD. Effects of soy isoflavones on intake, body weight, sex hormones, antioxidant performance, and semen quality in Xinong Saanen goats. Journal of Applied Animal Research. 2021;**49**(1):125-132. DOI: 10.1080/09712119.2021.1901716

[67] Xu D, Hu MJ, Wang YQ, Cui YL. Antioxidant activities of quercetin and its complexes for medicinal application. Molecules. 2019;**24**(6):1123. DOI: 10.3390/MOLECULES24061123

[68] Shimizu T, Nakanishi Y, Nakahara M, et al. Structure effect on antioxidant activity of Catecholamines toward singlet oxygen and other reactive oxygen species in vitro. Journal of Clinical Biochemistry and Nutrition. 2010;**47**(3):181-190. DOI: 10.3164/JCBN.09-112

[69] Lin JK, Chen PC, Ho CT, Lin-Shiau SY. Inhibition of xanthine oxidase and suppression of intracellular reactive oxygen species in HL-60 cells by theaflavin-3,3′-digallate, (−)-epigallocatechin-3-gallate, and propyl gallate. Journal of Agricultural and Food Chemistry. 2022;**48**(7):2736-2743. DOI: 10.1021/jf000066d

[70] Trujillo J, Chirino YI, Molina-Jijón E, Andérica-Romero AC, Tapia E, Pedraza-Chaverrí J. Renoprotective effect of the antioxidant curcumin: Recent findings. Redox Biology. 2013;**1**(1):448-456. DOI: 10.1016/J.REDOX.2013.09.003

[71] Kono Y, Shibata H, Kodama Y, Ueda A, Sawa Y. Chlorogenic acid as a natural scavenger for hypochlorous acid. Biochemical and Biophysical Research Communications. 1995;**217**(3):972-978. DOI: 10.1006/BBRC.1995.2865

[72] Gao J, Feng Z, Wang X, et al. SIRT3/SOD2 maintains osteoblast differentiation and bone formation by regulating mitochondrial stress. Cell Death and Differentiation. 2018;**25**(2):229-240. DOI: 10.1038/CDD.2017.144

[73] Koh E, Carmieli R, Mor A, Fluhr R. Singlet oxygen-induced membrane disruption and serpin-protease balance in vacuolar-driven cell death. Plant Physiology. 2016;**171**(3):1616-1625. DOI: 10.1104/PP.15.02026

[74] Wagner KH, Elmadfa I. Biological relevance of terpenoids. Overview focusing on mono-, di- and tetraterpenes. Annals of Nutrition & Metabolism. 2003;**47**(3-4):95-106. DOI: 10.1159/000070030

[75] Humbert PG, Haftek M, Creidi P, et al. Topical ascorbic acid on photoaged skin. Clinical, topographical and ultrastructural evaluation: Double-blind study vs. placebo. Experimental Dermatology. 2003;**12**(3):237-244. DOI: 10.1034/J.1600-0625.2003.00008.X

[76] Borodina I, Kenny LC, McCarthy CM, et al. The biology of ergothioneine, an antioxidant nutraceutical. Nutrition Research Reviews. 2020;**33**(2):190-217. DOI: 10.1017/S0954422419000301

[77] Larson RA. The antioxidants of higher plants. Phytochemistry. 1988;**27**(4):969-978. DOI: 10.1016/0031-9422(88)80254-1

[78] Blokhina O, Virolainen E, Fagerstedt K v. Antioxidants, oxidative damage and oxygen deprivation stress: A review. Annals of Botany 2003;**91** Spec No(2):179-194. DOI:10.1093/AOB/MCF118

[79] Yoshimoto S, Yoshida M, Ando H, Ichihashi M. Establishment of Photoaging In vitro by repetitive UVA irradiation: Induction of characteristic markers of senescence and its prevention by PAPLAL with potent catalase activity. Photochemistry and Photobiology. 2018;**94**(3):438-444. DOI: 10.1111/PHP.12871

[80] Akanmu D, Cecchini R, Aruoma OI, Halliwell B. The antioxidant action of ergothioneine. Archives of Biochemistry

and Biophysics. 1991;**288**(1):10-16. DOI: 10.1016/0003-9861(91)90158-F

[81] Obayashi K, Kurihara K, Okano Y, Masaki H, Yarosh DB. L-Ergothioneine scavenges superoxide and singlet oxygen and suppresses TNF-alpha and MMP-1 expression in UV-irradiated human dermal fibroblasts. Journal of Cosmetic Science. 2005;**56**(1):17-27. DOI: 10.1111/j.0142-5463.2005.00265_2.x

[82] Eli M, Li DS, Zhang WW, et al. Decreased blood riboflavin levels are correlated with defective expression of RFT2 gene in gastric cancer. World Journal of Gastroenterology. 2012;**18**(24):3112-3118. DOI: 10.3748/WJG.V18.I24.3112

[83] Aili A, Hasim A, Kelimu A, et al. Association of the plasma and tissue riboflavin levels with C20orf54 expression in cervical lesions and its relationship to HPV16 infection. PLoS One. 2013;**8**(11):e79937. DOI: 10.1371/JOURNAL.PONE.0079937

[84] Eichner ER, Hillman RS. Effect of alcohol on serum folate level. The Journal of Clinical Investigation. 1973;**52**(3):584-591. DOI: 10.1172/JCI107219

[85] Pichler R, Fritz J, Heidegger I, et al. Predictive and prognostic role of serum neopterin and tryptophan breakdown in prostate cancer. Cancer Science. 2017;**108**(4):663-670. DOI: 10.1111/CAS.13171

[86] Nakajima A, Tahara M, Yoshimura Y, Nakazawa H. Study of compounds suppressing free radical generation from UV-exposed ketoprofen. Chemical and Pharmaceutical Bulletin (Tokyo). 2007;**55**(10):1431-1438. DOI: 10.1248/CPB.55.1431

[87] Zhu Y, Tchkonia T, Pirtskhalava T, et al. The Achilles' heel of senescent cells: From transcriptome to senolytic drugs. Aging Cell. 2015;**14**(4):644-658. DOI: 10.1111/ACEL.12344

[88] Yoon JE, Kim Y, Kwon S, et al. Senescent fibroblasts drive ageing pigmentation: A potential therapeutic target for senile lentigo. Theranostics. 2018;**8**(17):4620-4632. DOI: 10.7150/THNO.26975

Chapter 10

Mollusk Shell Waste as Composite Photocatalyst for Methylene Blue Removal

Hazlini Dzinun, Nurul Amila Diyanah Kamarul Bharin, Norsyazwani Binti Md Nizam, Hui Ean Lim, Ahmad Danish Haikal Bin Sha'ari and Muhammad Amjad Bin Noor Azlan

Abstract

Mollusk shell is abundant in particular areas and frequently deposited in landfills, contributing to environmental pollution. However, mollusk shell waste has been proven as an absorbent that has a high possibility of acting as a photocatalyst when integrated with metal support in composite form due to the synergistic effect. Therefore, in this study, mussel and cockle shells as agricultural wastes were selected to be innovated as support for photocatalysts. The solid-state dispersion (SSD) method was used to prepare a composite photocatalyst where mussel and cockle were integrated with titanium dioxide (TiO_2) nanoparticles at a ratio of 9:1. In total, 100% of mussel and cockle were used as control samples. The prepared composite photocatalyst was evaluated with methylene blue (MB) removal in the suspension system. The result reveals that mussel/TiO_2 and cockle/TiO_2 composite photocatalyst show 25.92 and 24.08% for MB removal within 2 hours. It is due to the prepared composite photocatalyst particle sizes, where mussel/TiO_2 and cockle/TiO_2 were 259 and 268 nm, respectively. It is interesting to note that the prepared composite photocatalyst particle size should be in nanosize, where it can enhance the photocatalytic performance. Overall, agricultural waste should be utilized to ensure a clean environment for future generations.

Keywords: mollusk shell waste, composite photocatalyst, methylene blue removal

1. Introduction

Industrial development is a significant contributor to a country's financial status, including using dyes to produce an eye-catching product that meets the consumers' demand [1]. Estimated that 10–15% of dyes used in textile processing were lost in the wastewater. Coincidentally, methylene blue (MB), a sulfur-containing heterocyclic aromatic dye, is mostly applied in textiles industries for dyeing cotton, silk, and wool [2, 3].

 IntechOpen

In addition, other applications in chemistry, biology, and medical science also use this basic cation dye in the treatment of methemoglobinemia and cyanide poisoning [4].

Biological and chemical precipitation is the standard dye removal treatment [5] while several conventional wastewater treatment methods have also been tested for the same purpose. Coagulation or flocculation, ozonation, chemical oxidation, and electrochemical treatment reported by Subki et al., [6], novel technology, such as membrane filtration and bio-sorption [7]. Among all of the possible techniques, the photocatalysis process has proven to be the most favorable technology in textile wastewater treatment, where the most organic matter can be oxidized to water, carbon dioxide, and simple inorganics materials using light radiation and selected catalysts [8].

Due to the expensive production of chemical photocatalyst compounds, utilizing waste materials as adsorbents integrated with metal oxides to form composite photocatalysts has become the main focus of researchers nowadays in maintaining the sustainability of both the treatment and the environment. Mollusk shells catch attention as a potential derived catalyst in dye removal due to the high content of calcium carbonate ($CaCO_3$) in raw material to produce calcium oxide (CaO) as the most promising heterogeneous alkali catalyst obtained *via* the calcination process [9]. As the most heavily traded bivalve ample mollusk shell, cockle shells have been recorded to have a high percentage of CaO contained in the natural compound of the shell, which is 99.17% of $CaCO_3$ before calcined [10] while mussel shells came in the second place with 98.37% of $CaCO_3$ [11].

In order to get the small microstrain of CaO from $CaCO_3$ through the calcination process, Sari et al., [12] investigated the effect of calcination temperature on the crystallization of CaO from green mussel shells. They found out that the calcined CaO at 950°C obtained a small microstrain compared to the samples calcined at other temperatures. Besides, CaO calcined at 950°C exhibited the largest crystallite size, meaning it had high crystallinity and a shortened amorphous phase. The CaO calcined at 950°C showed a small microstrain compared to the other samples, meaning the crystal defects in the sample were small.

The particle sizes of composite photocatalysts are also one of the factors for enhanced photocatalytic performances. Based on previous researchers, they used various methods for synthesizing mollusk shells with various particle sizes produced. Among cockles, scallops, oysters, pyramidella, green mussels, razor clams, golden apple snails, and snail shells, the highest CaO is produced from cockle and green mussel shells [11]. Commonly, the synthesized waste shell involves the same steps, which are cleaning, drying, crushing, grinding, sieving, and calcination process. Mostly, the particle sizes of CaO are produced in micro-size as prepared by Buasri with his group research starting 2013 [11] until 2015 [13] with various types of shells. Interesting to note that Gbadeyan et al., [14] was successfully synthesized snail waste shells in nanosized *via* dry and wet ball milling methods.

The ratio of adsorbent with metal oxides to form a composite photocatalyst needs to be considered in order to get the highest removal of pollutants. Dzinun et al. [15] found that the optimum ratio for the highest adsorption and MB photocatalytic degradation was achieved by using a (1/9) ratio of TiO_2/eggshell. Therefore, in this study, 9:1 ratio of mollusk shell with TiO_2 was investigated. In total 100% of mollusk and cockle shell was used as the control sample with indicated ratios of 10:0. The composite photocatalyst was prepared by solid-state dispersion (SSD) method for MB removal in the suspension system.

2. Experimental

2.1 Materials

This study used the raw mussel and cockle shell waste then synthesized into powder form. TiO_2 (Sigma-Aldrich, 21 nm, 95% purity) was used and integrated with prepared powder of mussel and cockle shell to be a composite photocatalyst. Ethanol (95%) was used from QRëc.

2.2 Synthesis of cockle and mussel shell in powder form

Wasted mussel and cockle shells were washed and brushed several times with tap water to remove any foreign matter and impurities present. The cleaned shells were dried until no moisture was left then proceeded to crush into smaller sizes. Dry-milled both shells for 45 minutes at 500 rpm in batch mode. Then, the crushed shells were sieved to a particle size of ≤75 μm using Cooper's TG-0411/0412 Sieve Shaker for an average of 10 minutes. The powdered cockle shells then were mixed with 50 mL of ethanol in a 250 mL stainless steel jar and were wet-milled at 500 rpm for 4 hours equivalent to 240 minutes in a clockwise direction. The same condition was set for powdered mussel shells except for the volume of ethanol used is 30 mL. Dry and wet milling processes were operated in a planetary ball mill (Pritsch Pulverisette 6) to gain nanosize powder [14]. Both shells were purified *via* the decantation method four times every 20 minutes before filtration. Then, the fine particles were dried in a drying oven at 40°C for 3 days corresponding to 72 hours. After cockle and mussel shells dried, the samples were calculated in a muffle furnace (Protherm-Furnaces) of 950°C with a heating rate of 10°C/min for a whole 4 hours. The obtained CaO powder was cooled before being stored in a sealed zipper bag.

2.3 Synthesis of composite photocatalyst

The composite photocatalyst of mussel shell/TiO_2 and cockle shell/TiO_2 was prepared using the solid-state dispersion (SSD) method. There were two different samples of mussel shell/TiO_2 and cockle shell/TiO_2 with a ratio of 9:1 with a 10 g basis and also two control samples consisting of 100% of mussel shell and cockle shell. The samples were evaporated for 24 hours in a 35°C oven. Finally, the samples were calcined in a muffled furnace for 5 hours at 450°C.

2.4 Characterization of composite photocatalyst

The functional group was characterized using Fourier Transform Infrared (Agilent Cary 630 FT-IR). UV–Vis Spectrophotometer (HITACHI U-3900H) was used for the determination of photocatalytic degradation performance. The structural morphology of composite photocatalyst was investigated using scanning electron microscopy (SEM, HITACHI TM3)00).

2.5 Methylene blue removal testing

The setup of MB removal testing was prepared as in the previous study [15]. 10 mg/L of MB solution was prepared in a 200 mL beaker. 0.8 g of the composite photocatalyst was added into a beaker. The suspension was first stirred without a UV lamp using a magnetic stirrer for 30 min for adsorption equilibrium. After taking the sample for adsorption, the UV lamp and the stirrer were turned on for photocatalytic activity testing. The experiment was conducted for 150 min and for each 10 min interval, the aliquot samples were taken and centrifuged at 6000 rpm [16] for 5 minutes to measure the concentration of MB with a UV–Vis spectrophotometer at 665 nm as the maximum absorbance [17]. The samples were analyzed by a UV–Vis spectrophotometer.

The calibration curve is a method for determining the sample concentration that must be completed before the photocatalytic suspension test. The absorbance value of each sample from each experiment run was determined using a UV–Vis spectrophotometer. The sample absorption value can be translated into concentration using the linear equation obtained from the calibration curve (ppm). After normalizing the measurement, the removal of MB can be computed using the formula:

$$MB\,removal\% = \frac{(C_o - C_t)}{C_o} \times 100\% \tag{1}$$

where R is the removal efficiency of MB in percentage, C_o (mg/L) as initiate al concentration of dye solution, and C_t (mg/L) as the concentration of dye solution at equilibrium [18].

3. Results and discussion

3.1 Morphology and physiochemical properties of composite photocatalyst

Figure 1 shows the synthesizing of cockle and mussel shells into powder form. Raw waste shells were crushed into smaller sizes that are allowable for the milling process, where $CaCO_3$ powder was obtained, as shown in **Figure 1a** and **b**, respectively. After 4 hours of calcination at 950°C, the composite powder converted into CaO powder as shown in **Figure 1c**, and was used as an adsorbent for MB removal. The difference in color shows each shell has its appearance throughout the whole process. An investigation of mass reduction on both samples after calcination was performed, and

Figure 1.
(a) Raw waste shells, (b) powder shells before, and (c) after the calcination process to produce CaO.

it was discovered that the mass for CCS reduced to 5.56 g from 10.06 g, while CMS depleted from 10.08 to 5.12 g, both shells proved to lose half of its mass after conversion of CaO. A similar finding was observed in a study of $MgCO_3$ conversion to MgO, the mass loss was 50% approximately at a calcination temperature equal to or higher than 700°C, independent of the duration [19].

FT-IR analysis was performed on both shells to identify the composite formed before and after calcination as shown in **Figure 2**. Uncalcined cockle shell as proved in **Figure 2a** has a peak of C-O bond at 1438 cm^{-1} wavelength that comes in a group with a few bands at 1088, 861, and 714 cm^{-1} representing carbonate ion, CO_3^{2-} with aragonite microstructure [20] indicating the $CaCO_3$ presence before calcination. This result slightly corresponds to Ref. [21] with sharp bands found around 1450, 1080, 858, and 712 cm^{-1}.

Situated at the peak of 939 and 1438 cm^{-1} in the calcined cockle shell band (**Figure 2b**), the C-O bonds seem to lose their former strength as the

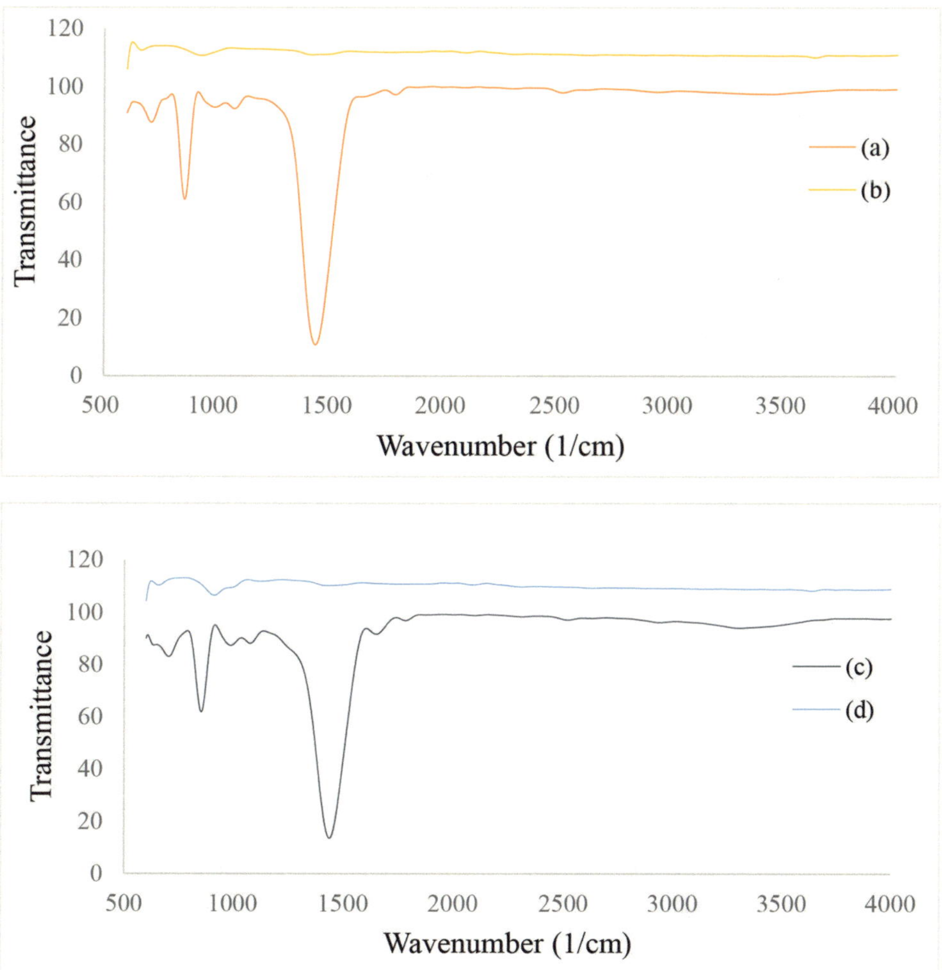

Figure 2.
FT-IR spectrum of (a) uncalcined and (b) calcined cockle shell, (c) uncalcined, and (d) calcined mussel shell.

CO_3^{2-} presence is gradually lost in the calcination process, which causes the shifting in the mentioned peak. This discovery occurred due to dissipation in the reduced mass of the functional group associated with CO_3^{2-} ion [22]. The same theory applied to uncalcined mussel shells concerning that both shells are one of the few types of mollusk shells.

As for the uncalcined mussel shell, as shown in **Figure 2c**, the C-O bonds were observed at 1442, 1077, 853, and 711 cm^{-1} peaks except there is a slight difference in the intensity of these peaks compared to **Figure 2a**, which have a stronger intensity of C-O content. The calcined cockle shell lost its C-O functional group, where the peak intensity of 924 and 1449 cm^{-1} is weakening (**Figure 2d**). According to Sari et al., [12], the functional group of CaO was formed starting at a temperature of 750°C, which is agreeable with this study as the CaO bond was formed at 656 cm^{-1} still in the range of 667.32 cm^{-1} from their characterization results.

The morphological structures of a mussel shell, mussel/TiO_2, cockle shell, and cockle/TiO_2 composite photocatalyst were examined by SEM as shown in **Figure 3**. TiO_2 appeared as fine particles (**Figure 3c** and **d**), whereas the cockle and mussel shell is more prominent due to the micro-sized particles and exhibits irregular shape and size (**Figure 3a** and **b**). The particle sizes of mussel and cockle shells were bigger than mussel/TiO_2 and cockle/TiO_2 composite photocatalysts due to the integration of TiO_2 in nanoparticles size. Besides, it has been established that the TiO_2 and mollusk shell particles bonded together and formed a composite photocatalyst following the calcination process.

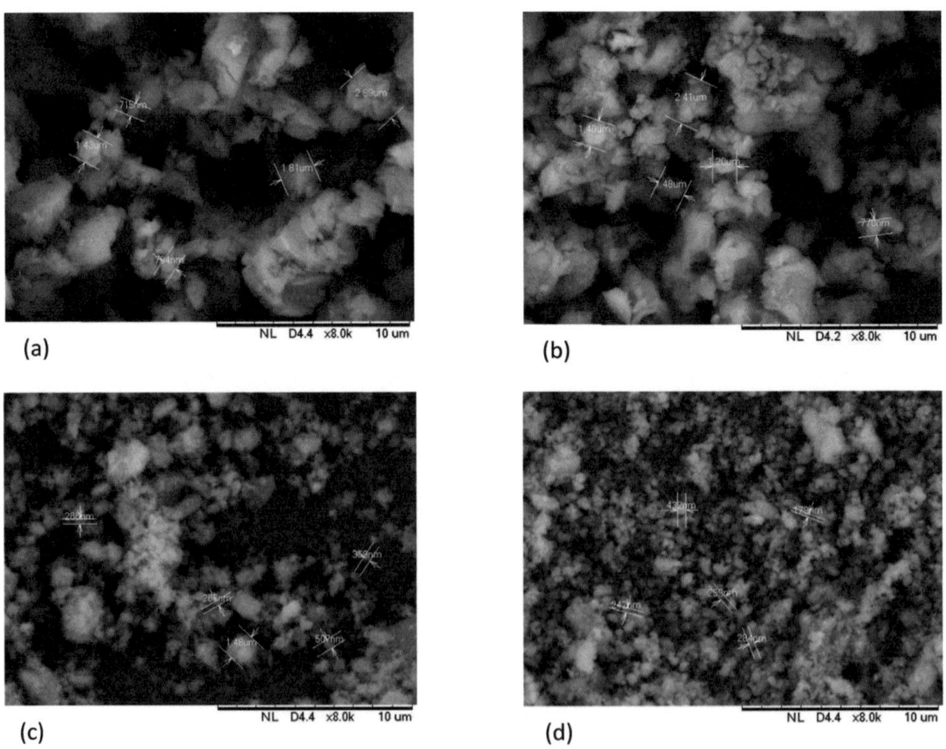

Figure 3.
SEM analysis for (a) cockle shell, (b) mussel shell (c) cockle shell/TiO_2, and (d) mussel shell/TiO_2 composite photocatalyst.

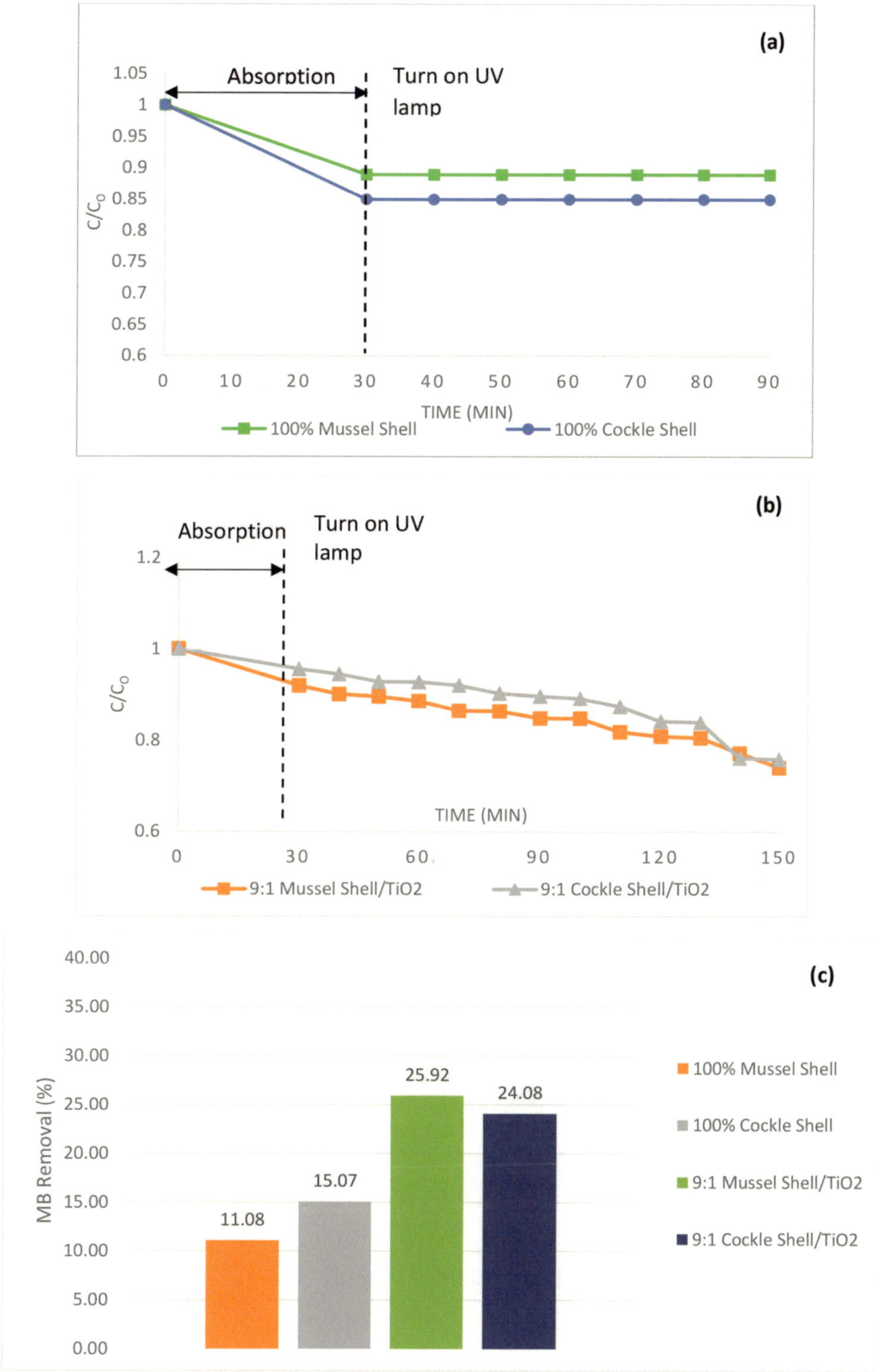

Figure 4.
(a) Normalized absorption and photocatalytic activity for (a) mussel and cockle shell, (b) for composite photocatalyst, and (c) percentage MB removal.

3.2 Performance of methylene blue removal

The removal of MB was carried out under UV illumination for 150 min, the photocatalytic efficiency of the composite photocatalyst was determined and the result is shown in **Figure 4**. When the cockle and mussel shell was used as an adsorbent, there was no photocatalytic degradation of the MB solution, which was as expected (**Figure 4a**). By comparison of the composite photocatalyst, mussel shell/TiO_2 exhibited the highest adsorption and achieved 25.92% of MB photocatalytic degradation compared to cockle shell/TiO_2 (**Figure 4b** and **c**). It is interesting to note that the particle size of the mussel shell/TiO_2 composite photocatalyst is smaller than cockle shell/TiO_2, which gives a higher surface area for adsorption and photocatalytic activity takes place.

4. Conclusion

By using the solid-state dispersion (SSD) method, photocatalyst composites made of cockle shell/TiO_2 and mussel shell/TiO_2 were successfully prepared. Using a suspension system and methylene blue as a model pollutant, we investigated the adsorption and photocatalytic activity of the material. The mussel shell/TiO_2 composite photocatalyst with the smallest particle size gives the highest MB removal when compared to the cockleshell/TiO_2 photocatalyst. Therefore, mussel shells have the potential to be utilized as useful materials for the removal of reactive dyes compared to cockle shells. In order to increase the MB removal, it would be suggested that the particle sizes or mussel shells should be nanosized and integrated with TiO_2 to form nanocomposite photocatalysts.

Acknowledgements

The authors would also like to thank the Center for Diploma Studies, Universiti Tun Hussein Onn Malaysia for its support.

Author details

Hazlini Dzinun*, Nurul Amila Diyanah Kamarul Bharin, Norsyazwani Binti Md Nizam, Hui Ean Lim, Ahmad Danish Haikal Bin Sha'ari and Muhammad Amjad Bin Noor Azlan
Department of Science and Mathematics, Center for Diploma Studies, University Tun Hussein Onn, Johor, Malaysia

*Address all correspondence to: hazlini@uthm.edu.my

References

[1] Carmen Z, Daniela S. Textile Organic Dyes-Characteristics, Polluting Effects and Separation/elimination Procedures from Industrial Effluents, a Critical Overview. InTech; 2011

[2] Arora C, Soni S, Sahu S, Mittal J, Kumar P, Bajpai P. Iron based metal organic framework for efficient removal of methylene blue dye from industrial waste. Journal of Molecular Liquids. 2019;**284**:345-352

[3] Mengqing H, Lou H, Yan X, Xiaoyan H, Feng R, Zhou M. In-situ fabrication of ZIF-8 decorated layered double oxides for adsorption and photocatalytic degradation of methylene blue. Microporous and Mesoporous Materials. 2018;**271**:68-72

[4] Khuluk RH, Rahmat A. Removal of methylene blue by adsorption onto activated carbon from coconut Shell (Cocous Nucifera L.). Indonesian Journal of Science & Technology. 2019;**4**(2):229-240

[5] Subramaniam R, Kumar S. Novel adsorbent from agricultural waste (cashew nut shell) for methylene blue dye removal: Optimization by response surface methodology. Water Resources and Industry. 2015;**11**:64-70

[6] Subki NS, Han LY. A preliminary study on the potential dye adsorbance by activated meretrix lamarckii (M. lamarckii) shells. In: IIOP Conference Series: Earth and Environmental Science. 2020;**494**:012008

[7] Geed SR, Samal K, Tagade A. Development of adsorption-biodegradation hybrid process for removal of methylene blue. Journal of Environmental Chemical Engineering. 2019;7(6):103439

[8] Chairungsri W, Subkomkaew A, Kijjanapanich P, Chimupala Y. Direct dye wastewater photocatalysis using immobilized titanium dioxide on fixed substrate. Chemosphere. 2022;**286**:131762

[9] Yang H, Yan N. Transformation of seafood wastes into chemicals and materials. In: Green Chemistry and Chemical Engineering. Encyclopedia of Sustainability Science and Technology. 2019:461-462

[10] Muhammd Mailafiya M, Abubakar K, Danmaigoro AM, Chiroma SB, Abdul Rahim E, Moklas AMM, et al. Cockle shell-derived calcium carbonate (Aragonite) nanoparticles: A dynamite to nanomedicine. Applied Sciences. 2019;**9**:2897

[11] Buasri A, Chaiyut N, Loryuenyong V, Worawanitchaphong P, Trongyong S. Calcium oxide derived from waste shells of mussel, cockle, and scallop as the heteregenous catalyst for biodiesel production. The Scientific World Journal. 2013;**2013**:460923

[12] Sari M, Yusuf Y. Synthesis and Characterization of Hydroxyapatite Based Green on Mussel Shells (Perna Viridis) with Calcination Temperature Variation using the Precipitation Method. International Journal of Nanoelectronics and Materials. 2017;**11**(3):357-370

[13] Buasri A, Loryuenyong V. The new green catalysts derived from waste razor and surf clam shells for biodiesel production in a continuous reactor. Green Processing and Synthesis. 2015;**4**(5):389-397

[14] Gbadeyan OJ, Adali S, Bright G, Sithole B, Onwubu S. Optimization of milling procedures for synthesizing

Nano-$CaCO_3$ from Achatina fulica shell through mechanochemical techniques. Journal of Nanomaterials. 2020;**2020**:4370172

[15] Dzinun H, Abd Khalid NH, Hairom NHH. Photocatalytic performance of TiO_2/eggshell composite for wastewater treatment. Materials Today: Proceedings. 2022;**65**:3000-3006

[16] Ganta U, Gargi B, Susmita D, Kalyan A. Adsorptive removal of methylene blue dye from simulated wastewater using shale: Experiment and modelling. Journal- Indian Chemical Society. 2017;**94**:1-12

[17] Ahsanulhaq Q, Zhonghai Z, Waqas AM. Synthesis and photocatalytic properties of α-Fe2O3 nanoellipsoids. Solid State Sciences. 2010;**12**:1516-1519

[18] Waheed M et al. Eggshell calcium: A cheap alternative to expensive supplements. Trends in Food Science and Technology. 2019;**91**:219-230

[19] Krähenbühl M, Etter B, Udert K. Pretreated magnesite as a source of low-cost magnesium for producing struvite from urine in Nepal. The Science of the Total Environment. 2015;**542**:1155-1161

[20] Islam KN et al. Characterisation of calcium carbonate and its polymorphs from cockle shells (Anadara granosa). Powder Technology. 2011;**213**:188-191

[21] Haslinda Sharifuddin J, Chee Yean W, Shariah Ghazali S. Investigating the catalytic properties of calcium compounds derived from marine based shell waste for wastewater treatment. Materials Today: Proceedings. 2018;**5**(10):21718-21727

[22] Joshi G, Rawat DS, Lamba BY, Bisht KK, Kumar P, Kumar N, et al. Energy Convers. Manage. 2015;**96**:258-267

Chapter 11

Excited States of Six Anthocyanidin Variants with Different Solvents as Dye Sensitizers for Photocatalysis

Diana Barraza-Jiménez, Hugo Iván Flores-Hidalgo, Sandra Iliana Torres-Herrera, Raúl Armando Olvera-Corral and Manuel Alberto Flores-Hidalgo

Abstract

Anthocyanidins in the gas phase and under the effects of solvents such as water, ethanol, n-hexane, and methanol have been studied using DFT and TDDFT electronic structure calculations for applications as natural dyes in photocatalysis. The results include HOMO and LUMO orbitals, HOMO-LUMO gap, chemical properties, reorganization energies, and excited states. Malvidin presented the lower HOMO-LUMO gap energy. After the inclusion of solvents, HOMO-LUMO gap energy increased in all cases, presenting malvidin with n-hexane as the narrower gap energy. Conceptual DFT results showed that cyanidin, malvidin, and pelargonidin present good charge transfer properties. Cyanidin presented a lower electron reorganization energy (λ_e) when water is used as the solvent. TDDFT has been used for excited states calculation and absorption data show the main peaks in a wavelength between 479.1 and 536.4 nm. The UV–Vis absorption spectra were generated and the solvent effects in each case are discussed. In consequence, pigments selected in this attempt are suitable to work in the visible part of the electromagnetic spectrum and display the main peak in the green region. These pigments are found as good options for photocatalysis applications, and the best choices for dye sensitization are cyanidin, malvidin, and petunidin after including the more common anthocyanidins in the analysis.

Keywords: anthocyanidins, Dyes, DSSC, TDDFT, conceptual DFT

1. Introduction

Dye-sensitized solar cell (DSSC) is a promising photovoltaic technology and part of several green technologies used for environmental remediation based on taking advantage of sunlight as the energy source. The number of researchers dedicated to working on the development of this technology has increased exponentially in late years [1]. An important part of this technology relates to dyes which are used to sensitize the semiconductor in DSSCs. The development of new more efficient dyes is part of the research trends to improve this technology [1–3]. Natural pigments represent one of the more important choices to have improved dyes in DSSCs.

DSSCs discovery by Michael Grätzel and Brian O'Regan is dated in 1991 [4]. This device is a photoelectrochemical cell that imitates the photosynthesis process in plants. The cell consists of a semiconductor-based photoanode covered with a dye layer, a summarized functioning of the device is described as follows to understand the dye's role and importance in a DSSC [5, 6]. Dye photoexcitation provides an electron injection into the semiconductor conduction band from the dye LUMO which is caused by energy bands overlap. Next, the oxidized dye is regenerated when an electron is given up from the redox electrolyte. Electrolyte species reduction is completed with the addition of an electron at the platinum-coated transparent conducting oxide (TCO). The remainder of the semiconductor Fermi level and the electrolyte redox potential is equivalent to the open-circuit voltage [7, 8].

The idea of using the reactions of photosynthesis to convert sunlight into electrical power was published in 1974 by Melvin Calvin and became a common technique in solar technology [9–11]. Solar cells started with silicon devices, but technology has advanced, and new materials and devices were created, this progress includes DSSCs as part of an emerging third-generation photovoltaic concept in which stands out the use of synthetic or natural dyes as light-harvesting pigments [5, 6]. DSSCs components require more research and development to reach higher efficiencies [12–14]. Photosensitizers based on natural pigments are more desirable in DSSCs than dyes from metal complexes and may reach similar performances and stability [9]. Our interest within this work relates to natural pigments' electronic structure and will be focused on anthocyanidins.

Selected pigments are among the more commonly found anthocyanidins in nature, six different aglycones or anthocyanidins are included within this work and its common name with its distribution in fruits and vegetables is as follows: cyanidin 50%, pelargonidin 12%, delphinidin 12%, peonidin 12%, petunidin 7%, and malvidin 7% [15–18]. Hydroxyl and methoxy groups differentiate these molecules by the number and position of their B-ring [19–21]. Prior work related to anthocyanidins by our research group has been published elsewhere and includes cyanidin, malvidin, and peonidin [22]. The methodology from such work was reproduced with an upgraded version of the Gaussian program as part of the continuity work by our research group and we developed further work with additional pigments and calculations which were included in the present work. This work is considered a deeper study because all calculations were re-done using Gaussian 16 [23] (prior work was developed with G09), in addition, this work includes three additional pigments which enrich twice the options to make the best choice among the more used anthocyanidins for photocatalysis applications. The new calculations not included in the prior work were developed using Cis-TDDFT which enabled an analysis/discussion of emission data and the respective spectra. After increasing to six anthocyanidins in the study, the lower value for gap energy is malvidin in its gas phase and remains as a general behavior that with the addition of solvents gap energy increases in all cases except for malvidin with n-hexane because it had the narrower gap followed by petunidin also with n-hexane. For charge transfer, based on conceptual DFT results, cyanidin, malvidin, and pelargonidin present the best results. Water as solvent followed by ethanol and methanol applied in cyanidin displayed the lower values for electron reorganization energy (λ_e). Also, TDDFT calculations were carried out to calculate absorption properties for each pigment. After increasing the sample from three molecules up to six, it was again cyanidin, malvidin, and petunidin the pigments with the best performance indices for dye sensitization.

2. Theory and computational details

Theoretical calculations were performed in Gaussian16 (G16) programs suite [23]. Calculations include four solvents (water, ethanol, n-hexane, and methanol) in addition to the gas phase. Selection criteria are mainly based on how often the solvent is used in the laboratory to obtain pigments. The solvation model was PCM (polarizable continuum solvation model) as implemented in G16 program suite. B3LYP/6–311 + g (d,p) is the theoretical method used during geometry relaxation. Open-source databases were used to obtain the first geometry version and then our theoretical methodology was applied to optimize geometric parameters. Functional B3LYP is a widely accepted approach for this kind of molecule, and it was selected for this study mainly for that reason [24]. Basis set 6–311 + g(d,p) as implemented in the Gaussian16 program package [23] complements B3LYP very well according to preliminary calculations. 6–311 + g(d,p) was tested by running a set of calculations with different organic molecules with more than acceptable results. The literature considers B3LYP/6–311 + g(d,p) a theoretical method that provides a good level of accuracy for similar molecules [25–29]. A local minimum needs to be reached at the geometric optimization and it was confirmed with the calculation of harmonic vibrational frequencies. The zero-point vibrational energy (ZPVE) scaling was performed as a thermal correction (TC) at 298.15 K. Complementing geometry and frequency calculations, neutral energy, and adiabatic energies were obtained. Thereupon, chemical properties (HOMO, LUMO, gap, ionization potential (IP), electronic affinity (EA), electrophilicity (ω), electronegativity (χ), and hardness (η)) were computed based on the chemical reactivity indexes obtained in energy calculations. The sequence followed during the calculations was: first gas phase and then different solvents, one by one were included such as water, ethanol, n-hexane, and methanol.

Data reported by other research teams were included to compare with our results. So, a good idea is provided on performance against other theoretical methodologies or results obtained experimentally. Discussion is made on whether these molecules may be good dye sensitizers with TiO_2 [25–29] for future work. Excited states were calculated (over 10 states) but only the first excited states will be discussed in this document. TDDFT calculations were carried on with B3LYP/6-311 g + (d,p) for consistency with energy calculations. Useful energy graphs and excited states spectra diagrams considering the six more common anthocyanidin variants in the same figure are included in this work for comparison of results under similar theoretical methods. Chemissian code [30] was used to develop most energy graphs included in the results section.

3. Results and discussion

3.1 Molecular structure

Anthocyanidins are based on the flavylium ion or 2-phenylchromenylium (chromenylium may be referred to as benzopyrylium). These natural pigments are derivatives of 2-phenylchromenylium cation or flavylium cation. A relevant feature for this structure is the capability to carry different substituents in the phenyl group at 2-position. Another particularity to note is anthocyanidins differ from other flavonoids because of a positive charge. Molecule substituents and main features are displayed in **Table 1** which includes a molecular scheme in **Figure 1** shown next to the table so the reader may have a good view for a general interpretation of structural differences between anthocyanidin variants. Anthocyanidins have a 15-carbon atoms main structure arranged in two aromatic rings (A and B) as shown in **Figure 1**. A third ring (C) provides the positive charge from an oxygen atom contained in this ring. Two C=C bonds in the C ring differentiate anthocyanidins among the flavonoid family and it is responsible for a positive charge in this molecule, therefore, it is a cation (flavylium) when it is at the stable form at low pH [31].

The phenylbenzopyrylium core of anthocyanins may be modified by the addition of a wide range of chemical groups using hydroxylation, acylation, and methylation.

Geometric parameters are summarized in **Table 2**. The phenylbenzopyrylium is normally combined with a wide range of chemical groups using hydroxylation, acylation, and methylation.

The C—C bond length found within this work has a similar length or nearly enough to 140 pm (C—C bond length average size) which is the typical bond length

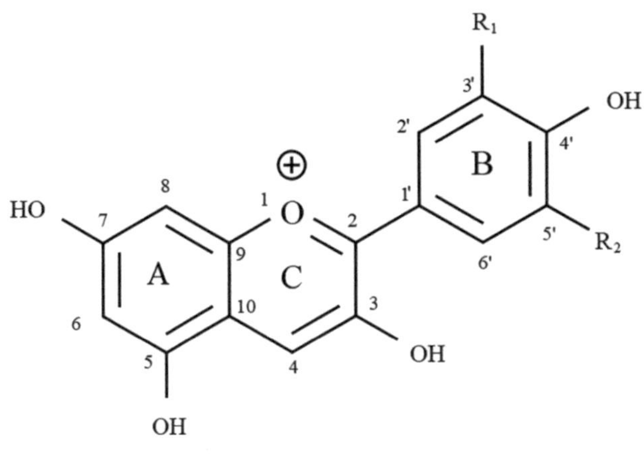

Figure 1.
General chemical structure of anthocyanidins according to ***Table 1.***

		Substitution pattern		
Name	**Chemical formula**	**R1**	**R2**	**Color**
Cyanidin	$(C_{15}H_{11}O_6)^+$	OH	OH	Orange-red
Delphinidin	$(C_{16}H_{11}O_7)^+$	OH	OH	Blue-red
Malvidin	$(C_{15}H_{13}O_5)^+$	OCH_3	OCH_3	Blue-red
Pelargonidin	$(C_{15}H_{11}O_5)^+$	H	H	Orange
Peonidin	$(C_{15}H_{13}O_6)^+$	OCH_3	H	Orange-red
Petunidin	$(C_{15}H_{12}O_6)^+$	OCH_3	OH	Blue-red

Table 1.
More known anthocyanidins structure and substitution patterns.

DOI: http://dx.doi.org/10.5772/intechopen.108158

Parameter	Cyanidin	Delphinidin	Malvidin	Pelargonidin	Peonidin	Petunidin
O(1)-C(2)	1.350	1.344	1.347	1.345	1.345	1.346
O(1)-C(9)	1.358	1.358	1.359	1.359	1.358	1.359
C(2)-C(3)	1.420	1.403	1.407	1.404	1.404	1.406
C(2)-C(1′)	1.436	1.448	1.444	1.445	1.447	1.446
C(3)-C(4)	1.382	1.390	1.388	1.389	1.390	1.390
C(4)-C(10)	1.403	1.397	1.399	1.397	1.397	1.398
C(5)-C(6)	1.376	1.375	1.376	1.375	1.375	1.375
C(5)-C(10)	1.427	1.436	1.435	1.436	1.436	1.436
C(6)-C(7)	1.412	1.409	1.408	1.409	1.409	1.409
C(7)-C(8)	1.395	1.399	1.398	1.399	1.398	1.398
C(8)-C(9)	1.386	1.380	1.382	1.381	1.381	1.382
C(9)-C(10)	1.409	1.423	1.421	1.422	1.423	1.422
C(1′)-C(2′)	1.422	1.406	1.414	1.415	1.409	1.414
C(1′)-C(6′)	1.414	1.416	1.407	1.413	1.411	1.410
C(2′)-C(3′)	1.377	1.389	1.381	1.377	1.387	1.385
C(3′)-C(4′)	1.422	1.404	1.418	1.405	1.420	1.409
C(4′)-C(5′)	1.396	1.403	1.407	1.403	1.399	1.401
C(5′)-C(6′)	1.383	1.386	1.395	1.383	1.384	1.389
O(1)-C(2)-C(1′)-C(6′)	180	150.1	151.4	150.5	150.1	150.6
C(3)-C(2)-C(1′)-C(2′)	180	148.5	149.3	149.8	149.1	148.3
O-C(3′)-C(4′)-C(5′)	180	179.9	179.3	180.0	177.4	179.9
H-C(5′)-C(4′)-C(3′)	180	178.2	175.9	177.9	178.5	178.2
O-C(4′)-C(3′)-C(2′)	180	179.7	177.0	179.9	176.7	179.8
O-C(4′)-C(5′)-C(6′)	180	179.2	178.0	179.3	177.7	179.4
C(8)-C(9)-C(10)-C(4)	180	175.7	176.3	175.6	175.9	176.2
O(1)-C(9)-C(10)-C(5)	180	178.6	178.8	178.6	178.7	178.8
C(8)-C(9)-O(1)-C(2)	180	179.0	179.9	179.1	179.4	179.6
C(5)-C(10)-C(4)-C(3)	180	179.7	179.3	179.7	179.5	179.4
C(9)-O(1)-C(2)-C(1′)	180	179.4	179.1	179.0	179.0	179.3

Data regenerated by our research group with a similar theoretical method reported elsewhere [22].

Table 2.
Selected anthocyanidins geometric parameters summary including bond length and bond angles in Å and °, respectively.

for benzene. An average of 154.0 pm and 134 pm were found for single and double C—C bonds length, respectively. C—C bond lengths for benzene are customarily accepted around 139 pm in the literature which is near to our findings. The discrepancy is minor near to 0.1 Å in average for C—C bonds of selected pigments considering an average length between 1.346 and 1.444 Å.

Methodologies such as B3LYP/6-31 g(d), B3LYP/6–31 + g(d,p) have been reported in the literature for similar molecules [32–36] and will be included a few selected data from some of these sources to enrich the discussion in this work. At his point, one can say B3LYP reaches accurate results for these molecules' geometries and may be expected good results for similar organic molecules as well. Thus, results in this work for C—C bond lengths comply with the reported data.

The planarity in a structure is related to dihedral angles. For anthocyanidins within this work, the planarity among the three rings forming these molecules skeleton within each anthocyanidin represents an important feature that differentiates one from another. In the literature, the parameter reported is the torsion angle instead of dihedral angles and this value may be in the same way a factor that characterizes an anthocyanidins and influences its electronic structure behavior [32]. Cyanidin is considered a planar molecule because its dihedrals vary by less than 1° from a perfectly planar structure. Delphinidin and petunidin have similar planarity between them but their torsion angle causes the molecules to have the lower planarity level. Peonidin has more dihedrals different than 180° but only a couple of them differ more than 5°. Then, the analysis indicate there are differences in the dihedrals but only a couple cases deviate significantly from a perfect planarity. However, despite the numeric difference is small, it is such differences in planarity that determine most of the molecule character and its chemical properties. Put it in other words, it may be seen that few dihedrals correspond with a nonplanar structure, in such a way that there is a direct relationship with the relative angle or torsion angle between rings, and it represents the main difference observed in the B ring compared with the rest of the structure. All selected structures fall into the torsion angle and planarity concepts mentioned, with exception of cyanidin which has an almost perfectly planar structure confirmed by its dihedral values.

3.2 Electronic structure

Energy calculations were executed using B3LYP/6311 + g(d,p) method for the gas phase and four solvents (water, ethanol, n-hexane, and methanol). The reader may see HOMO and LUMO molecular orbitals numeric results in **Table 3**. An idea to analyze from these results is how these molecules energy orbitals may overlap with a semiconductor energy orbital for DSSCs and photocatalytic applications.

This procedure consists in reproducing a process where an electron is photo-induced in the molecular system by being transferred from the dye-excited state to the semiconductor. The process takes place at HOMO and LUMO energy orbitals. Therefore, a dye sensitizer should have HOMO and LUMO energy levels that mate with electrolyte redox potential and the semiconductor conduction band [25]. Pigments included in this work well with the electrolyte redox level (−4.85 eV) and the conduction band edge for TiO_2 (−4.00 eV), considering values reported in the literature [25–29].

Calculations include molecular orbitals for all variants in the gas phase and with solvents four different solvents. LUMO results are between −6.856 and − 6.624 eV for the gas phase, which is relevant because LUMO molecular orbital may be beneficial for the application as dye sensitizers. An expected condition is dye molecular orbitals overlapping semiconductor band gap in some way so it can take place an easier charge transfer process.

There is a shift around 3 eV in HOMO and LUMO for gas phase results if compared to results when added solvents like water and ethanol. This shift is evidenced in

Pigment	Solvent	H-L	HOMO	LUMO	λ_e	EEP	λ_h	HEP
$(C_{15}H_{11}O_6)^+$	Gas phase	2.664	−9.288	−6.624	0.318	5.525	0.344	10.361
	Water	2.824	−6.452	−3.628	0.262	4.064	0.284	6.038
	Ethanol	2.816	−6.528	−3.712	0.264	4.102	0.288	6.155
	n-hexane	2.712	−7.916	−5.204	0.295	4.818	0.324	8.284
	Methanol	2.818	−6.501	−3.683	0.263	4.089	0.267	6.115
$(C_{16}H_{11}O_7)^+$	Gas phase	2.667	−9.513	−6.846	0.346	5.753	0.364	10.572
	Water	2.844	−6.577	−3.732	0.292	4.191	0.320	6.136
	Ethanol	2.835	−6.658	−3.823	0.294	4.235	0.321	6.260
	n-hexane	2.726	−8.114	−5.388	0.325	5.013	0.346	8.467
	Methanol	2.838	−6.630	−3.792	0.293	4.220	0.320	6.217
$(C_{15}H_{13}O_5)^+$	Gas phase	2.539	−9.24	−6.701	0.371	5.666	0.452	10.162
	Water	2.823	−6.532	−3.709	0.294	4.172	0.460	5.946
	Ethanol	2.810	−6.610	−3.800	0.295	4.216	0.462	6.066
	n-hexane	2.657	−7.975	−5.318	0.335	4.968	0.479	8.169
	Methanol	2.815	−6.583	−3.768	0.294	4.201	0.461	6.024
$(C_{15}H_{11}O_5)^+$	Gas phase	2.881	−9.737	−6.856	0.333	5.732	0.348	10.820
	Water	2.994	−6.693	−3.699	0.290	4.156	0.321	6.250
	Ethanol	2.989	−6.778	−3.789	0.292	4.200	0.322	6.379
	n-hexane	2.918	−8.290	−5.372	0.319	4.984	0.335	8.658
	Methanol	2.990	−6.748	−3.758	0.291	4.185	0.321	6.334
$(C_{15}H_{13}O_6)^+$	Gas phase	2.691	−9.465	−6.774	0.364	5.703	0.498	10.371
	Water	2.955	−6.668	−3.713	0.293	4.173	0.527	6.019
	Ethanol	2.945	−6.748	−3.803	0.294	4.217	0.527	6.142
	n-hexane	2.815	−8.166	−5.351	0.328	4.980	0.533	8.316
	Methanol	2.948	−6.720	−3.772	0.294	4.202	0.527	6.100
$(C_{15}H_{12}O_6)^+$	Gas phase	2.604	−9.379	−6.775	0.341	5.693	0.366	10.411
	Water	2.796	−6.511	−3.715	0.289	4.173	0.318	6.066
	Ethanol	2.785	−6.590	−3.805	0.291	4.216	0.320	6.188
	n-hexane	2.664	−8.014	−5.35	0.318	4.976	0.348	8.353
	Methanol	2.789	−6.562	−3.773	0.290	4.201	0.319	6.145

**Data regenerated by our research group with a similar theoretical method reported elsewhere [22].*

Table 3.
Energy results for selected molecules: H-L is the HOMO-LUMO gap or energy band. All units in eV.

HOMO magnitude by around 3 eV. A shift of less than 1.5 eV in HOMO and LUMO is estimated for n-hexane molecular orbitals calculations. The HOMO and LUMO molecular orbitals are shown in **Figures 2** and **3**.

The difference between HOMO-LUMO is generally accepted as a similar value to the band gap. Results for HOMO-LUMO gap were between 2.539 and 2.881 eV in the

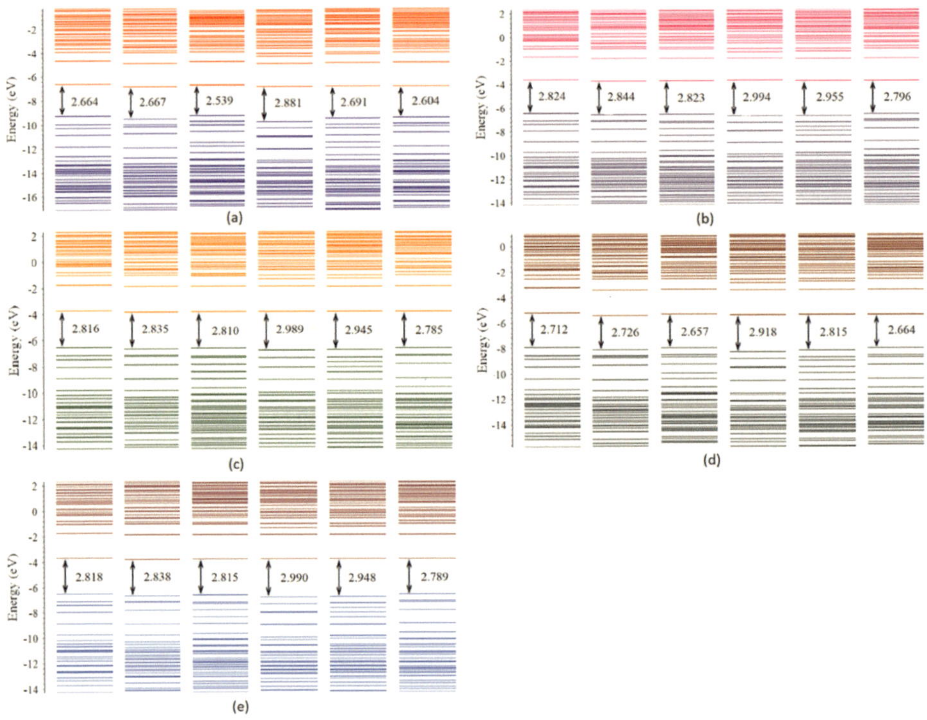

Figure 2.
Molecular orbitals for selected anthocyanidins cyanidin [22], delphinidin, malvidin [22], pelargonidin, peonidin [22], and petunidin corresponding to (a) gas phase, (b) water solvent, (c) ethanol solvent, (d) n-hexane solvent and (c) methanol solvent. H-L gap energy units are in eV.

gas phase with malvidin having the narrower gap and pelargonidin with the wider gap among the six pigments.

Therefore, solvents are responsible for a slight shift of HOMO-LUMO values in all cases. Malvidin in the gas phase has a lower value for the HOMO-LUMO energy. With the addition of solvents, these gaps increase in all cases with n-hexane as the narrower value, followed by petunidin also with n-hexane. Ethanol and methanol solvents have a slighter effect than water. In general, effects in HOMO-LUMO are small, which means there is a similarity in magnitude on the results when used any solvent water, ethanol, n-hexane, or methanol. The HOMO-LUMO gap varies in all cases by less than 10% if compared with the HOMO-LUMO values in the gas phase. The greater shift was 11 and 10% corresponding to malvidin and peonidin, respectively, however, n-hexane effect on malvidin shifted only 5%. Then, when water is used a bigger shift in HOMO-LUMO is observed and, in contrast, n-hexane caused the smaller shift. The energy gap had a small variation with ~0.3 eV on average considering all variants. Cyanidin and delphinidin have alike energy gap values despite the solvent and despite their differences in geometric parameters and constituents. HOMO-LUMO gap energy seems almost unaffected by planarity and relative angles, which means the effects of the main geometric parameters in gap energy are considered small.

There is an amount of energy needed so a molecule can become ionized, which means if one charge is lost it becomes a cation and if one charge is gained it becomes an anion. Such energy was calculated using intramolecular reorganization energies.

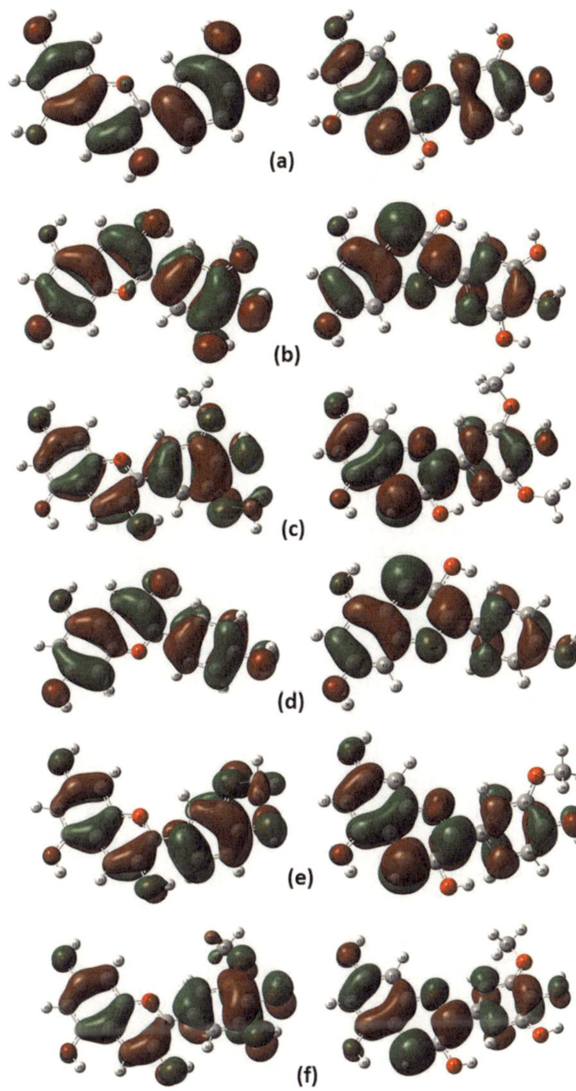

Figure 3.
Molecular orbitals charge distribution using B3LYP/6–311 + g(d,p), corresponding to: (a) cyanidin [22], (b) delphinidin, (c) malvidin [22], (d) pelargonidin, (e) peonidin [22] and (f) petunidin.

When the ionized molecule becomes neutral, then these two processes relate to the charge transfer process. The energy needs to be available for charge transfer for that reason reorganization energy (λ) values are low to prevent wasting energy in reorganization processes. Then, the reason why λ is low is to maximize the use of solar energy instead of using sunlight during the energy transfer process. Water, ethanol, and methanol solvent addition cause a decrease in λ. Solvent n-hexane also decreases λ but slightly. The lower electron reorganization energy (λ_e) was cyanidin with water but with similarities when used ethanol and methanol.

Cyanidin lower hole reorganization energy (λ_h) was obtained when used solvent methanol followed by water with similar values but not as close as in the case of λ_e.

The results for hole extraction potential (HEP) and electron extraction potential (EEP) present higher values for the gas phase, and a decrease is observed when any of the solvents is used.

Water, ethanol, and methanol have a similar effect in HEP and EEP, and it is bigger than n-hexane in all cases. Higher values for HEP were observed in pelargonidin in the gas phase, in general, for gas-phase HEP results are around 10 eV.

With solvent n-hexane, HEP goes around 8 eV and with solvents such as water, ethanol, and methanol its value is nearly 6 eV. EEP values are near 1 eV for solvents water, ethanol, and methanol and go down to 0.5 eV with n-hexane. Higher EEP was observed in delphinidin in the gas phase as expected because of the OH radical present in its molecular structure, but the rest of the selected anthocyanidins had similar values in the gas phase with 0.1 eV variation. Reorganization energies show malvidin is the best choice followed by petunidin.

Cyanidin with methanol produces the best electron reorganization energy λ_e followed by water and ethanol. Cyanidin is more suitable for hole energy λ_h with the same solvents. Petunidin is the next more suitable but with a modest advantage by 0.05 eV over cyanidin. It is possible that λ values performance relates with molecule planarity. The effect of solvents in EEP and HEP is unclear in contrast with λ. Malvidin with water is the best choice from EEP and HEP viewpoint but the variation is minor considering the same solvent is used in other molecules.

3.3 Chemical reactivity properties

Chemical reactivity properties were calculated with conceptual DFT. These properties are shown in **Table 4**.

Ionization potential (IP) is associated with the electronic cloud stiffness. In terms of reactivity, the cloud is wary to become a participant in electron transfer. Therefore, a lower ionization potential is enticing to have a larger molecular potential so electron donation boosts. Malvidin presents the lower IP in its gas phase and decreases further with solvent addition. A similar effect in IP magnitude was caused by water, ethanol, and methanol but the lower IP value was when water is used as a solvent in cyanidin among all variants.

In the gas phase, IP was near 11 eV and when used water, ethanol, and methanol IP decreased to values near 6 eV. IP values also had a reducing trend with solvent n-hexane with results around 8 eV. The lower IP was observed in cyanidin with water and methanol meanwhile for malvidin and petunidin their lower values were observed with these two solvents.

For molecules in its gas phase, EA results were around 5 eV and with solvents water, ethanol, and methanol a reducing trend was observed with results around 3 eV, and n-hexane effect on EA also was a reducing trend to values around 4 eV. Delphinidin in n-hexane has the higher EA but its EA values are only slightly higher than those for pelargonidin, petunidin, peonidin, and malvidin, all with n-hexane.

Attracting electron pairs may be measured with electronegativity (χ). For a better suitability to act as a charge acceptor, a high electronegativity (χ) is desirable. Pelargonidin displayed the highest χ value in the gas phase, in general χ results are near 8 eV and have a decreasing trend with values near 5 eV when solvents such as water, ethanol, and methanol are used. For n-hexane solvent, results are near 6 eV. Pelargonidin with n-hexane presents the higher value but it is slightly over the rest of the molecules using n-hexane as well.

Pigment	Solvent	IP	EA	χ	η	ω	S
$(C_{15}H_{11}O_6)^+$	Gas phase	10.642	5.154	7.898	2.744	11.439	0.364
	Water	6.322	3.802	5.062	1.26	10.165	0.793
	Ethanol	6.443	3.838	5.141	1.302	10.147	0.768
	n-hexane	8.608	4.522	6.565	2.043	10.549	0.490
	Methanol	6.382	3.825	5.104	1.278	10.189	0.782
$(C_{16}H_{11}O_7)^+$	Gas phase	10.876	5.353	8.114	2.761	11.923	0.362
	Water	6.456	3.899	5.177	1.278	10.485	0.782
	Ethanol	6.582	3.941	5.261	1.320	10.484	0.757
	n-hexane	8.813	4.688	6.751	2.062	11.048	0.485
	Methanol	6.537	3.927	5.232	1.305	10.485	0.766
$(C_{15}H_{13}O_5)^+$	Gas phase	10.614	5.296	7.955	2.659	11.899	0.376
	Water	6.406	3.878	5.142	1.264	10.462	0.791
	Ethanol	6.528	3.921	5.224	1.304	10.469	0.767
	n-hexane	8.647	4.633	6.640	2.007	10.983	0.498
	Methanol	6.486	3.906	5.196	1.29	10.466	0.775
$(C_{15}H_{11}O_5)^+$	Gas phase	11.181	5.399	8.29	2.891	11.886	0.346
	Water	6.571	3.866	5.219	1.352	10.07	0.739
	Ethanol	6.701	3.908	5.305	1.396	10.077	0.716
	n-hexane	8.992	4.665	6.829	2.164	10.775	0.462
	Methanol	6.656	3.894	5.275	1.381	10.074	0.724
$(C_{15}H_{13}O_6)^+$	Gas phase	10.869	5.34	8.105	2.765	11.879	0.362
	Water	6.545	3.881	5.213	1.332	10.199	0.751
	Ethanol	6.670	3.922	5.296	1.374	10.209	0.728
	n-hexane	8.850	4.652	6.751	2.099	10.859	0.477
	Methanol	6.627	3.908	5.267	1.359	10.205	0.736
$(C_{15}H_{12}O_6)^+$	Gas phase	10.777	5.352	8.064	2.713	11.987	0.369
	Water	6.384	3.884	5.134	1.25	10.541	0.8
	Ethanol	6.508	3.925	5.217	1.291	10.536	0.774
	n-hexane	8.701	4.658	6.679	2.022	11.034	0.495
	Methanol	6.465	3.911	5.188	1.277	10.537	0.783

Data regenerated by our research group with similar theoretical method reported elsewhere [22].

Table 4.
Chemical reactivity of selected anthocyanidins. Properties displayed are ionization potential (IP), electron affinity (EA), electronegativity (χ), chemical hardness (η), electrophilicity index (ω), and chemical softness (S), units are eV.

Therefore, chemical properties show similarity among resulting values which may be induced by the molecular resemblances including the torsion angle, and the small structural differences may be responsible for the main differences as well as their molecule constituents.

3.4 Excited states with TDDFT

TDDFT excited states were computed with B3LYP/6311 + g(d,p) methodology in Gaussian16. The literature supports that B3LYP is a suitable hybrid functional [25–29, 36] for this kind of computations and has been successful in similar molecules.

A good match between the absorption spectrum and the solar irradiation spectrum in DSSCs benefits its performance. An indicator of the light-harvesting effectiveness may be data related to the dye's absorption and such data become relevant for the performance of the DSSCs [37–41] as a whole. Our results are in acceptable accord with experimental values and the differences may be caused by solvent effects and variation added by measuring methodologies [36, 42–44]. For ΔE, a low value is desirable so the first excited state may need as low energy as possible. In the gas phase, malvidin presents the lower value for ΔE and, with solvent addition, the lower value was malvidin with n-hexane closely followed by petunidin with n-hexane. The literature reports two main regions in anthocyanidins UV–Vis spectra. The first one at 260–280 nm and the second one at the visible region between 490 and 550 nm. There is a third peak at 310–360 nm [43], but we will focus on the main peak located in the visible region.

This group of anthocyanidins in the gas phase had absorption wavelengths between 479.1 and 536.4 nm. These molecules work in the visible with both cyanidin and pelargonidin working in the blue region. Pelargonidin and malvidin are the lower and higher values while cyanidin presents a similar value with pelargonidin results which may be related to the fact that both have a small relative angle at the B ring and, they are the simplest molecules regarding their constituents. Addition of solvent shifts absorption spectra by increasing its wavelength by less than 5 nm in the case of water, ethanol, and methanol. When used n-hexane, absorption spectra shift by nearly 10 nm. TDDFT excited states absorption data are shown in **Table 5** and absorption spectra are shown in **Figure 4**. Photon-to-current conversion relies on the visible and near UV regions results and based on these results one can attain microscopic information related to electronic transitions and MO properties.

A goal of TDDFT excited states was to calculate absorption data and our numeric results are shown in **Table 5** and absorption spectra are shown in **Figure 4**.

Light harvesting energy (LHE) index was calculated due to its importance in electronic transfer.

The light-harvesting energy (LHE) index was calculated due to its importance in electronic transfer. In a dye sensitizer, a high LHE maximizes photo-current response, and it can be calculated with equation (1):

$$LHE = 1 - 10^{-f} \tag{1}$$

where f is the oscillator strength of the dye associated with the wavelength corresponding to the peak absorbance through intramolecular charge transfer [45, 46]. Singlet-to-singlet transitions of the absorption bands with maximum wavelength and oscillator strength were obtained for all selected anthocyanidins. In the gas-phase cyanidin had a higher LHE followed by petunidin and on the other hand, malvidin had a lower LHE value. After the addition of solvents, there is an increase in LHE in all cases but with malvidin, the effect of the solvent is more noticeable especially when methanol is used. Cyanidin, petunidin, and malvidin have higher LHE values after solvent addition. The lowest energy absorption in these molecules is due

Molecule	Solvent	State	ΔE (eV)	λ (nm)	Transition	Contribution	f	LHE
$(C_{15}H_{11}O_6)^+$	Gas Phase	1	2.546	487.1 (522*)	H - > L	67%	0.507	0.689
	Water	1	2.524	491.2	H - > L	68%	0.619	0.760
					H-1 - > L	17%		
					H-2 - > L	12%		
	Ethanol	1	2.528	490.4	H - > L	68%	0.629	0.765
					H-1 - > L	15%		
					H-2 - > L	12%		
	n-hexane	1	2.473	501.4	H - > L	69%	0.686	0.794
	Methanol	1	2.524	491.3	H - > L	68%	0.622	0.761
					H-1 - > L	17%		
					H-2 - > L	12%		
$(C_{16}H_{11}O_7)^+$	Gas Phase	1	2.448	506.4 (534*)	H - > L	61%	0.324	0.526
					H-1 - > L	30%		
					H-2 - > L	18%		
	Water	1	2.498	496.4	H - > L	68%	0.579	0.736
					H-2 - > L	17%		
	Ethanol	1	2.498	496.4	H - > L	68%	0.565	0.728
					H-2 - > L	17%		
	n-hexane	1	2.421	512.17	H-2 - > L	16%	0.574	0.733
					H - > L	68%		
	Methanol	1	2.495	496.9	H - > L	68%	0.578	0.736
					H-2 - > L	17%		
$(C_{15}H_{13}O_5)^+$	Gas Phase	1	2.312	536.4 (542*)	H - > L	60%	0.240	0.425
	Water	1	2.434	509.3	H-1 - > L	30%	0.604	0.751
	Ethanol	1	2.481	499.8	H - > L	68%	0.591	0.744
					H-2 - > L	17%		
	n-hexane	1	2.376	521.9	H - > L	70%	0.627	0.764
	Methanol	1	2.431	510.1	H - > L	61%	0.601	0.749
$(C_{15}H_{11}O_5)^+$	Gas Phase	1	2.588	479.1 (513*)	H - > L	61%	0.325	0.527
					H-1 - > L	35%		
	Water	1	2.591	478.6	H - > L	67%	0.493	0.679
					H-1 - > L	21%		
	Ethanol	1	2.592	478.4	H - > L	67%	0.472	0.663
					H-1 - > L	22%		
	n-hexane	1	2.537	488.7	H-1 - > L	24%	0.487	0.674
					H - > L	66%		
	Methanol	1	2.589	478.9	H - > L	67%	0.491	0.677
					H-1 - > L	21%		

Molecule	Solvent	State	ΔE (eV)	λ (nm)	Transition	Contribution	f	LHE
$(C_{15}H_{13}O_6)^+$	Gas Phase	1	2.401	516.3 (532*)	H - > L	67%	0.288	0.485
					H-1 - > L	11%		
	Water	1	2.509	494.2	H - > L	69%	0.530	0.705
	Ethanol	1	2.564	483.6	H - > L	67%	0.515	0.695
	n-hexane	1	2.465	503	H - > L	69%	0.535	0.708
	Methanol	1	2.505	494.9	H - > L	69%	0.527	0.703
$(C_{15}H_{12}O_6)^+$	Gas Phase	1	2.423	511.7 (543*)	H - > L	65%	0.410	0.611
	Water	1	2.460	503.9	H - > L	69%	0.600	0.749
	Ethanol	1	2.757	449.7	H-2 - > L	23%	0.682	1.792
					H - > L	66%		
	n-hexane	1	2.382	520.6	H-2 - > L	13%	0.606	0.752
					H - > L	69%		
	Methanol	1	2.458	504.5	H - > L	69%	0.599	0.748

Cyanidin, malvidin, and peonidin data has been regenerated for this work with very similar results, considering there is a set of our own results obtained with different methodology that were published previously elsewhere [22].
**Experimental data from the literature [43].*

Table 5.
TD-DFT excited states absorption data for selected molecules.

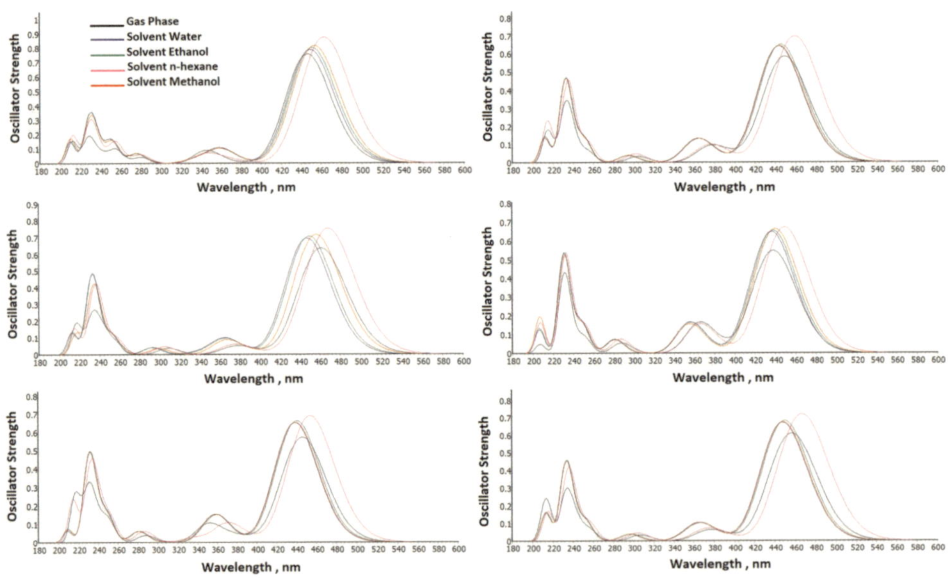

Figure 4.
Absorption spectra using TD-DFT for gas phase and solvents water, ethane, n-hexane, and methane for: (a) cyanidin [22], (b) delphinidin, (c) malvidin [22], (d) pelargonidin, (e) peonidin [22], and (f) petunidin.

to the transition from HOMO to LUMO with the largest oscillator strength resulting in an enhanced LHE, this approach emphasizes the parameters recommended in the literature to identify the best choice [41, 47, 48].

To obtain an effect where the absorption spectrum overlaps with the solar spectrum, the energy gap will have to reduce. Such action could be possible with the inclusion of a co-absorber of appropriate properties. Among the dyes studied, anthocyanidins with the higher LHE may work well with solar energy and may be recommended as the better suited for use as a potential sensitizer for DSSC.

3.5 Excited states using CIS-TDDFT

Cis-TDDFT methodology was used to calculate excited states with the scheme implemented in Gaussian16 [23]. Emission wavelength values increase after solvents are included in most cases if compared with results obtained for emission in anthocyanidins gas phase, respectively, for each variant. Emission spectra for each selected anthocyanidin are shown in **Figure 5**. According to experimental data, DFT calculations underestimate wavelength values by approximately 8%. Cyanidin, delphinidin, malvidin, and pelargonidin have similar effects with each solvent presenting slightly increased emission wavelength values for water, ethanol, and methanol and a slight decrease in wavelength value for n-hexane solvent. Peonidin and petunidin have similar effects when water and ethanol are used with a slight increase in wavelength value and are similar to the effect when used solvent n-hexane with a slight decrease but there is a different effect on these two anthocyanidins when methanol is used since, in these two variants, wavelength presents a slight decrease.

Overall, the effects of solvents in these six anthocyanidins are similar, maybe petunidin presents stronger effects in the wavelength with solvents than the others but the effects can be considered small even for this case.

Oscillator strength for the selected anthocyanidins has a particular effect for each variant that may be related to the solvent. In all cases, the gas phase displayed the lower oscillator strength value except for malvidin which presents the lower oscillator

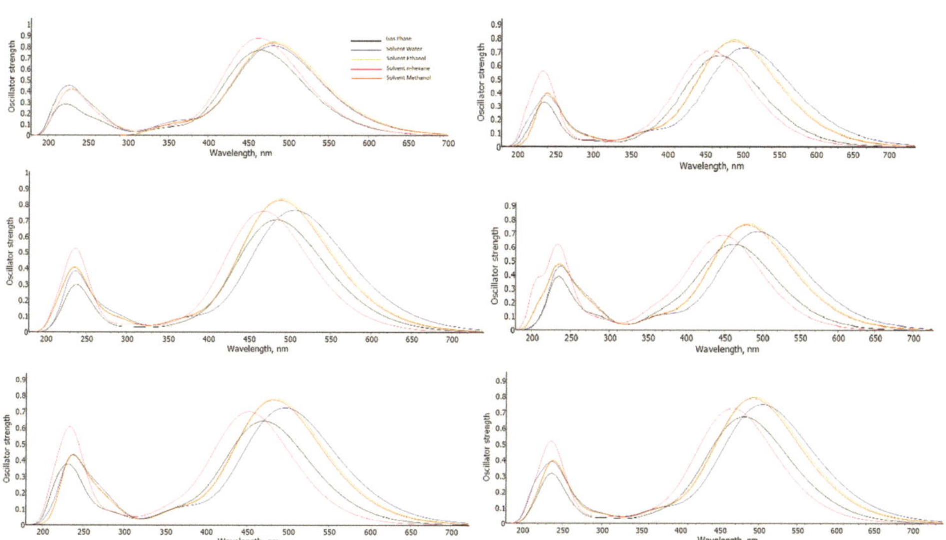

Figure 5.
Emission spectra data using TD-DFT excited states for gas phase and solvents water, ethane, n-hexane, and methane applied in the next molecules: (a) cyanidin, (b) delphinidin, (c) malvidin, (d) pelargonidin, (e) peonidin, and (f) petunidin.

strength value for the solvent methanol. In all cases, the higher oscillator strength value was observed when used solvent ethanol except for petunidin since it was with solvent water where the higher value was observed. Oscillator strength and data obtained for excited states calculations are shown in **Table 6**.

Cyanidin presented the higher oscillator strength values if compared with each anthocyanidin variant, and this behavior is maintained when the solvent is included. For cyanidin, when solvents are added the oscillator strength values increase if

Molecule	Solvent	State	ΔE (eV)	λ (nm)	Transition	Contribution	f	LHE
$(C_{15}H_{11}O_6)^+$	Gas Phase	1	2.665	465.2 (522*)	H-1 - > L	15%	0.769	1.830
					H - > L	68%		
	Water	1	2.576	481.2	H-1 - > L	13%	0.808	1.845
					H - > L	69%		
	Ethanol	1	2.569	482.7	H - > L	68%	0.877	1.867
	n-hexane	1	2.677	463.1	H - > L	68%	0.877	1.867
	Methanol	1	2.578	480.9	H-1- > L	13%	0.832	1.853
					H - > L	69%		
$(C_{16}H_{11}O_7)^+$	Gas Phase	1	2.633	470.8 (534*)	H-2 - > L	22%	0.662	1.782
					H - > L	66%		
	Water	1	2.460	503.9	H-1 - > L	13%	0.729	1.813
					H - > L	69%		
	Ethanol	1	2.532	489.6	H - > L	69%	0.785	1.836
	n-hexane	1	2.700	459.3	H-2 - > L	24%	0.699	1.800
					H - > L	65%		
	Methanol	1	2.540	488.2	H - > L	69%	0.776	1.832
$(C_{15}H_{13}O_5)^+$	Gas Phase	1	2.561	484.2 (542*)	H - > L	67%	0.702	1.802
	Water	1	2.451	505.8	H-1 - > L	13%	0.764	1.828
					H - > L	69%		
	Ethanol	1	2.531	489.9	H-1 - > L	14%	0.835	1.854
					H - > L	69%		
	n-hexane	1	2.648	468.2	H - > L	66%	0.755	1.824
	Methanol	1	2.537	488.7	H-1 - > L	14%	0.601	1.749
					H - > L	69%		
$(C_{15}H_{11}O_5)^+$	Gas Phase	1	2.672	464.0 (513*)	H-1 - > L	18%	0.612	1.755
					H - > L	68%		
	Water	1	2.509	494.2	H - > L	70%	0.712	1.806
	Ethanol	1	2.577	481.2	H - > L	69%	0.769	1.830
	n-hexane	1	2.760	449.3	H-1 - > L	18%	0.674	1.788
					H - > L	68%		
	Methanol	1	2.586	479.5	H - > L	69%	0.760	1.826
$(C_{15}H_{13}O_6)^+$	Gas Phase	1	2.632	471.1 (532*)	H - > L	67%	0.637	1.769

Molecule	Solvent	State	ΔE (eV)	λ (nm)	Transition	Contribution	f	LHE
	Water	1	2.500	496.0	H - > L	69%	0.724	1.811
	Ethanol	1	2.560	484.4	H - > L	69%	0.779	1.834
	n-hexane	1	2.734	453.5	H - > L	66%	0.689	1.796
	Methanol	1	2.734	453.5	H - > L	66%	0.689	1.796
$(C_{15}H_{12}O_6)^+$	Gas Phase	1	2.574	481.7 (543*)	H - > L	67%	0.669	1.786
	Water	1	2.455	505.0	H - > L	69%	0.749	1.822
	Ethanol	1	2.757	449.7	H-2 - > L	23%	0.682	1.792
					H - > L	66%		
	n-hexane	1	2.657	466.7	H-2 - > L	23%	0.720	1.809
					H - > L	66%		
	Methanol	1	2.657	466.7	H-2 - > L	23%	0.720	1.748
					H - > L	66%		

Experimental data from the literature [43].

Table 6.
Excited states emission results for selected anthocyanidins using CIS-TD-DFT.

compared with its gas phase value and the higher oscillator strength values were observed with ethanol and n-hexane.

Malvidin resembles cyanidin with slightly smaller values for oscillator strength but a similar trend and in all cases when added a solvent oscillator strength, values increase aside from methanol which can be considered the only exception. Delphinidin, pelargonidin, and peonidin resemble cyanidin as well but with smaller values than malvidin so the difference with cyanidin is bigger in these cases.

Cyanidin presented the higher oscillator strength values if compared with each anthocyanidin variant and this behavior is maintained when solvent is included. For cyanidin addition of solvents cause an increase in the oscillator strength values from the gas phase value with ethanol and n-hexane having the higher oscillator strength values.

Malvidin resembles cyanidin but with smaller values for oscillator strength with a similar trend and in all cases when added solvent oscillator strength values increase with the only exception of methanol. Delphinidin, pelargonidin, and peonidin resemble cyanidin as well but with smaller values than malvidin so the difference with cyanidin is bigger in these cases. Also, in these cases, all values obtained for oscillator strength when added solvents were smaller than gas-phase oscillator strengths. For petunidin, oscillator strength changes after solvent addition are moderate, the bigger change was with the addition of water solvent.

The transition energy for the selected anthocyanidins has a similar trend among all selected after the addition of the different solvents. For cyanidin, delphinidin, malvidin, and pelargonidin transition energy for gas-phase decreases for all solvents except for n-hexane which is the only solvent where transition energy increases. For petunidin and peonidin, all the prior effects occur with the only difference that using methanol also presents an increase in the transition energy values if compared with the gas phase with similar values to those observed in n-hexane.

Also, in these cases all values obtained for oscillator strength when added solvents were smaller than gas phase oscillator strengths.

For petunidin, oscillator strength changes after solvent addition are moderate, the bigger change was with the addition of water solvent.

Transition energy for the selected anthocyanidins has a similar trend among all selected after the addition of the different solvents. For cyanidin, delphinidin, malvidin, and pelargonidin transition energy for the gas phase decreases for all solvents except for n-hexane which is the only solvent where transition energy increases. For petunidin and peonidin, all prior effects occur with the only difference that methanol also presents an increase in the transition energy values if compared with gas phase with similar values to those observed in n-hexane.

4. Conclusions

The molecule's structural geometry was analyzed a generally accepted methodology with a different basis set. Parameters such as relative angles, dihedrals, main features of individual rings, and a discussion on each molecule planarity were included as part of the discussion to relate the main geometry parameters with the molecule behavior and chemical features. Molecules functionalization with COH_3 is highlighted as an important feature for the structural and energy gap differences in each molecule geometry structure definition and is directly related with molecular orbital distribution with a direct effect in gap energy.

MOs and spectra results show there is a good fit with TiO_2 and concluded these pigments may be good dye sensitizers. Malvidin in its gas phase may be a good option from a gap energy perspective. Solvents increase their gap energy in all cases except with n-hexane which is the narrower followed by petunidin also with n-hexane. A good charge transfer feature is important as well and it was assessed with conceptual DFT. Results show cyanidin, malvidin, and pelargonidin may have a better charge transfer. A lower as possible electron reorganization energy (λ_e) and a high LHE are desirable since this would benefit charge transfer. Cyanidin has the smaller λ_e with water, but ethanol and methanol λ_e resulting values were nearly different. For LHE, the highest were cyanidin, malvidin, and petunidin with similar values between solvents. Based on our analysis of absorption capabilities for the selected pigments, it is corroborated cyanidin, malvidin, and petunidin may be acceptable dye sensitizers for DSSCs and photocatalysis applications.

Acknowledgments

Special acknowledgment to the Scientific Computational Laboratory at FCQ-UJED for computational resources. Our deep appreciation to Academic Group UJED-CA-129 for the technical discussions.

Funding statement

Authors declare that our research and publication of this chapter was partially financed by CONACyT (Mexican Science and Technology National Council) through 2015 CONACyT SEP-CB (Basic Science-Public Education Ministry) project fund

258,553/CONACyT/CB-2015-01 and by SEP through DSA/103.5/15/7028 project fund within the 2015 National PRODEP call.

Conflicts of interest exemption statement

The authors declare that this research was completed without any conflicts of interest related to the authoring, technical data, and received funding related to this work.

Author details

Diana Barraza-Jiménez[1], Hugo Iván Flores-Hidalgo[2], Sandra Iliana Torres-Herrera[3], Raúl Armando Olvera-Corral[1] and Manuel Alberto Flores-Hidalgo[1*]

1 Scientific Computing Laboratory, Faculty of Chemistry Science, Juarez University of Durango State, Av. Veterinaria S/N, Circuito Universitario, Durango, México

2 Department of Protected and Sustainable Agriculture, Technology University of Camargo Campus Jimenez, Chihuahua, Mexico

3 Juarez University of Durango State, Faculty of Forestry Science Durango, México

*Address all correspondence to: manuel.flores@ujed.mx

References

[1] Clement RC, Prasanth R. A critical review of recent developments in nanomaterials for photoelectrodes in dye sensitized solar cells. Journal of Power Sources. 2016;**317**:120-132. DOI: 10.1016/j.jpowsour.2016.03.016

[2] Dhonde M, Sahu K, Das M, Yadav A, Ghosh P, Murty VVS. Review—recent advancements in dye-sensitized solar cells; From photoelectrode to counter electrode. Journal of The Electrochemical Society. 2022;**169**:066507

[3] Green MA et al. Solar cell efficiency tables (Version 44). Progress in Photovoltaics: Research and Applications. 2014;**22**(7):701-710. DOI: 10.1002/pip.2525

[4] O'Regan B, Grätzel M. A low-cost, high-efficiency solar cell based on dye-sensitized colloidal TiO2 films. Nature. Oct. 1991;**353**(6346):737-740. DOI: 10.1038/353737a0

[5] Dwivedi G, Munjal G, Bhaskarwar AN, Chaudhary A. Dye-sensitized solar cells with polyaniline: A review. Inorganic Chemistry Communications. 2022;**135**:109087

[6] Cole JM, Pepe G, Al Bahri OK, Cooper CB. Cosensitization in dye-sensitized solar cells. Chemical Reviews. 2019;**119**(12):7279-7327. DOI: 10.1021/acs.chemrev.8b00632

[7] Mathew S et al. Dye-sensitized solar cells with 13% efficiency achieved through the molecular engineering of porphyrin sensitizers. Nature Chemistry. 2014;**6**(3):242-247. DOI: 10.1038/nchem.1861

[8] Roy P et al. TiO2 nanotubes and their application in dye-sensitized solar cells. Nanoscale. 2010;**2**(1):45-59. DOI: 10.1039/b9nr00131j

[9] Hug H et al. Biophotovoltaics: Natural pigments in dye-sensitized solar cells. Applied Energy. Feb. 2014;**115**:216-225. DOI: 10.1016/j.apenergy.2013.10.055

[10] Calvin M. Solar energy by photosynthesis. Kagaku to Seibutsu. 1974;**12**(7):481-498. DOI: 10.1271/kagakutoseibutsu1962.12.481

[11] Mohr H, Schopfer P, Hans M, Schopfer P. Wirkungen ultravioletter strahlung. Lehrbuch Der Pflanzenphysiologie. 1978:346-353. DOI: 10.1007/978-3-642-96453-4_28

[12] Lee MM et al. Efficient hybrid solar cells based on meso-superstructured organometal halide perovskites. Science. 2012;**338**(6107):643-647. DOI: 10.1126/science.1228604

[13] Norris DJ, Aydil ES. Getting moore from solar cells. Science. 2012;**338**(6107):625-626. DOI: 10.1126/science.1230283

[14] Hagfeldt A. Brief overview of dye-sensitized solar cells. AMBIO. 2012;**41**(S2):151-155. DOI: 10.1007/s13280-012-0272-7

[15] Woodford JN. A DFT investigation of anthocyanidins. Chemical Physics Letters. 2005;**410**(4):182-187. DOI: 10.1016/j.cplett.2005.05.067

[16] Estévez L, Mosquera RA. First characterization of the formation of anthocyanin–Ge and anthocyanin–B complexes through UV–Vis spectroscopy and density functional theory quantum chemical calculations. Journal of Agricultural and Food Chemistry. 2021;

69(4):1272-1282. DOI: 10.1021/acs.jafc.0c06827

[17] Castañeda-Ovando A, Pacheco-Hernández Md L, Páez-Hernández ME, Rodríguez JA, Galán-Vidal CA. Chemical studies of anthocyanins: A review. Food Chemistry. 2009;**113**:859-871

[18] Cavalcanti RN et al. Non-thermal stabilization mechanisms of anthocyanins in model and food systems —An overview. Food Research International. 2011;**44**(2):499-509. DOI: 10.1016/j.foodres.2010.12.007

[19] Francis F. Chapter 12 - Food colorings. In: Douglas B, editor. Published in Book: Color in Food. CRC Press: MacDougall. Woodhead Publishing; 2002. p. 297

[20] Kalt W et al. Anthocyanin content and profile within and among blueberry species. Canadian Journal of Plant Science. 1999;**79**(4):617-623. DOI: 10.4141/p99-009

[21] Davies KM. Chapter: Modifying anthocyanin production in flowers, published in book: Anthocyanins: Biosynthesis, Functions, and Applications. In: Gould K, Davies K, Winefield C, editors. Science+Business Media LLC. 2009

[22] Barraza-Jiménez D, Flores-Hidalgo MA, et al. Solvent effects on dye sensitizers derived from anthocyanidins for applications in photocatalysis. In: Glossman-Mitnik MD, Maciejewska M, editors. Chapter Published in Book: Solvents, Ionic Liquids and Solvent Effects. London, UK: IntechOpen; 2019

[23] Frisch MJ, Fox DJ, et al. Gaussian 16, Revision C.01. Wallingford CT: Gaussian, Inc.; 2016

[24] Becke AD. Density-functional thermochemistry III. The role of exact exchange. The Journal of Chemical Physics. 1993;**98**:5648

[25] Terranova U, Bowler DR. Δ Self-consistent field method for natural anthocyanidin dyes. Journal of Chemical Theory and Computation. 2013;**9**(7): 3181-3188. DOI: 10.1021/ct400356k

[26] Fan W, Deng W. Incorporation of thiadiazole derivatives as π-spacer to construct efficient metal-free organic dye sensitizers for dye-sensitized solar cells: A theoretical study. Communications in Computers & Chemistry. 2013;**1**:152-170. DOI: 10.4208/cicc.2013.v1.n2.6

[27] Armas R, Miguel M, Oviedo J, Sanz JF. Coumarin derivatives for dye sensitized solar cells: A TD-DFT study. Physical Chemistry Chemical Physics. 2012;**14**:225. DOI: 10.1039/C1CP22058F

[28] Baldenebro-López J et al. Density functional theory (DFT) study of triphenylamine-based dyes for their use as sensitizers in molecular photovoltaics. International Journal of Molecular Sciences. 2012;**13**(4, 1):4418-4432. DOI: 10.3390/ijms13044418

[29] Xu J et al. DFT Studies on the electronic structures of indoline dyes for dye-sensitized solar cells. Journal of the Serbian Chemical Society. 2010;**75**:259

[30] Leonid S. Chemissian. V4.43. pp. 2005-2016

[31] Ge X, Xiaochuan, et al. Accurate and inexpensive prediction of the color optical properties of anthocyanins in solution. The Journal of Physical Chemistry A. 2015;**119**(16):3816-3822. DOI: 10.1021/acs.jpca.5b01272

[32] Woodford JN, Li W. Recent progress of thin-film photovoltaics for indoor application. Chinese Chemical Letters. 2020;**31**:643

[33] Buseta PB, Colleter JC, Gadret M. Structure du chlorure d'apigéninidine monohydrate. Acta Crystallographica B. 1974;**30**:1448

[34] Ueno K, Saito N, Ueno K, Saito N. Cyanidin bromide monohydrate (3,5,7,3',4'-pentahydroxyflavylium bromide monohydrate). Acta Crystallographica Section B: Structural Crystallography and Crystal Chemistry. 1977;**33**(1):114-116. DOI: 10.1107/S0567740877002702

[35] Meyer M. Ab initio study of flavonoids. International Journal of Quantum Chemistry. 2000;**76**(6):724-732

[36] Sanchez-Bojorge NA et al. Theoretical calculation of the maximum absorption wavelength for Cyanidin molecules with several methodologies. Computational and Theory of Chemistry. 2015;**1067**:129-134

[37] Liu Z. Theoretical studies of natural pigments relevant to dye-sensitized solar cells. Journal of Molecular Structure: THEOCHEM. 2008;**862**:44. DOI: 10.1016/j.theochem.2008.04.022

[38] Sánchez de Armas R, San-Miguel MA, Oviedo J, Sanz J. Direct vs. indirect mechanisms for electron injection in DSSC: Catechol and alizarin. Computational Theory of Chemistry. 2011;**975**:99

[39] Glossman-Mitnik D. Computational molecular characterization of Coumarin-102. Journal of Molecular Structure: THEOCHEM. 2009;**911**:105

[40] Mohammadi N, Wang F. First-principles study of Carbz-PAHTDDT dye sensitizer and two carbz-derived dyes for dye sensitized solar cells. Journal of Molecular Modeling. 2014;**20**: 2177

[41] Megala M, Rajkumar BJM. Theoretical study of anthoxanthin dyes for dye sensitized solar cells (DSSCs). Journal of Computational Electronics. 2016;**15**(557). DOI: 10.1007/s10825-016-0791-8

[42] Harborne JB, Harborne JB. Comparative biochemistry of flavonoids. The Biochemical Journal. 1958;**V70**:22

[43] Brouillard R. The Mechanism of Co-Pigmentation of Anthocyanins in Aqueous Solutions. London: Chapman and Hall Ltd.; 1988. p. 525

[44] El Kouari Y, Migalska-Zalas A, Arof AK, Sahrao B. Computations of absorption spectra and nonlinear optical properties of molecules based on anthocyanidin structure. Optical and Quantum Electronics. 2015;**47**:1091-1099

[45] Roy P, Kumar Sinha NK, Tiwari S, Khare A. A review on perovskite solar cells: Evolution of architecture, fabrication techniques, commercialization issues and status. Solar Energy. 2020;**198**:665

[46] Zhang CR et al. The role of the conjugate bridge in electronic structures and related properties of tetrahydroquinoline for dye sensitized solar cells. H.S. Chen. International Journal of Molecular Sciences. 2013;**14**: 5461

[47] Abdullah MI, Janjua MRSA, Mahmood A, Ali S, Ali M, Abdullah B, et al. Quantum chemical designing of efficient sensitizers for dye sensitized solar cells. Bulletin of the Korean Chemical Society. 2013;**34**(7):2093-2098. DOI: 10.5012/bkcs.2013.34.7.2093

[48] Chien CY, Hsu BD. Optimization of the dye-sensitized solar cell with anthocyanin as photosensitizer. Solar Energy. 2013;**98**:203-211

Chapter 12

Utilizing Photocatalysts in Reducing Moisture Absorption in Composites of Natural Fibers

Mohammed Mohammed and Rozyanty Rahman

Abstract

Due to growing environmental consciousness and the depletion of oil supplies, numerous efforts have been made to replace synthetic fibers in fiber-reinforced composites with natural fibers (NFr). The low cost and abundance of NFr and its biodegradability and low density have encouraged researchers worldwide to study their potential applications in several industrial sectors. However, NFr has several disadvantages: excessive moisture absorption and subsequent swelling and degradation, low chemical and fire resistance, and insufficient interfacial interactions with polymers. Consequently, there is great interest in modifying the surface of NFr using a variety of methods. This chapter presents an overview of the NFr, its characterization, the problems associated with adding NFr to polymer composites. This literature survey suggests an in-depth review of photocatalysis by utilizing photocatalysts nanoparticle (PHNPs) aimed at increasing the hydrophobicity and interfacial bonding between the NFr and the matrix Using a photo-induced oxidation mechanism to disassemble water molecules, pollutants, and bacteria in a wet environment. Additionally, we reviewed the effects of these PHNPs on the moisture absorption, mechanical characteristics, and dimensional stability of NFr composites. As a result, this review article may make a valuable contribution to researchers interested in coating and treating NFr to further enhance their surface characteristics.

Keywords: Photocatalysts, natural fiber, surface treatment, moisture absorption

1. Introduction

Environmental concerns and escalating social demand for the adoption of less hazardous composites materials have prompted a paradigm shift toward employing NFr as a replacement for synthetic and non-renewable reinforcements. This has led to the development of what are usually known as "Green Composites," or composites made from biodegradable or renewable raw materials with lower environmental effect [1]. The application of NFr in polymer composites is expanding in many sectors such as automotive, furniture, packaging and construction. The utilization of NFr is based on its advantages including decreased tool wear, inexpensive, light weight, high toughness, good specific strength characteristics, ease of separation, improved energy recovery, carbon dioxide sequestration, and biodegradability, as well as their sustainable renewable characteristics, and having competitive mechanical performance are just some of the significant properties that make them suitable for us. NFr are available in a broad multitude of types and are often used as fillers and reinforcement [2].

However, NFr reinforced composites have a number of disadvantages, including limited wettability, poor fiber/polymer matrix performance, and the ability to absorb more water and moisture [3]. For the majority of NFr users, the influence of moisture absorption properties on NFr-reinforced composites became a warning signal. In reality, this absorption destroys the adhesion force between the fiber and matrix in an enormous manner. Subsequently, it diminishes the composite's strength in a manner proportional to the absorption duration or nature. In addition, it results in inadequate stress transfer and ultimately specimen fracture over time. The moisture absorption involves the following mechanisms: at first, the water molecule penetrates the micro gaps that are present within the chains, then the penetrated water diffuses deep within the chain via the capillary transport mechanism, and finally, the fibers swell due to the micro gaps [4]. The hydrophilic nature of NFr and the presence of components such as wax and pectin preclude an efficient reaction with the thermoset or thermoplastic polymer, therefore preventing the reinforcement and matrix from bonding better. Improved bonding between NFr and resin is critical for stress transmission from the matrix to the reinforcement. As a result, a weak interface reduces the physical and mechanical characteristics of composites [5].

To resolve this concern, chemical treatments of NFr are utilized to enhance the physical, mechanical, and thermal characteristics of the composites. According to Edeerozey et al. and Baiardo et al., chemical treatments can increase interfacial interaction between NFr and polymer matrix by diminishing the hydrophilicity of NFr, cleaning the surface of NFr, augmenting fiber roughness, and lowering moisture content of NFr [6, 7]. The numerous chemical processes were conducted for NFr, such as alkaline, silane, and stearic acid. Further studies have shown that the chemically treated NFr composite materials have superior tensile, bending, impact, and interlaminar shear strengths and hardness levels compared to untreated composites.

According to Njuguna et al., PHNPs are a promising filler material for improving the mechanical and physical characteristics of NFr polymer composites while lowering moisture absorption [8]. Since nanoscale fillers are often defect-free, their applications in polymer composite area setup can overcome the limitations of the traditional/conventional micrometer scale. Owing to the uniform and homogenous dispersion of PHNPs, a large matrix-PHNPs interfacial area is formed, which influences relaxation behavior and ensues the mechanical, molecular mobility, and thermal characteristics [9, 10]. PHNPs are often found in the minor zone., whereas just a few microparticles contribute to the deformation of the plastic area. This enables PHNPs to enhance the fracture and mechanical characteristics of fragile matrixes. PHNPs with a higher surface area are of particular interest because they provide superior reinforcement for the production of nanocomposites [9]. PHNPs are typically included as a percentage of the total weight of the composite for nanocomposite advancement [11]. The particular surface area of PHNPs has a significant effect on the composite characteristics, which is continuous. The basic principle behind nanocomposites is to make a high interface between the PHNPs and the matrix. The homogenous dispersion of PHNPs is frequently problematic [12]. Nanocomposites are a type of high-performance material that has exceptional characteristics, combinations, and design options [13].

Polymer/ NFr-nanocomposites	Result of integration of Nano particles into NFr	Reference
Epoxy/banana fiber/ nano clay	(a) The nanoclay-infused banana fiber composites enhanced compressive yield stress, modulus, and strength by 17%, 2 times, and 43%, respectively (b) Microscopy Examination demonstrated that nanoclay infusion in fiber enhanced fiber-matrix interfacial and load transfer characteristics, resulting in a higher level of property enhancement.	[15]
unsaturated polyester resin/ sisal fiber/ Multi-walled carbon nanotubes	The produced composite displayed appreciable mechanical and acoustical characteristics	[16]
Epoxy/Flax fiber/ nano titanium dioxide	composite material projected best flexural strength, impact strength and the hardness value with an increase of 25.05%, 24.29% and 16.01% respectively.	[17]
Polypropylene/wheat straw fiber/nano titanium dioxide	(a)Composite exhibited excellent UV-shielding properties in the range of 100–200 nm(b) exhibited the highest flexural (29.27 MPa), tensile (14.38 MPa), and impact (4.55 kJ/m^2) strengths.	[18]
Epoxy/Coccinia Indica fiber/nano clay	(a) The experimental evidence shows that the increase in weight percentage of nanoclay enhances the tensile, flexural, impact, compression properties of the treated *Coccinia Indica* fiber reinforced epoxy composites (b) the presence of 3 wt.% nanoclay provides better interfacial adhesion between fibers and matrix	[19]

Table 1.
The implications of integration of PHNPs into NFr.

Despite the fact that review papers and even books have been published on the totality impact of surface treatment on NFr in terms of moisture absorption, mechanical performance, and morphology [2, 14], the researchers believe that a precise book or review papers on the overall surface treatment with PHNPs is an essential first step in providing immunity to NFr polymer composites against quick and unnecessary degradation, resulting in the production of NFr polymer composites with high water repellency and strength properties, has not yet been published; nevertheless, it is anticipated that such a book will be of considerable importance to the composite scientific community. In the published literature, we found no review or book has been written to evaluate the impact of surface treatment with inorganic PHNPs on the hydrophobicity of NFr and relate that with improving the mechanical, physical, and chemical attributes of its composite. In order to achieve this goal, we tabulated in **Table 1**, the effects of the incorporation of PHNPs into NFr on moisture absorption and mechanical properties. The findings will serve as a scientific benchmark for developing high-performance polymer composites incorporating NFr as reinforcement, particularly for use in building applications such as floor decking, door and window frames.

2. Compositions and natural Fibers properties

To acquire a thorough grasp of the degrading issues, it is vital to first fathom the nature and compositions of NFr. NFr are categorized according to whether they

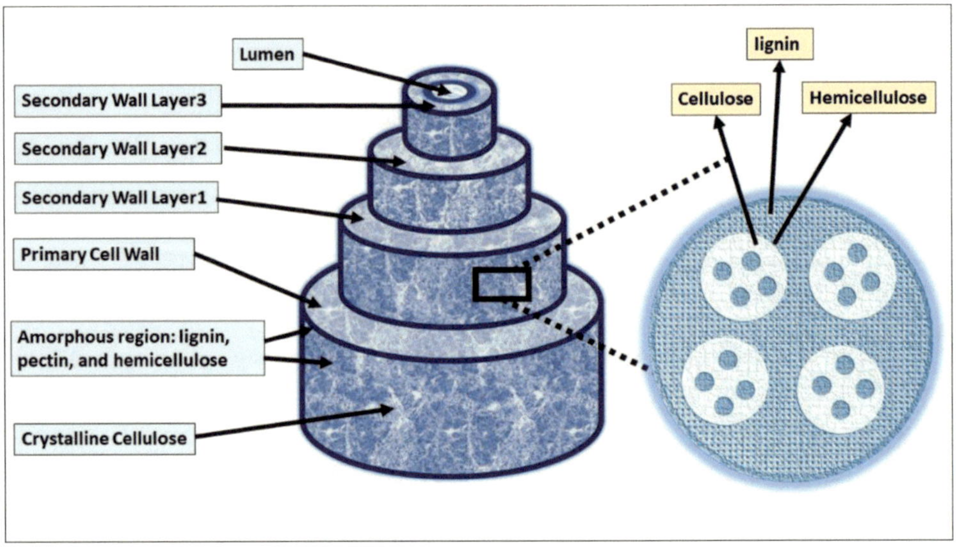

Figure 1.
Schematic representation of structure of natural fiber.

originated from plants, animals, or minerals. The major component of NFr generated from plants is cellulose fibrils embedded in lignin matrix. **Figure 1** demonstrates the structure of a biofiber. Three secondary cell walls and a primary cell wall make up the complicated layered structure of each fiber. The thick middle layer of secondary cell walls governs the mechanical characteristics of fiber. It is composed of a series of helical microfibrils generated from long-chain cellulose molecules. Each cell wall consists of three primary elements: cellulose, hemicelluloses, and lignin. Lignin-hemicelluloses serve as the matrix, whereas microfibrils (composed of cellulose molecules) serve as the fibers [20, 21]. Other constituents consist of pectins, oil, and waxes [21, 22]. The most important structural element of many NFr is cellulose, a natural polymer with three hydroxyl groups per repeating unit. Cellulose is resistant to hydrlolysis, strong alkali and oxidizing agents. However, when exposed to chemical and solution treatments, cellulose degrades to some extent. Hemicelluloses are polysaccharides with a reduced molecular weight that act as a cementing matrix between cellulose microfibrils and are the fundamental structural component of the fiber cell. Because it is hydrophilic, weak acids and bases can quickly hydrolyze it. Lignin is a complex hydrocarbon polymer that provides plants their stiffness and aids in water transport. It is hydrophobic, resistant to acid hydrolysis and the assaults of most microbes, soluble in hot alkali, rapidly oxidized, and readily condensable with phenol [23].

3. Challenges for natural fiber as a reinforcement in polymer composites

3.1 Moisture absorption

Humidity aging is well acknowledged as a significant contributor to the long-term degradation of an organic matrix subjected to the environment or in contact with an aqueous medium. Moreover, humidity aging has a detrimental influence on the mechanical characteristics and dimensional stability of composites [24].

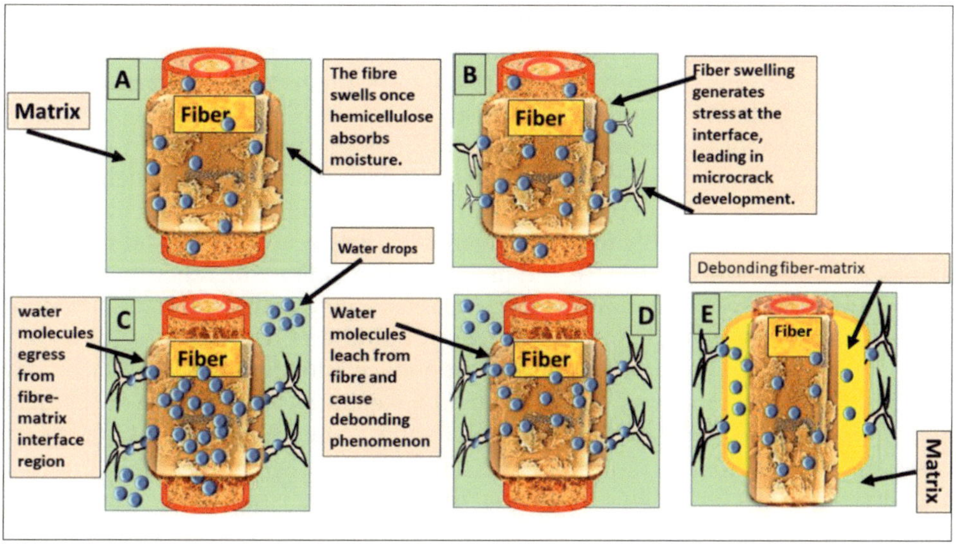

Figure 2.
Schematic representation of impact of moisture on Fiber-Matrix Interface.

Consequently, it is critical to research the moisture absorption behavior of NFr composites to determine the possible impacts of the absorbed water and the durability of the NFr composites utilized in the humidity environment [25, 26]. The majority of studies believe that moisture penetrates composite materials through three distinct mechanisms. The primary process is due to water molecules diffusing into the microgaps between polymer chains. The second mechanism includes capillary transportation through the holes and defects caused by insufficient wettability at the interfaces between the fiber and matrix. The third mechanism is the transfer through microcracks formed in the matrix during the combining process with fiber or due to fiber swelling [26, 27]. Moisture quickly hydrolyzes matrixes such as polyester, resulting in a reduce in molecular weight of polymer [24, 28]. Temperature and humidity have an impact on the rate of decomposition of NFr [28]. In a humid climate, this would significantly lose mechanical characteristics for NFr [29]. The loading of NFr has a significant impact on the moisture absorption characteristics. IAS et al. [30] evaluated the impact of fiber loading (10–40 wt%) on the moisture absorption capacity of a kenaf/polyester composite at ambient temperature. Moisture absorption increased with increasing fiber volume in composites, it was reported. The water uptake process of a NFr is shown in **Figure 2**.

4. Impact of surface treatment of natural fiber on moisture absorption

Numerous drawbacks of NFr include decreased strength, decreased interfacial adhesion, poor moisture absorption resistance, and reduced durability and dimensional stability. To overcome these constraints, different methods for NFr modification have been proposed [31]. Four different approaches are used for treating NFr, physical processes, chemical methods, biological methods, and nano enhance photocatalysis treatment. These revised techniques have varying degrees of efficacy in enhancing the mechanical properties of fibers, adhesion between matrix and fiber, and improving various characteristics of the end products [32]. However, moisture absorption leads to debonding between NFr and the matrix, because of the high susceptibility of NFr to moisture, leading to a significant decrease in the mechanical characteristics of the NFr composite [27]. This is because NFr in nature is polar and hydrophilic due to the existence of non-cellulosic components (i.e., pectin, lignin, and hemicelluloses), because it makes moisture absorption possible through the availability of functional groups such as hydroxyl (OH) and carboxylic acid [33]. Low fiber bonding with a polymer matrix generates empty areas surrounding organic composite fibers, resulting in increased moisture absorption. Fiber modification through alkalization will minimize moisture content. Potassium hydroxide or sodium hydroxide is frequently employed to remove open groups of hydroxyls that bind to water molecules [22]. Additionally, it removes hemicelluloses and reduces the capacity of fiber to absorb moisture. This combination results in the formation of chemical and hydrogen bonds, which retards the degradation of the NFr-matrix interface caused by humidity [34].

There are certain limitations to utilizing the physical, chemical, and biological treatments, such as obtaining optimal chemical concentrations to avoid additional delignification of the NFr because greater concentrations might weaken and deteriorate the NFr, as reported by Alnaid et al. [35] and Dahham et al. [36]. The chemical compounds that are employed in treating NFr will continue to damage the fibers after exposure to weathering if they are not thoroughly cleaned after-treatment process, eventually leading to severe fiber swelling and disintegration. Furthermore, the NFr composites treated with these techniques have a low UV resistance, resulting in UV deterioration of composite components, such condition contributes to more interlaminar delamination and more brittleness of the composites when placed under high loads [23].

5. Mechanism of repellence moisture via photocatalysts nanoparticle

Photocatalysis is a viable environmental protection process because it can oxidize small quantities of organic contaminants into benign compounds [37, 38]. Photocatalysis employs PHNPs to perform a photo-induced oxidation rection in order to degrade organic pollutants, inactivate microorganisms, and break down water molecules [39, 40]. **Figure 3** depicts the photocatalysis process. When photons with energies higher than the band gap energy of PHNPs are absorbed, the valence band (VB) electrons in PHNPs are excited to the conduction band, which opens up a variety of potential photoreactions. With sufficient photo energy, the photocatalytic surface produces a positive hole (h+) in the valence band and an electron (e) in the conduction band (CB). The positive hole may either directly oxidize organic contaminants or generate highly reactive hydroxyl radicals (OH•). The major oxidants in the photocatalytic system [41] are the hydroxyl radicals (OH•), which oxidize organics. The electron in the conduction band decreases the amount of oxygen adsorbed by PHNPs. TiO_2, ZnO, and Fe_2O_3 are only a few of the semiconductors that could be used as PHNPs. The band gap energy plays a crucial role in the photocatalytic process. The bandgap is defined as the void area that extends from the uppermost portion of the filled valence band to the lowest part of the unoccupied conduction band and is determined to be around 3.3 eV for ZnO nanoparticles in its crystalline state [42]. Thus, the light of a specific wavelength (i.e., ultraviolet radiation (UV)) contains enough energy to promote electrons (e-) to the conduction band while leaving an

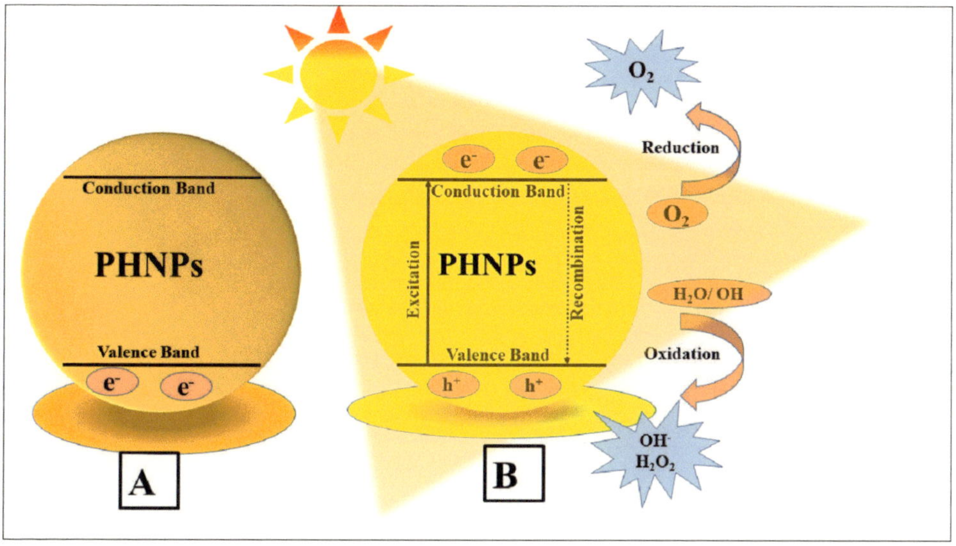

Figure 3.
Schematic representation of the photocatalytic mechanism of ZnONPs (a) before exposure to UV, (b) after exposure to UV.

electron–hole (h+) behind. Both the hole and electron can typically be recombined very quickly [43]. They can also migrate to the surface of the PHNPs as a result of their reaction with the adsorbed elements, initiating (i) the reaction between the electron and the oxygen and (ii) the reaction between the hole and the hydroxyl ion or H_2O molecule to form superoxide and hydroxyl radicals, as shown in **Figure 3** the photocatalysis process of PHNPs.

The majority of experts concur that there are five distinct mechanisms involved in the moisture-repelling action via photocatalyst nanoparticles. The first mechanism involves PHNPs were able to penetrate through the cell wall of the NFr and deposited mainly in the cell lumens and partly in the cell walls of the NFr. It is expected that the physical and mechanical characteristics of the resultant NFr would be altered to varying degrees since the PHNPs are present in both cell walls and lumens [44, 45]. Hygroscopicity of the modified NFr was reduced because the inorganic PHNPs were integrated into the cell wall and were likely occupying empty space (micropores) inside the cell wall that would otherwise be accessible to water molecules [46, 47]. By establishing hydrogen bonds with inorganic PHNPs, the hydroxyl groups of NFr cell wall components, which are principally responsible for water absorption, may have been inhibited, hence lowering the hygroscopicity of NFr. As similar study conducted by Mohammed et al. [48] showed that integration of zinc nano particle is proposed; the water repellent capability is good enough to achieve the high performance of the composite with proper system formulation during the modification process.

The second mechanism involves the incorporation inorganic PHNPs into cell walls of the fiber decreases the capacity of the cell wall to swell owing to bulking, therefore increasing the dimensional stability of the impregnated NFr. Indeed, only inorganic PHNPs integrated into the cell wall are predicted to affect the hydrophilicity and dimensional stability of NFr significantly. In contrast, those deposited in the cell lumen are anticipated to have a negligible effect [49]. Moreover, by incorporating inorganic PHNPs into the NFr cell wall, the cell wall elements are expected to be coated with water-repellent inorganic PHNPs, which may stop water molecules

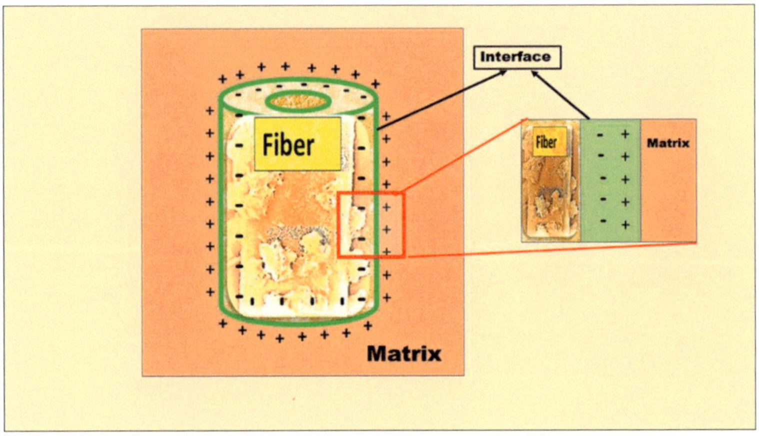

Figure 4.
Schematic representation of electrostatic adhesion mechanisms between Fiber and matrix.

and moisture from reaching them, thus slowing swelling of NFr and improving NFr dimensional stability [48, 50, 51].

Due to the presence of inorganic PHNPs in the cell wall and on the fiber surface, the third mechanism includes strong static electric forces that may attract the nonpolar polymer surface to the NFr. As a result of the compatibility between the NFr and the polymer matrix, the mechanical characteristics of the composites will be improved and their moisture absorption will be reduced. Electrostatic adhesion is a common bonding mechanism in which opposite charges (cathode and anode) are formed on the fiber and matrix surfaces. This results in two layers of opposite charges at the interface, which enhances the attraction between the fiber and matrix of the composite [52, 53]. Electrostatic discharge treatments were applied to the surfaces of polymer and electrostatic fibers to promote electrostatic adhesion at the interface region between matrix and NFr, thereby improving the transfer the stress from matrix to NFr leading to enhancing performance of the NFr composites [54]. In other words, the incorporation of the inorganic PHNPs into the micropore structure of the fiber cell walls can diminish the microvoid volumes in the fiber, as well as decrease air bubble formation during the composite manufacturing process [55]. **Figure 4** illustrates schematic representation of interfacial electrostatic bonding mechanisms between fiber and matrix.

According to Stark [56], moisture penetration increases as the number of moisture channels within NFr composites increases; consequently, the fourth mechanism refers to the addition of inorganic PHNPs as an additive to close these pathways and reduce moisture absorption. However, greater density NFr composites have lower moisture content and less swelling than untreated NFr composites [57]. This is because the inorganic PHNPs enhance the density of the composite. Therefore, by incorporating inorganic PHNPs, it is possible to lower the loading of NFr present in composites while simultaneously rising the tensile strength of the NFr composites [45, 58]. According to specific research, reducing the fiber loading improves the tensile strength of composites, whereas rising the fiber loading results in increased moisture absorption of composites depend on NFr as ingredients [59, 60].

PHNPs play two key roles in the water repellence mechanism of NFr composites via the fifth mechanism. **Figure 5** shows that when NFr absorbs a water molecule, the critical radius of PHNP atomic structures increases, pushing NFr molecules to press outward and be released to the composite surface, which is warmed by UV absorption [44]. Second,

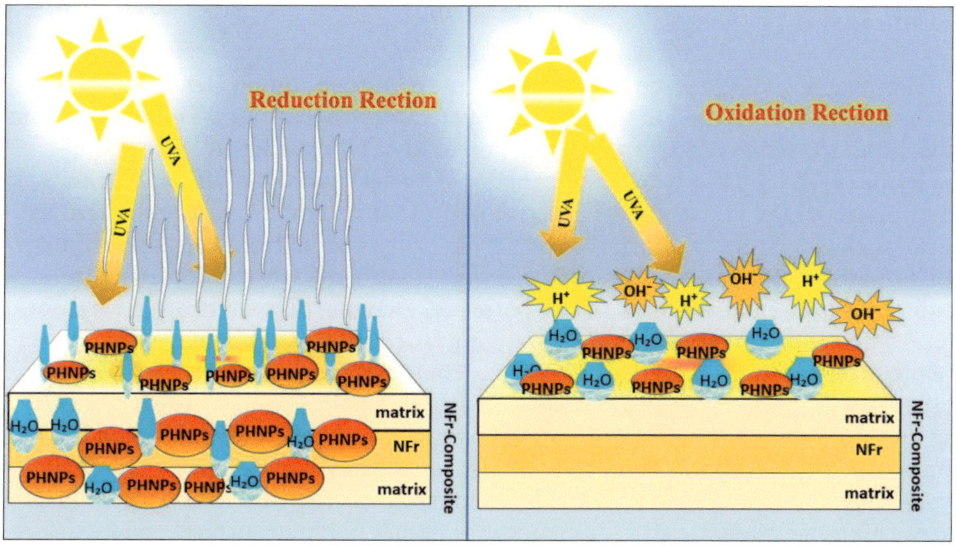

Figure 5.
Illustration proposed reduction and oxidation reactions by PHNPs.

suppose the water molecule is straight or near the composite surface. In that case, PHNPs will initiate an oxidation process with the assistance of radiated energy, resulting in the breakdown of the water molecule into H+ and OH- [44].

Raghupathi et al. [61] revealed that enhanced Active Oxygen Species (ROS) generation from PHNPs during UV exposure resulted in higher PHNPs activity, which regenerated the NFr water repellency mechanisms. These reactive species include anion of superoxide (O_2), peroxide of hydrogen (H_2O_2), and hydroxide (OH-). The formation of ROS during the dark has been noticed by Hirota et al. [62] who tested ZnO-NPs toward E. Coli. They observed that activity might occur in the dark, which is consistent with the findings of Jones et al. [63] such unambiguous findings imply that new methods for producing reactive species in the absence of light and the dark are likely to be discovered in the future.

The activation of PHNPs by UV light can be represented by the following steps:

$$\text{PHNPs} + hv \rightarrow e^{-} + h^{+} \tag{1}$$

$$e^{-} + O_2 \rightarrow O_2^{-} \tag{2}$$

In this reaction, h^+ and e- are powerful oxidizing and reductive agents respectively. The oxidative and reductive reaction steps are expressed as:

Oxidative reaction:

$$h^{+} + Organic \rightarrow CO_2 \tag{3}$$

$$h^{+} + H_2O \rightarrow OH^{-} + H^{+} \tag{4}$$

Reductive reaction:

$$OH + Organic \rightarrow CO_2 \tag{5}$$

6. Conclusions

This chapter reviewed the research on NFr-reinforced composites and the constraints that have emerged in their development and serviceability. These constraints must be addressed in order to utilize the potential of NFr successfully. To begin, a suitable surface treatment of the fiber should be performed. Additionally, the study discusses the literature on NFr treatments that researchers have employed to decrease the moisture absorption of NFr composites and increase the compatibility of NFr with the polymer matrix. Many surface treatments are available, such as sodium chlorite, methacrylate, silane, peroxide, enzyme, plasma, ozone treatment, etc. Furthermore, the study describes how surface treatments-based photocatalysts nanoparticle could increase NFr adhesion to their matrix and reduce NFr moisture absorption. Therefore, surface treatments with PHNPs can be regarded for modifying the characteristics of NFr. Moisture resistance and fiber-matrix adhesion can be improved by including specific PHNPs, such as Zinc oxide, titanium dioxide, and copper oxide, inside the cell wall and on the surface of NFr. In addition, these PHNPs exert a high electric static attraction on the nonpolar polymer surface, enhancing the compatibility of the fibers with the polymer matrix. Water-repelling inorganic PHNPs are projected to cover the cell wall components, which could prevent water molecules from accessing the wall components, delaying their swelling and decreasing moisture absorption rate of NFr. The review is given here will aid researchers in better comprehending the many characteristics of NFr composites, therefore facilitating the development of novel green materials with enhanced performance. However, research in this field is extensive, and the current literature discusses using photocatalysts nanoparticle in reducing the moisture uptake by NFr. As a result, it is essential to do a thorough analysis of the current literature to understand the pros and cons of photocatalysis treatment for the specific requirements. In light of this review, it is advised that more research be conducted to determine the impact of matrix modification and fiber length on the characteristics of NFr composite materials.

Acknowledgements

The author like to express their appreciation to the Faculty of Chemical Engineering Technology, Universiti Malaysia Perlis (UniMAP), Special thanks to those who contributed to this project directly or indirectly.

Notes/thanks/other declarations

I would like to thank my colleagues specifically Dr. Bashir Betar, Dr. Tijjani Adam, Dr. Aeshah Mohammed, Dr. Azlin Osman, and Sara Debeuc, your continued support is the one thing I can always count on. You all showed me patience, guided and motivated me whenever the situation arose. Thank you all for supporting me wholeheartedly.

Author details

Mohammed Mohammed* and Rozyanty Rahman
Center of Excellence Geopolymer and Green Technology, School of Materials Engineering, Universiti Malaysia Perlis, Kangar, Perlis, Malaysia

*Address all correspondence to: hmn7575@yahoo.com

References

[1] Gholampour A, Ozbakkaloglu T. A review of natural fiber composites: Properties, modification and processing techniques, characterization, applications. Journal of Materials Science. 2020;**55**:829-892. DOI: 10.1007/s10853-019-03990-y

[2] Negawo TA, Polat Y, Buyuknalcaci FN, Kilic A, Saba N, Jawaid M. Mechanical, morphological, structural and dynamic mechanical properties of alkali treated Ensete stem fibers reinforced unsaturated polyester composites. Composite Structures. 2019;**207**:589-597. DOI: 10.1016/j.compstruct.2018.09.043

[3] Saba N, Tahir PM, Jawaid M. A review on potentiality of nano filler/natural fiber filled polymer hybrid composites. Polymers. 2014;**6**:2247-2273. DOI: 10.3390/polym6082247

[4] Srinivasan T, Suresh G, Ramu P, Ram VG, Giresh M, Arjun K. Effect of water absorption of the mechanical behavior of banana fiber reinforced IPN natural composites. Materials Today: Proceedings. 2021;**45**:1334-1337. DOI: 10.1016/j.matpr.2020.06.024

[5] Lu N, Oza S, Tajabadi MG. Surface modification of natural fibers for reinforcement in polymeric composites. Surface Modification of Biopolymers. 2015;**1**:224-237. DOI: 10.1002/9781119044901.ch9

[6] Edeerozey AM, Akil HM, Azhar AB, Ariffin MZ. Chemical modification of kenaf fibers. Materials Letters. 2007;**61**:2023-2025. DOI: 10.1016/j.matlet.2006.08.006

[7] Baiardo M, Frisoni G, Scandola M, Licciardello A. Surface chemical modification of natural cellulose fibers. Journal of Applied Polymer Science. 2002;**83**:38-45. DOI: 10.1002/app.2229

[8] Njuguna J, Pielichowski K, Desai S. Nanofiller-reinforced polymer nanocomposites. Polymers for Advanced Technologies. 2008;**19**:947-959. DOI: 10.1002/pat.1074

[9] De Azeredo HM. Nanocomposites for food packaging applications. Food Research International. 2009;**42**:1240-1253. DOI: 10.1016/j.foodres.2009.03.019

[10] Schadler LS, Brinson LC, Sawyer WG. Polymer nanocomposites: A small part of the story. Journal of Metals. 2007;**59**:53-60. DOI: 10.1007/s11837-007-0040-5

[11] Haq M, Burgueño R, Mohanty AK, Misra M. Hybrid bio-based composites from blends of unsaturated polyester and soybean oil reinforced with nanoclay and natural fibers. Composites Science and Technology. 2008;**68**:3344-3351. DOI: 10.1016/j.compscitech.2008.09.007

[12] Mohammed M, Rozyanty R, Mohammed AM, Osman AF, Adam T, Dahham OS, et al. Fabrication and characterization of zinc oxide nanoparticle-treated kenaf polymer composites for weather resistance based on a solar UV radiation. BioResources. 2018;**13**:6480-6496. DOI: 10.15376/biores.13.3.6480-6496

[13] Camargo PH, Satyanarayana KG, Wypych F. Nanocomposites: Synthesis, structure, properties and new application opportunities. Materials Research. 2009;**12**:1-39. DOI: 10.1590/s1516-14392009000100002

[14] Musa L, Rozyanty AR, Zhafer SF. Effect of modification time of kenaf

bast fiber with maleic anhydride on tensile properties of kenaf-glass hybrid fiber unsaturated polyester composites. In: Solid State Phenomena. Vol. 280. Switzerland: Trans Tech Publications Ltd.; 2018. pp. 353-360. DOI: 10.4028/www.scientific.net/ssp.280.353

[15] Mohan TP, Kanny K. Compressive characteristics of unmodified and nanoclay treated banana fiber reinforced epoxy composite cylinders. Composites Part B: Engineering. 2019;**169**:118-125. DOI: 10.1016/j.compositesb.2019.03.071

[16] Abdel-Hakim A, Awad EH, El-Nemr KF, El-Basheer TM. Impact of gamma radiation and multi-walled carbon nanotubes on the mechanical and acoustical properties of reinforced sisal fiber/polyester resin composites. Radiation Physics and Chemistry. 2021;**189**:109768. DOI: 10.1016/j.radphyschem.2021.109768

[17] Prasad V, Joseph MA, Sekar K, Ali M. Flexural and impact properties of flax fibre reinforced epoxy composite with nano TiO_2 addition. Materials Today: Proceedings. 2018;**5**:24862-24870. DOI: 10.1016/j.matpr.2018.10.285

[18] Wang D, Xuan L, Han G, Wong AH, Wang Q, Cheng W. Preparation and characterization of foamed wheat straw fiber/polypropylene composites based on modified nano-TiO_2 particles. Composites Part A: Applied Science and Manufacturing. 2020;**128**:105674. DOI: 10.1016/j.compositesa.2019.105674

[19] Mylsamy B, Palaniappan SK, Subramani SP, Pal SK, Aruchamy K. Impact of nanoclay on mechanical and structural properties of treated Coccinia indica fibre reinforced epoxy composites. Journal of Materials Research and Technology. 2019;**8**:6021-6028. DOI: 10.1016/j.jmrt.2019.09.076

[20] Dittenber DB, GangaRao HV. Critical review of recent publications on use of natural composites in infrastructure. Composites Part A: Applied Science and Manufacturing. 2012;**43**:1419-1429. DOI: 10.1016/j.compositesa.2011.11.019

[21] John MJ, Thomas S. Biofibres and biocomposites. Carbohydrate Polymers. 2008;**71**:343-364. DOI: 10.1016/j.carbpol.2007.05.040

[22] Wong KJ, Yousif BF, Low KO. The effects of alkali treatment on the interfacial adhesion of bamboo fibres. Proceedings of the Institution of Mechanical Engineers, Part L: Journal of Materials: Design and Applications. 2010;**224**:139-148. DOI: 10.1243/14644207jmda304

[23] Azwa ZN, Yousif BF, Manalo AC, Karunasena W. A review on the degradability of polymeric composites based on natural fibres. Materials & Design. 2013;**47**:424-442. DOI: 10.1016/j.matdes.2012.11.025

[24] Yew GH, Yusof AM, Ishak ZM, Ishiaku US. Water absorption and enzymatic degradation of poly (lactic acid)/rice starch composites. Polymer Degradation and Stability. 2005;**90**:488-500. DOI: 10.1016/j.polymdegradstab. 2005.04.006

[25] Merdas I, Thominette F, Tcharkhtchi A, Verdu J. Factors governing water absorption by composite matrices. Composites Science and Technology. 2002;**62**:487-492. DOI: 10.1016/s0266-3538(01)00138-5

[26] Kim HJ. Effect of water absorption fatigue on mechanical properties of sisal textile-reinforced composites. International Journal of Fatigue.

2006;**28**:1307-1314. DOI: 10.1016/j.ijfatigue.2006.02.018

[27] Dhakal HN, Zhang ZA, Richardson MO. Effect of water absorption on the mechanical properties of hemp fibre reinforced unsaturated polyester composites. Composites Science and Technology. 2007;**67**:1674-1683. DOI: 10.1016/j.compscitech.2006.06.019

[28] Garlotta D. A literature review of poly (lactic acid). Journal of Polymers and the Environment. 2001;**9**:63-84. DOI: 10.1023/A:1020200822435

[29] Paul A, Joseph K, Thomas S. Effect of surface treatments on the electrical properties of low-density polyethylene composites reinforced with short sisal fibers. Composites Science and Technology. 1997;**57**:67-79. DOI: 10.1016/s0266-3538(96)00109-1

[30] Salem IA, Rozyanty AR, Betar BO, Adam T, Mohammed M, Mohammed AM. Study of the effect of surface treatment of kenaf fiber on chemical structure and water absorption of kenaf filled unsaturated polyester composite. In: Proceedings of the International Conference on Applications and Design in Mechanical Engineering (ICADME 2017); 21-22 August 2017; Penang. Malaysia: In Journal of Physics: Conference Series. 2017. p. 012001

[31] Mohammed M, Rozyanty AR, Beta BO, Adam T, Osman AF, Salem IA, et al. The weathering effect in natural environment on kenaf blast and unsaturated polyester composite. In: Proceedings of the International Conference on Applications and Design in Mechanical Engineering (ICADME 2017); 21-22 August 2017; Penang. Malaysia: In Journal of Physics: Conference Series. 2017. p. 012003

[32] Mohammed M, Rozyanty AR, Adam T, Betar BO. Study of the weathering effect in a natural environment on the hybrid kenaf bast/glass fibre-filled unsaturated polyester composite. In: Proceedings of the 3rd Electronic and Green Materials International Conference 2017 (EGM 2017); 29-30 April 2017; Krabi. Thailand: AIP Conference Proceedings. 2017. p. 020201

[33] Akil HM, Cheng LW, Ishak ZM, Bakar AA, Abd Rahman MA. Water absorption study on pultruded jute fibre reinforced unsaturated polyester composites. Composites Science and Technology. 2009;**69**:1942-1948. DOI: 10.1016/j.compscitech.2009.04.014

[34] Salem IA, Rozyanty AR, Betar BO, Adam T, Mohammed M, Mohammed AM. Study of the effect of surface treatment of kenaf fibre on mechanical properties of kenaf filled unsaturated polyester composite. In: Proceedings of the International Conference on Applications and Design in Mechanical Engineering (ICADME 2017); 21-22 August 2017; Penang. Malaysia: In Journal of Physics: Conference Series. 2017. p. 012002

[35] Alnaid A, Noriman NZ, Dahham OS, Hamzah R, Sudin S, Idrus SS, et al. Effects of sodium hydroxide treatment on LLDPE/DS composites: Tensile properties and morphology. In: Proceedings of the International Conference on Green Design and Manufacture (IConGDM 04); 29-30 Apr 2018; Saigon. Vietnam: Journal of Physics. 2018. p. 012061

[36] Dahham OS, Noriman NZ, Hamzah R, Al-Samarrai MN, Idrus SS, Shayfull Z, et al. The influences naoh treatment on polypropylene/cyperus odoratus (pp/cy) composites: Tensile and morphology. In: Proceedings of

the International Conference on Green Design and Manufacture (IConGDM 04); 29-30 Apr 2018; Saigon. Vietnam: Journal of Physics. 2018. p. 012057

[37] Srinivasan SS, Wade J, Stefanakos EK, Goswami Y. Synergistic effects of sulfation and co-doping on the visible light photocatalysis of TiO2. Journal of Alloys and Compounds. 2006;**424**:322-326. DOI: 10.1016/j.jallcom.2005.12.064

[38] Vijayaraghavan S, Goswami DY. Photocatalytic oxidation of toluene in water from an algae pond with high dissolved oxygen content. Journal of Solar Energy Engineering. 2003;**125**:230-232. DOI: 10.1115/sed2002-1061

[39] Goswami DY, Trivedi DM, Block SS. Photocatalytic disinfection of indoor air. Journal of Solar Energy Engineering, Transactions of the ASME. 1997;**119**:92-96. DOI: 10.1115/1.2871871

[40] Rehman S, Ullah R, Butt A, Gohar ND. Strategies of making TiO_2 and ZnO visible light active. Journal of Hazardous Materials. 2009;**170**:560-569. DOI: 10.1016/j.jhazmat.2009.05.064

[41] Ahmed S, Rasul MG, Martens WN, Brown R, Hashib MA. Heterogeneous photocatalytic degradation of phenols in wastewater: A review on current status and developments. Desalination. 2010;**261**:3-18. DOI: 10.1016/j.desal.2010.04.062

[42] Lany S, Osorio-Guillén J, Zunger A. Origins of the doping asymmetry in oxides: Hole doping in NiO versus electron doping in ZnO. Physical Review B. 2007;**75**:241203. DOI: 10.1103/physrevb.75.241203

[43] Rasmussen JW, Martinez E, Louka P, Wingett DG. Zinc oxide nanoparticles for selective destruction of tumor cells and potential for drug delivery applications. Expert Opinion on Drug Delivery. 2010;**7**:1063-1077. DOI: 10.1517/17425247.2010.502560

[44] Mohammed M, Betar BO, Rahman R, Mohammed AM, Osman AF, Jaafar M, et al. Zinc oxide nano particles integrated kenaf/unsaturated polyester biocomposite. Journal of Renewable Materials. 2019;**7**:967-982. DOI: 10.32604/jrm.2019.07562

[45] Wang X, Liu J, Chai Y. Thermal, mechanical, and moisture absorption properties of wood-TiO_2 composites prepared by a sol-gel process. BioResources. 2012;**7**:0893-0901. DOI: 10.15376/biores.7.1.0893-0901

[46] Soltani M, Najafi A, Yousefian S, Naji HR, Bakar ES. Water repellent effect and dimension stability of beech wood impregnated with nano-zinc oxide. BioResources. 2013;**8**:6280-6287. DOI: 10.15376/biores.8.4.6280-6287

[47] Kathirvelu S, D'souza L, Dhurai B. A comparative study of multifunctional finishing of cotton and P/C blended fabrics treated with titanium dioxide/zinc oxide nanoparticles. Indian Journal of Science and Technology. 2008;**1**:1-12. DOI: 10.17485/ijst/2008/v1i7/29597

[48] Mohammed M, Rozyanty AR, Osman AF, Adam T, Hashim U, Mohammed AM, et al. The weathering effect in natural environment on KenafBast filled unsaturated polyester composite and integration of Nano zinc particle for water repellent. Micro and Nanosystems. 2017;**9**:16-27. DOI: 10.2174/1876402909666170531075138

[49] Tshabalala MA, Libert R, Schaller CM. Photostability and moisture uptake properties of wood veneers coated with a combination of thin sol-gel films

and light stabilizers. Holzforschung. 2011;*65*:215-220. DOI: 10.1515/hf.2011.022

[50] Li Z, Wang Y, Cheng L, Guo W, Wu G. Effect of Nano-$CaCO_3$ on the structure and properties of holocellulose-fiber/polypropylene biomass composites. Journal of Wood Chemistry and Technology. 2017;**37**:62-74. DOI: 10.1080/02773813.2016.1235586

[51] Ibrahim ID, Jamiru T, Sadiku ER, Kupolati WK, Agwuncha SC. Impact of surface modification and nanoparticle on sisal fiber reinforced polypropylene nanocomposites. Journal of Nanotechnology. 2016;**2016**:1-9. DOI: 10.1155/2016/4235975

[52] Heath CJ, Bond IP, Potter KD. Integrating electrostatic adhesion to composite structures. In: Industrial and Commercial Applications of Smart Structures Technologies 2015. USA: SPIE; 2015;**9433**:95-105. DOI: 10.1117/12.2084073

[53] Heath CJ, Bond IP, Potter KD. Electrostatic adhesion for added functionality of composite structures. Smart Materials and Structures. 2016;**25**:025016. DOI: 10.1088/0964-1726/25/2/025016

[54] Yamamoto T, Uematsu K, Irisawa T, Tanabe Y. Controlling of the interfacial shear strength between thermoplastic resin and carbon fiber by adsorbing polymer particles on carbon fiber using electrophoresis. Composites Part A: Applied Science and Manufacturing. 2016;**88**:75-78. DOI: 10.1016/j.compositesa.2016.05.021

[55] Mai C, Militz H. Modification of wood with silicon compounds. Inorganic silicon compounds and sol-gel systems: A review. Wood Science and Technology. 2004;**37**:339-348. DOI: 10.1007/s00226-003-0205-5

[56] Stark NM. Effect of weathering cycle and manufacturing method on performance of wood flour and high-density polyethylene composites. Journal of Applied Polymer Science. 2006;**100**:3131-3140. DOI: 10.1002/app.23035

[57] Tuduce Trăistaru AA, Crina Anca Sandu I, Cristina Timar M, Lucia Dumitrescu G, Sandu I. SEM-EDX, water absorption, and wetting capability studies on evaluation of the influence of nano-zinc oxide as additive to paraloid B72 solutions used for wooden artifacts consolidation. Microscopy Research and Technique. 2013;**76**:209-218. DOI: 10.1002/jemt.22155

[58] Farahani MR, Banikarim F. Effect of nano-zinc oxide on decay resistance of wood-plastic composites. BioResources. 2013;**8**:5715-5720. DOI: 10.15376/biores.8.4.5715-5720

[59] Li X, Tabil LG, Panigrahi S. Chemical treatments of natural fiber for use in natural fiber-reinforced composites: A review. Journal of Polymers and the Environment. 2007;**15**:25-33. DOI: 10.1007/s10924-006-0042-3

[60] Stark N. Influence of moisture absorption on mechanical properties of wood flour-polypropylene composites. Journal of Thermoplastic Composite Materials. 2001;**14**:421-432. DOI: 10.1106/udky-0403-626e-1h4p

[61] Raghupathi KR, Koodali RT, Manna AC. Size-dependent bacterial growth inhibition and mechanism of antibacterial activity of zinc oxide nanoparticles. Langmuir. 2011;**27**:4020-4028. DOI: 10.1021/la104825u

[62] Hirota K, Sugimoto M, Kato M, Tsukagoshi K, Tanigawa T, Sugimoto H. Preparation of zinc oxide ceramics with a sustainable antibacterial activity under dark conditions. Ceramics International. 2010;**36**:497-506. DOI: 10.1016/j.ceramint.2009.09.026

[63] Jones N, Ray B, Ranjit KT, Manna AC. Antibacterial activity of ZnO nanoparticle suspensions on a broad spectrum of microorganisms. FEMS Microbiology Letters. 2008;**279**:71-76. DOI: 10.1111/j.1574-6968.2007.01012.x

Chapter 13

New Strategy to Improve Photocatalytic Activity and Mechanistic Aspect for Water Splitting

Tanu Mittal

Abstract

Photocatalytic water splitting under irradiation of light is an ultimate system to make available renewable energy sources. However, significant efforts have been made to fabricate advanced nanocomposites; the major challenge persists, which is low efficiency and selectivity towards hydrogen evolution under the illumination of solar energy. In this chapter, the recent developments in photocatalysts, and heterojunction fabrication factors influencing the photocatalytic process for the dynamic production of hydrogen have been discussed. The improvements in photocatalyst systems have been classified as strategies to improve different factors of photocatalytic splitting such as Z-scheme systems and the influence of operating parameters such as band gap, morphology, temperature, oxygen vacancies, intensity, and pH. In addition, the thermodynamics of selective photocatalysts is conferred critically. The future research aspect of photocatalytic water splitting has been also discussed.

Keywords: water splitting, nanocomposite, photocatalyst

1. Introduction

Global energy consumption has remarkably increased due to the continual increase in world population and lifestyle standards [1]. Transportation and industrial sectors lead to high emissions of greenhouse gases including carbon monoxide, and carbon dioxide, 90% of the global energy and fossil fuel supply amounting to substantial depletion of the environment [2, 3]. Eventual depletion of energy supplies, market uncertainty, and mitigating the consequences of fossil fuel burning demands the development of clean and renewable resources of energy [4–6]. There are several alternative energy sources, including geothermal energy, hydropower, wind power, and solar energy, which are relatively cleaner and more sustainable than fossil fuels. However, the substitution of natural resources with alternative resources is much more challenging due to the limitations of each resource. Such as we cannot store the electricity which is produced by wind turbines. Hydropower undergoes dam construction limitations due to possible adverse environmental effects and high costs. Geothermal energy is a costly operation because it is a continuous source and is limited in its lifetime [7]. Renewable and free solar energy is unlimited and capable of producing electricity without any maintenance or requirement of having turbines. Half an hour of solar radiation on the earth's surface can be used for 1 year [8].

Though, sunlight is an intermittent energy source that limits the ratio of solar radiation due to its dependence on the geographical season, day, time, and position [9, 10]. Another major disadvantage of solar energy is its low density per unit area of the earth's surface [11]. Therefore, to meet the global energy demand it is very necessary to develop a source of energy that is continuous, storable, and renewable. In this line of research, hydrogen is a profitable fuel for being profuse from different sustainable sources of water and biomass, high energy yield and efficiency, eco-friendly and capable of storage, thus it is considered an ideal alternative source of energy for nonrenewable energy source [12, 13]. Photocatalytic water splitting has attracted considerable attention as a potential renewable energy resource with limited use of fossil fuel and no CO_2 emission [1–4, 14]. Recent development in photocatalytic systems for photocatalytic water splitting can be divided into two main approaches. In the first approach, water is split into hydrogen and oxygen by visible light irradiation on the photocatalyst. In this type of system, the photocatalyst should have the suitable thermodynamic potential for photocatalytic water splitting. The narrow band gap of the photocatalyst system harvests visible light photons and provides stability against photo corrosion. Due to the rigorous requirement of photocatalyst, one-step water splitting is limited [5, 6]. The second approach is to apply a two-way mechanism by using two different types of photocatalysts [7]. These mechanisms were inspired by the natural photosynthesis phenomenon in green plants and are called the Z-scheme. The advantage of water splitting under the Z scheme is that a wider range of visible light is available because a change in Gibbs free energy is required for the one-step water splitting for the separation of water molecules and oxygen. Use of semiconductors used for either reduction or oxidation potential for one side of the system. For example, use of metal oxides such as WO_3 & $BiVO_4$ act as good oxygen evolution photocatalysts in two-way water splitting systems by using a suitable redox mediator, even though they are unable to reduce water [8, 9]. Overall water splitting via two-step photoexitation by the use of visible light and with different combinations of photo catalysts has been successfully reported [8–13, 15, 16]. However, there are several challenges in the promotion of electron transfer between two semiconductors. In addition, the photocatalytic activity of water splitting is strongly dependent on the physiochemical properties of photocatalysts such as the nature of active sites and the reaction conditions [2, 3, 14]. In past decade there are a number of material have been reported as a visible light active photocatalyst, which produces both oxygen and hydrogen under visible light irradiations [1, 4–17]. However, a number of photocatalysts have successfully achieved water splitting without any reagent. Significant progress has been made in the development of cocatalysts and the interpretation of reaction mechanisms. This perception of the mechanism highlights some important aspects of the recent development in water-splitting research.

2. Mechanistic aspect for water splitting

In the development of overall photocatalytic water splitting under visible light a number of photocatalytic materials and preparation methods have been studied. A number of studies have focused on the material development of materials that are suitable for overall water splitting and their visible light absorption properties, crystallographic quality, phase purity, band edge position, and particle morphology. However, it is very challenging to identify the factor which dominates the net photocatalytic activity based on physical properties. Photocatalytic reactions proceed through an intricate sequence of competing for multistep processes. This multistep process establishes the kinetics and dynamics of photocatalytic water-splitting reactions for future applications [18]. The rate of water splitting with the modified photocatalyst was proportional to light intensity under solar irradiations. Excess loading of the co-catalyst did not improve the photocatalytic water splitting rate. The experimental results define the shortage of photo-exited carriers available for surface redox reactions under visible irradiations, this indicates that the balance between the rates of redox reactions on the surface of photocatalyst and charge carrier recombination in the bulk photocatalyst determines the charge concentration in the photocatalyst. The kinetic model of photocatalytic water splitting also determines the rate of reaction and probability of photoexcited holes for oxygen evolution and recombination. It was easy to assume two different co-catalyst distinctly facilitate hydrogen and oxygen evolution thereby stimulating overall water splitting in harmony. As stated earlier, visible light-responsive photocatalyst has been devoted to the development of active sites on photocatalyst and elucidating reaction mechanisms, which leads to significant progress in the field of heterogeneous photocatalytic water splitting [18–31]. Conversion of solar energy most efficiently can achieve overall water splitting under longer wavelength irradiation; this is because the number of accessible photons in the solar spectrum increased with an increased wavelength of the solar spectrum.

In this line of research, the development of a photocatalyst having a wider absorption band is highly desirable for overall water splitting. Although, a photocatalyst with an absorption edge of 600 nm would be optimal for the activation of the surface barrier and so surface reactions that can produce hydrogen and oxygen. The most promising candidates in this category have already developed, such as $LaTiO_2$, Ta_3N_5, and $Ti_2S_2O_5$ with a band gap of 2 eV which means they have an absorption edge near 600 nm [2]. However, the photocatalytic activities of these materials are not sufficient to achieve overall water splitting. Recent progress has been also made in material chemistry towards reducing the density defects [13, 31]. Consequently, it is very important to study the nature of defects, which can facilitate the undesirable electron–hole pair recombination in the photocatalytic system. In a two-step, a water-splitting system for $BaTaO_2N_{12}$ and Ta_3N_5 the absorption, the wavelength has been increased to 660 nm [16]. The research in this area is underway in direction of both photo catalyst preparations and mechanistic aspects of water splitting processing in harmony.

3. Band structure and surface properties for photocatalytic water splitting

Photocatalytic activity of water splitting reactions has been affected by various factors such as surface structure chemical functionality deactivating destructive surface state, band edge position, and extract of carriers to improve catalytic activity [32, 33]. In spite of surface area properties responsible for the photocatalytic activity, there is an adjustment between charge diffusion length and light absorption. An increase in surface area may lead to a decrease in photovoltage and an increase in the surface recombination process. Therefore, before surface modification, it is necessary to study the loss mechanism. It can be observed that sheet-like structures exhibited higher light absorption capacity in comparison with spherical morphologies. The band gap shift towards lower energies with the reduction in crystallite size [34]. The most important example of this is $BiVO_4$ with a narrow band gap of 2.4 eV with control and desirable structure and morphology which is mandatory for the photocatalytic activity [35, 36]. Photocatalytic activity if $BiVO_4$ depends upon the facets which are exposed to irradiations [37]. Improved photocatalytic activity due to charge separation is reported for CdS/ZnO and CDs/TiO_2 heterostructure [38–41]. The incorporation of cations and anions into UV-active materials turned into visible light-active materials.

4. Photocatalytic water splitting modeling

Theoretical modeling studies of water splitting concern a number of aspects such as electron–hole transport/recombination, photocatalytic reaction, semiconductor band edge alignment, and photo redox chemistry [27–34]. DFT has been used extensively for the theoretical modeling of water splitting to predict the electronic structure of materials due to the modest cost and high accuracy reproducibility [35, 36, 42, 43]. However, the inaccurate predictions of band gaps are the major drawback of DFT. Hybrid materials have better performance for the prediction of band gap and the exited state position [44]. The band gap drawback and problems is tackled by perturbation theory which has a long-standing record of success [45, 46]. Computational methods are very helpful, especially in the prediction of impurities in photocatalysts induced by dopants in tuning with band gap for example TiO_2. Theoretical and computational tools are given an understanding and idea about various aspects of material and its state. For example, $BiVO_4$ in the band structures and density of states, migration energy profile of surface reactions, band structure, and density state, electron/hole pair generation comprehensive study. Also, electron–hole pairs driven to different crystal facets [47, 48]. All of these findings can obtain from comprehensive computational studies that are related to or compared with facet (011), (010) have lower absorption at 420 nm, better transport of electron–hole pair, and lower potential energy surfaces [49]. Theoretical studies are related to the improvement of band structure and morphology of the photocatalytic material.

5. Photocatalytic material description and design

As per earlier studies, a suitable photocatalyst should have a band gap of at least 1.23 eV for overall water splitting. High crystallinity and small particle size are major requirements to minimize the recombination of photo-generated electrons and holes. Metal oxide, nitrides, sulfides, phosphates, Groups I and II metals along with their lanthanides can also be used as photocatalytic material for overall water splitting. For the improvement of the efficiency of the photocatalyst, modification of material by doping some transition metal cations can help to increase the visible light response of the photocatalyst. Also, to exclude the energy backward reaction of water splitting and to increase the hydrogen production yields co-catalyst such as RuO_2, Au, Pt, and NiO can be used. This section of the chapter has focused on the heterogeneous photocatalyst including TiO_2 and metal oxides.

Fujishima and Honda first confirmed that TiO_2 was a potential photo-anode for UV light active photocatalytic water splitting. TiO_2 has been widely studied in a number of photocatalytic reactions due to its low cost, environmentally friendly nature, chemical stability, and tunable energy band gap [50–54]. A number of alternative methods have been reported to extend the photocatalytic activity of TiO_2 into the visible light region, such as by doping with metal ions e.g. Carbon nanotubes [55, 56]. However, altered mechanistic methodologies have been proposed to explain this enhancement of activity. There are three mechanisms that have been reported to describe the synergistic effect of carbon nanotubes on TiO_2, the first prospective mechanism is that carbon can act as an electron sink, which can effectively prevent the recombination process [57]. In another mechanism, carbon acts as a photosensitizer, which can pump electrons into the TiO_2 conduction band [58]. In addition to proposed mechanisms, carbon can also act as a template for the dispersion of TiO_2 nanoparticles to avoid agglomeration [59]. Further nonmetal ion doping and metallic dopants usually add additional energetic levels in the band gap, which will reduce the energy barrier and introduce a new absorption band gap [58, 59]. Doping of TiO_2 with other elements can help to change the optical properties of nanomaterial and reduce the charge carrier recombination sufficiently s. Piskunov et al. suggested improvement in water splitting of TiO_2 doped with Fe, where Fe^{2+} and Fe^{3+} act as centers for electron trapping and Fe^{3+} & Fe^{4+} act as a center for hole trapping. Lu0 et al. confirmed the doping of vanadium into the crystal lattice of TiO_2 that shifts the absorption band to the visible range and V^{4+} V^{5+} efficiently traps the holes and electrons. Further, anionic doping has been extensively reported for TiO_2 by different dopant materials such as C, N, F, S, B, and sCl [58, 59].

Further, other than TiO_2 a number of metal Oxides such as Cu_2O, Al_2O_3, CoO and ZrO_2, Fe_2O_3, and Ta_2O_2 have been widely studied due to their low cost and stability. However, metal oxides suffer from limitations due to their large band gaps which limit their absorption of visible light. Ionic bonded materials have a large band gap because, in a typical metal oxide, the valence band and the conduction band have O_2p. To overcome this shortcoming transition metal cation has been used with dn configuration for example Fe_2O_3 with the band gap value (2.0 eV) and Co_3O_4 91.3 eV [31]. This may lead to an increase in light absorption but a decrease inefficient charge carrier transport due to high resistivity. Using post-transition metals such as PbO (2.1 eV), SnO_2 (2.4 eV), and Bi_2O_3 (2.5 eV) leads to better charge carrier generation however they are indirect semiconductors; so the optical absorption band edge varies with the square root of photon energy and gives less efficient charge carrier extraction process Therefore ternary metal oxides have been suggested to overcome these issues, for example, $Bi_{20}TiO_{32}$, $SnNb_2O_6$ and $BiVO_4$ [31, 35]. Properties of n and p-type semiconductor properties have been found in BiO_4 and high photon to current conversion efficiency [59]. In addition, Fe_2O_3 as photocatalytic material has a band gap of 2.2 eV which allows photon absorption under the irradiation of visible light. Morales-Guio et al. have proposed a photocatalyst of amorphous iron-nickel oxide (FeNiOx) for the oxygen evolution reaction. Similarly, WO_3 has been considered a good photo anode material due to its suitable valance band position, which favors a high water oxidation potential.

In addition, Amer et al. have suggested ZrO_2 modification with the deposition of ZrN on ZrO_2 thin layers for the preparation of core-shell structures which are visible light active. However, Moniz et al. stated that the main drawback of WO_3 is its instability towards corrosion. Due to the low e.g. of these materials, these can be modified with doping with metal cations or by combining with other semiconductors to form heterojunctions. Sivula et al. have confirmed a WO_3/Fe_2O_3 heterojunction for better water oxidation due to its suitable band gap and proper alignment between WO_3

and Fe_2O_3 metal oxide which leads to better electron transfer at the host and guest interface. Ta_2O_3 (Tantalum Oxide) has been considering an attractive semiconductor for overall photocatalytic water splitting because of its wide band gap value (4 eV), further it is required to narrow the band gap by doping with some doping ions. It is also mentioned by Lu et al. that Ta_2O_5 nanowires as an active photocatalyst with a high rate of hydrogen generation. Recently, Zhu et al. reported Ta_2O_3 nanowires that were modified by an aluminum reduction for the improvement of electron density and photoelectrochemical overall water splitting of the material.

6. Conclusions

One of the most promising steps towards generating clean and renewable alternatives to fossil fuels is hydrogen production from solar energy by using photocatalytic materials. Different approaches have been employed in order to use solar energy more efficiently and to shift the photocatalytic activity in the visible range with stability and efficiency. TiO_2 has been used most widely as a photocatalyst however it has also some limitations such as:

- high hydrogen overpotential, wide band gap, and increased recombination of electron–hole pairs. Other metal oxides such as copper oxide, zinc oxide, and iron oxide have been also discussed as photocatalytic materials.

- Nitrides and composite materials have been also reviewed. Efficient water splitting depends on the innovative design of photocatalytic material. In addition, recent studies on heterojunction photocatalysts have also studied the nature of charge transfer.

- Carbon-based material heterojunctions are feasible future routes for the efficient design of photocatalysts. The architecture of the heterojunction influences photocatalytic activity and could potentially lead to great improvements.

- The future development in direction of photocatalytic water splitting is focused on the development of an efficient photoanode with a band edge with a redox potential of water and increased charge transfer under visible light with stability.

- Computational and theoretical modeling could help to understand the electronic density state and band structure. Therefore, the rational design of the photocatalyst leads to better overall water splitting's. Computational screening is an emerging field that will be used in the selection of material and design of junctions to yield optimized band structures.

Author details

Tanu Mittal
Faculty of Natural Sciences, GNA University, Phagwara, India

Address all correspondence to: drmittaltanu@gmail.com

References

[1] Pao H-T, Tsai C-M. CO_2 emissions, energy consumption and economic growth in BRIC countries. Energy Policy. 2010;**38**:7850-7860

[2] Davis SJ, Caldeira K. Consumption-based accounting of CO_2 emissions. Proceedings of the National Academy of Sciences of the United States of America. 2010;**107**:5687-5692

[3] Dodman D. Blaming cities for climate change? An analysis of urban greenhouse gas emissions inventories. Environment and Urbanization. 2009;**21**:185-201

[4] Byrne J, Hughes K, Rickerson W, Kurdgelashvili L. American policy conflict in the greenhouse: Divergent trends in federal, regional, state, and local green energy and climate change policy. Energy Policy. 2007;**35**:4555-4573

[5] Solomon S, Plattner G-KG, Knutti R, Friedlingstein P. Irreversible climate change due to carbon dioxide emissions. Proceedings of the National Academy of Sciences of the United States of America. 2009;**106**:1704-1709

[6] Di Primio R, Horsfield B, Guzman-Vega M. Determining the temperature of petroleum formation from the kinetic properties of petroleum asphaltenes. Nature. 2000;**406**:173-176

[7] Barbier E. Geothermal energy technology and current status: An overview. Renewable and Sustainable Energy Reviews. 2002;**6**:3-65

[8] Dincer I, Zamfirescu C, Dinçer İ, Zamfirescu C, Dincer I, Zamfirescu C. Sustainable Energy Systems and Applications. Vol. 6. New York, NY, USA: Springer Science & Business Media; 2011

[9] Parida B, Iniyan S, Goic R. A review of solar photovoltaic technologies. Renewable and Sustainable Energy Reviews. 2011;**15**:1625-1636

[10] Xie WT, Dai YJ, Wang RZ, Sumathy K. Concentrated solar energy applications using Fresnel lenses: A review. Renewable and Sustainable Energy Reviews. 2011;**15**:2588-2606

[11] Zamfirescu C, Dincer I, Naterer GF, Banica R. Quantum efficiency modeling and system scaling-up analysis of water splitting with Cd1-xZnxS solid-solution photocatalyst. Chemical Engineering Science. 2013;**97**:235-255

[12] Turner J. A sustainable hydrogen production. Science. 2004;**305**:972-974

[13] Zhu J, Zäch M. Nanostructured materials for photocatalytic hydrogen production. Current Opinion in Colloid & Interface Science. 2009;**14**:260-269

[14] Moniz S, Shevlin SA, Martin D, Guo Z, Tang J. Visible-light driven heterojunction photocatalysts for water splitting—A critical review. Energy & Environmental Science. 2015;**8**:731-759

[15] Melián EP, Díaz OG, Méndez AO, López CR, Suárez MN, Rodríguez JMD, et al. Efficient and affordable hydrogen production by water photo-splitting using TiO_2-based photocatalysts. International Journal of Hydrogen Energy. 2013;**38**:2144-2155

[16] Midilli A, Ay M, Dincer I, Rosen MA. On hydrogen and hydrogen energy strategies I: Current status and needs. Renewable and Sustainable Energy Reviews. 2005;**9**:255-271

[17] Chiarello GL, Aguirre MH, Selli E. Hydrogen production by photocatalytic steam reforming of methanol on noble metal-modified TiO_2. Journal of Catalysis. 2010;**273**:182-190

[18] Das D, Veziroglu TN. Advances in biological hydrogen production processes. International Journal of Hydrogen Energy. 2008;**33**:6046-6057

[19] Hou K, Hughes R. The kinetics of methane steam reforming over a Ni/α-Al_2O catalyst. Chemical Engineering Journal. 2001;**82**:311-328

[20] Lewis NS, Nocera DG. Powering the planet: Chemical challenges in solar energy utilization. Proceedings of the National Academy of Sciences of the United States of America. 2006;**103**:15729-15735

[21] Kudo A. Photocatalysis and solar hydrogen production. Pure and Applied Chemistry. 2007;**79**:1917-1927

[22] Department of Energy. Fuel cell technologies office multi-year research, development, and demonstration plan. Available from: http://energy.gov/eere/fuelcells/downloads/fuel-cell-technologies-office-multi-yearresearch-development-and-22. [Accessed on 9 May 2016]

[23] Sivula K, van de Krol R. Semiconducting materials for photoelectrochemical energy conversion. Nature Reviews Materials. 2016;**70**:15010:1-15010:7

[24] Guijarro N, Prévot MS, Sivula K. Surface modification of semiconductor photoelectrodes. Physical Chemistry Chemical Physics. 2015;**17**:15655-15674

[25] Meshram SP, Adhyapak PV, Mulik UP, Amalnerkar DP. Facile synthesis of CuO nanomorphs and their morphology dependent sunlight driven photocatalytic properties. Chemical Engineering Journal. 2012;**204-206**:158-168

[26] Lu Y, Shang H, Shi F, Chao C, Zhang X, Zhang B. Preparation and efficient visible light-induced photocatalytic activity of m-$BiVO_4$ with different morphologies. Journal of Physics and Chemistry of Solids. 2015;**85**:44-50

[27] Etacheri V, Di Valentin C, Schneider J, Bahnemann D, Pillai SC. Visible-light activation of TiO_2 photocatalysts: Advances in theory and experiments. Journal of Photochemistry and Photobiology C: Photochemistry Reviews. 2015;**25**:1-29

[28] Nazeeruddin MK, De Angelis F, Fantacci S, Selloni A, Viscardi G, Liska P, et al. Combined experimental and DFT-TDDFT computational study of photoelectrochemical cell ruthenium sensitizers. Journal of the American Chemical Society. 2005;**127**:16835-16847

[29] Nayak PK, Periasamy N. Calculation of electron affinity, ionization potential, transport gap, optical band gap and exciton binding energy of organic solids using "solvation" model and DFT. Organic Electronics. 2009;**10**:1396-1400

[30] Chai S, Wen S-H, Huang J-D, Han K-L. Density functional theory study on electron and hole transport properties of organic pentacene derivatives with electron-withdrawing substituent. Journal of Computational Chemistry. 2011;**32**:3218-3225

[31] De Angelis F, Fantacci S, Selloni A. Alignment of the dye's molecular levels with the TiO2 band edges in dye-sensitized solar cells: A DFT-TDDFT study. Nanotechnology. 2008;**19**:424002-424008

[32] Wendt S, Seitsonen A, Kim Y, Knapp M, Idriss H, Over H. Complex redox chemistry on the RuO2 (100) surface: Experiment and theory. Surface Science. 2002;**505**:137-152

[33] Chen J, Zhang H, Tomov IV, Wolfsberg M, Ding X, Rentzepis PM. Transient structures and kinetics of the ferrioxalate redox reaction studied by time-resolved EXAFS, optical spectroscopy, and DFT. The Journal of Physical Chemistry. A. 2007;**111**:9326-9335

[34] Kohn W, Sham LJ. Self-consistent equations including exchange and correlation effects. Physics Review. 1965;**140**:A1133-A1138

[35] Hohenberg P. Inhomogeneous electron gas. Physics Review. 1964;**136**:B864-B871

[36] Burke K. Perspective on density functional theory. The Journal of Chemical Physics. 2012;**136**:150901

[37] Wu Q, Han R, Chen P, Qi X, Yao W. Novel synthesis and photocatalytic performance of $BiVO_4$ with tunable morphologies and macroscopic structures. Materials Science in Semiconductor Processing. 2015;**38**:271-277

[38] Li J, Hoffmann MWG, Shen H, Fabrega C, Prades JD, Andreu T, et al. Enhanced photoelectrochemical activity of an excitonic staircase in CdS@ TiO_2 and CdS@anatase@rutile TiO_2 heterostructures. Journal of Materials Chemistry. 2012;**22**:20472-20476

[39] Li J, Cushing SK, Zheng P, Senty T, Meng F, Bristow AD, et al. Solar hydrogen generation by a CdS-Au-TiO_2 sandwich nanorod array enhanced with Au nanoparticle as electron relay and plasmonic photosensitizer. Journal of the American Chemical Society. 2014;**136**:8438-8449

[40] Liu X, Wang C, Xu J, Liu X, Zou R, Ouyang L, et al. Fabrication of ZnO/CdS/Cu2ZnSnS4 p-n heterostructure nanorod arrays via a solution-based route. CrystEngComm. 2013;**15**:1139-1145

[41] Wang X, Yin L, Liu G, Wang L, Saito R, Lu GQM, et al. Polar interface-induced improvement in high photocatalytic hydrogen evolution over ZnO-CdS heterostructures. Energy & Environmental Science. 2011;**4**:3976-3979

[42] Lejaeghere K, Bihlmayer G, Bjorkman T, Blaha P, Blugel S, Blum V, et al. Reproducibility in density functional theory calculations of solids. Science. 2016;**351**:145-152

[43] Yang J, Wang D, Zhou X, Li C. A theoretical study on the mechanism of photocatalytic oxygen evolution on $BiVO_4$ in aqueous solution. Chemistry. 2013;**19**:1320-1326

[44] Khakpash N, Simchi A, Jafari T. Adsorption and solar light activity of transition-metal doped TiO_2 nanoparticles as semiconductor photocatalyst. Journal of Materials Science: Materials in Electronics. 2012;**23**:659-667

[45] Miyauchi M, Irie H, Liu M, Qiu X, Yu H, Sunada K, et al. Visible-light-sensitive photocatalysts. Nanocluster-grafted titanium dioxide for indoor environmental remediation. Journal of Physical Chemistry Letters. 2016;7:75-84

[46] Wang S, Pan L, Song JJ, Mi W, Zou JJ, Wang L, et al. Titanium-defected undoped anatase TiO_2 with p-type conductivity, room-temperature ferromagnetism, and remarkable photocatalytic performance. Journal

of the American Chemical Society. 2015;**137**:2975-2983

[47] Fei T, Rongshu Z, Kelin S, Feng O, Gang C. Synergistic photocatalytic degradation of phenol using precious metal supported titanium dioxide with hydrogen peroxide. Environmental Engineering Science. 2016;**33**:185-192 Molecules 2016, 21

[48] Li B, Zhao J, Liu J, Shen X, Mo S, Tong H. Bio-templated synthesis of hierarchically ordered macro-mesoporous anatase titanium dioxide flakes with high photocatalytic activity. RSC Advances. 2015;**5**:15572-15578

[49] Wang H, Dong S, Chang Y, Faria JL. Enhancing the photocatalytic properties of TiO2 by coupling with carbon nanotubes and supporting gold. Journal of Hazardous Materials. 2012;**235-236**:230-236

[50] Li Z, Gao B, Chen GZ, Mokaya R, Sotiropoulos S, Li Puma G. Carbon nanotube/titanium dioxide (CNT/TiO_2) core-shell nanocomposites with tailored shell thickness, CNT content and photocatalytic/photoelectrocatalytic properties. Applied Catalysis B: Environmental. 2011;**110**:50-57

[51] Silva CG, Faria JL. Photocatalytic oxidation of benzene derivatives in aqueous suspensions: Synergic effect induced by the introduction of carbon nanotubes in a TiO_2 matrix. Applied Catalysis B: Environmental. 2010;**101**:81-89

[52] Babu VJ, Vempati S, Uyar T, Ramakrishna S. Review of one-dimensional and two-dimensional nanostructured materials for hydrogen generation. Physical Chemistry Chemical Physics. 2015;**17**:2960-2986

[53] Chen X, Shen S, Guo L, Mao SS. Semiconductor-based photocatalytic hydrogen generation. Chemical Reviews. 2010;**110**:6503-6570

[54] Sayama K, Nomura A, Zou Z, Abe R, Abe Y, Arakawa H. Photoelectrochemical decomposition of water on nanocrystalline $BiVO_4$ film electrodes under visible light. Chemical Communications. 2003;**23**:2908-2909

[55] Mishra M, Chun D-M. α-Fe_2O_3 as a photocatalytic material: A review. Applied Catalysis A: General. 2015;**498**:126-141

[56] Haghighat S, Dawlaty JM. Continuous representation of the proton and electron kinetic parameters in the pH-potential space for water oxidation on hematite. Journal of Physical Chemistry C. 2015;**119**:6619-6625

[57] Morales-Guio CG, Mayer MT, Yella A, Tilley SD, Grätzel M, Hu X. An optically transparent iron nickel oxide catalyst for solar water splitting. Journal of the American Chemical Society. 2015;**137**:9927-9936

[58] Lü X, Ding S, Lin T, Mou X, Hong Z, Huang F. Ta_2O_5 nanowires: A novel synthetic method and their solar energy utilization. Dalton Transactions. 2012;**41**:622-627

[59] Zhu G, Lin T, Cui H, Zhao W, Zhang H, Huang F. Gray Ta_2O_5 nanowires with greatly enhanced photocatalytic performance. ACS Applied Materials & Interfaces. 2016;**8**:122-127

Chapter 14

A Perspective on the Recent Amelioration of Co_3O_4 and MnO_2 Bifunctional Catalysts for Oxygen Electrode Reactions

Selvaraj Venkateshwaran,
Karuppiah Selvakumar, Velu Duraisamy
and Sakkarapalayam Murugesan Senthil Kumar

Abstract

Metal-air batteries with the aid of high theoretical energy density and affability are trusted as propitious energy storage systems in today's energy research. However, enforcement of the technology is still hindered by the sluggish kinetics of their electrode reactions, that is, oxygen evolution and oxygen reduction reaction (OER/ORR). Developing a catalyst with inherently greater bifunctional activity and durability is the finest solution to confront the aforementioned challenges. Transition metal oxides (TMOs) are the most appropriate choice of materials for that purpose since they are highly active, inexpensive, abundant and non-hazardous. Among the various transition metal oxides, MnO_2 and Co_3O_4 are gaining much attention due to their superior bifunctional performance and alkaline stability owing to their structural features and physicochemical properties. With the inspiration from promoted catalytic activity of MnO_2 and Co_3O_4, this chapter is fully devoted to these two catalysts. The activity structural relationship, recent developments and future directions of these materials for bifunctional catalysis have been discussed in more detail. Besides, the significant parameters judging the bifunctional activity, that is, phase, crystal facets, morphology, defects, strains and mixed metals oxide formations, have been illustrated with suitable evidence. In addition, the fundamentals of water oxidation and reduction reactions are explained with the mechanisms. Moreover, the physiochemical properties of MnO_2 and Co_3O_4 materials and their influence on the catalytic activity are related for a better understanding of bifunctional catalysis. This collective perception will be highly useful for the comprehension and designing of advanced metal oxide catalysts to further improve bifunctional catalysis.

Keywords: ORR, OER, water splitting, MnO_2, Co_3O_4, bifunctional catalyst

1. Introduction

As far the energy consumption is concerned in the current energy scenario, the role of fossil fuels is exceptional by delivering a rich source in comparison with other sources [1, 2]. The catastrophic changes to the earth's climate are unavoidable with these carbon-emitting energy supplies. As the depletion of fossil fuels alarms the scientific community to move forward for future fuels, the search is triggered with the application of renewable energy sources like solar, tidal and wind-assisted energy devices. Considering the electrochemical energy conversion devices such as batteries, alkaline fuel cells and solar water-splitting devices [3], these are the technological functions based on the electrochemical reactions such as oxygen evolution reaction (OER), oxygen reduction reaction (ORR) [4, 5], hydrogen evolution reaction (HER) [6] and hydrogen oxidation reaction (HOR) [7]. The chemistry behind the hydrogen-based oxidation and reduction is facile compared with the oxygen-based oxidation and reduction reactions. The OER and ORR reactions are uphill processes and thus drag more overpotentials, and hence, the voltage applied also increases [8]. In metal-air batteries (MABs), the activity of the device predominantly depends on OER and ORR taking place during the charge and discharge process, respectively. The most active catalysts are the expensive Pt (HER, ORR), IrO_2 and RuO_2 (OER), and the search for inexpensive catalysts is unavoidable for the commercial usage of MABs. Moreover, the use of two different catalysts for ORR and OER in MAB makes the device more complicated and expensive. Besides, the ORR being the most active catalyst, Pt is not a suitable candidate for OER owing to the formation of the oxide over the surface of the catalyst, which gives the catalytic ability of OER. Similarly, IrO_2 is a good candidate for OER, due to its high conductivity and stability, and is not preferable for ORR [9]. Therefore, the development of non-precious transition metal-based bifunctional catalysts, which can simultaneously catalyze both OER and ORR, becomes the utmost important parameter to enhance the efficiency of these electrode reactions. Even though there are catalysts with increased efficiency for the selective ORR-like carbon-derived metal catalysts and their oxides, the OER becomes a matter of concern as the conversion of carbon to their oxides occurs before the thermodynamic potential of OER. For OER, the chalcogenides and oxides of transition metals are found to be more active and stable catalysts in basic environment. Therefore, identifying electrocatalysts with significant activity in both OER and ORR is more important for applying them in the energy conversion MAB devices [10].

In addition to the bifunctional activity, the stability of the electrocatalyst in the electrolyte medium is also vital for commercializing the technology. The corrosive nature of acids during long-term usage restricts their commercial usage, and in such cases, alkaline electrolytes become preferable, which facilitates the sluggish oxygen electrode reactions (ORR/OER) with ease and thus provides relatively less corrosive to the transition metal-based materials. Additionally, transition metal-based catalysts have more advantages such as the ability to exist in multiple (two or more) and mixed oxidation states, different coordination environments, and flexibility to replace one transition metal by another that is highly desirable to further promote the bifunctional activity. Electrocatalysts derived from transition metals like transition metal oxides (TMOs), sulfides and hydroxides have been demonstrated as efficient electrocatalysts in oxygen electrode reactions [11]. Among all TMOs, particularly cobalt oxide (Co_3O_4, spinel oxide) and manganese oxides are found to be the superior bifunctional catalysts for oxygen electrode reactions [10].

Manganese oxides, with their highly affordable nature and handling, could also deliver enhanced ORR activity and moderate OER activity through the metal and lattice oxygen that act as active sites. The enhanced activity from manganese oxides is ascribed to the ease of conversion of Mn to higher oxidation states and its preferable four-electron transfer reduction pathway in ORR [12]. However, in spinel oxides like

Co_3O_4, Co^{2+} and Co^{3+}, ions assist the conversion of reversible adsorption-desorption of oxygen to facilitate the oxygen electrode reactions [9, 13]. Therefore, much effort has been dedicated based on these TMOs for the efficient conversion of both OER and ORR.

After evaluating the need for new and advanced energy conversion devices in brief, in this chapter, we have first introduced fundamental electrochemical parameters and requirements for the efficient ORR/OER electrocatalyst in basic electrolytes. The bifunctional behavior of the superior non-precious TMOs such as MnO_2 and Co_3O_4 is explained in more detail and its recent developments from the literature are reviewed in extreme depth [13]. The fundamental factors that can be used to enhance the bifunctional activity are summarized, which include morphology, phase, crystal facets, defects, mixed metal oxides and doping of metals on the metal oxide surface. From this chapter, one can easily find the fundamentals behind the formation of highly active, earth abundant transition metal-based catalysts for both OER and ORR processes with increased efficiency. In addition, the exact catalytic behavior of the bifunctional catalyst is not fully understood, which varies from material to material. To pinpoint the active sites and discover the mechanism of the particular catalyst, density field theory (DFT) calculation and in situ study by using specific analytical techniques such as Raman spectroscopy, Mossbauer spectroscopy, X-ray adsorption/diffraction and Fourier transform infrared spectroscopy are highly recommended. Theoretical studies combined with these in situ experiments are able to provide a deeper insight into structures and processes at the atomic level, which together with laboratory experiments could lead to a better understanding of the mechanistic steps involved in the reactions, and these can pave the way for the future directions of OER and ORR for commercial application. Finally, the existing challenges and much-needed effort for improving the catalytic activity of bifunctional catalysts are discussed as the future effort directions.

2. Fundamentals of oxygen electrode reactions (ORR/OER)

To provide good clarity of bifunctional electrocatalysis, first, the basics and fundamentals of oxygen electrode reactions and electrochemical parameters used in the evaluation of the performance of electrocatalysts are discussed in more detail. The intensive knowledge of the above-mentioned field is necessary to design a new non-precious transition metal-based material and to understand its behavior toward ORR/OER.

2.1 Mechanism of ORR

ORR is an electrochemically sluggish reaction with four-electron transfer in a multiple-step process, where the reaction begins with the diffusion of oxygen (O_2) toward the catalyst, adsorption of O_2 molecule on the active sites of the catalyst, transfer of electrons to the adsorbed O_2 molecules and passage through multiple steps. Finally, it is converted to products and desorbed from the active sites of the catalyst [14].

If we look at the ORR more deeply, there are two possible ways to adsorb O_2 molecules on the electrode surface such as associative (bidentate, side-on, two oxygens atoms of O_2 coordinates with the metal) O_2 adsorption and dissociative (monodentate, end-on, one oxygen atom of O_2 coordinates perpendicular to the metal) O_2 adsorption as shown in **Figure 1**.

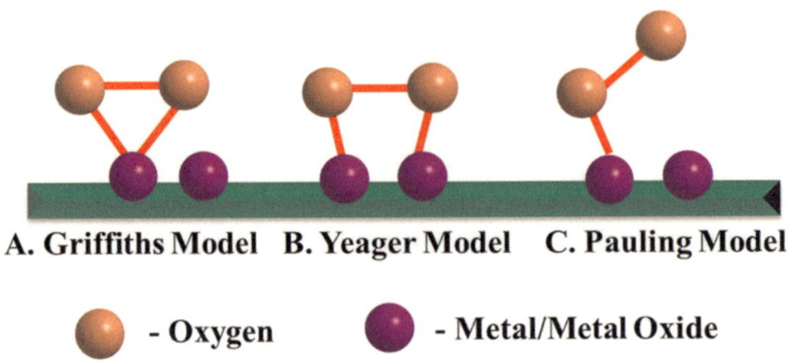

Figure 1.
The schematic representation of the adsorption of O_2 over the metal/oxide surface.

The adsorption mode of oxygen had the foremost influence and is crucially significant for the number of electrons transferred (n) during the ORR process. The associative adsorption mode of oxygen leads to the direct four-electron transfer reduction path and the dissociative adsorption mode of oxygen results in the two-electron transfer reduction path through the peroxide intermediate formation. Typically, noble metals predominantly follow four-electron transfer reduction, and carbon-based materials follow two-electron transfer reduction. On the other hand, non-precious metal oxides can follow either indirect two- or direct four-electron transfer path or both, influenced by the nature of metal oxide and overpotential region. The higher overpotential region offers highly desirable direct four-electron transfer reduction, whereas, in the lower overpotential region, two-electron transfer reduction with peroxide formation occurs predominantly. Therefore, the exact number of electron transfers and the mechanism of ORR on the metal oxide surface still remain unclear. However, the widely accepted mechanism of ORR is fairly complicated and involves oxygen-containing intermediates such as O*, OH* and OOH*. The mechanism of four-electron transfer pathway of ORR on a catalytic metal surface can be described as follows:

$$^{*} + O_2(g) \rightarrow O_2^{*} \tag{1}$$

$$1/2\,O_2^{*} \rightarrow O^{*} \tag{2}$$

$$O^{*} + H^{+}(aq) + e^{-} \rightarrow OH^{*} \tag{3}$$

$$OH^{*} + H^{+}(aq) + e^{-} \rightarrow H_2O^{*} \tag{4}$$

$$H_2O^{*} \rightarrow H_2O \tag{5}$$

Here, * represents the active sites of the catalyst under investigation [13, 15, 16]. The two-electron transfer mechanism of ORR is as follows:

$$O_2^{*} + H^{+}(aq) + e^{-} \rightarrow HOO^{*} \tag{6}$$

$$HOO^{*} + H^{+} + e^{-} \rightarrow H_2O_2 \tag{7}$$

In some cases, OOH* dissociation may be involved on the same catalytic surface as follows:

$$HOO^{*} \rightarrow O^{*} + HO^{*} \tag{8}$$

All possible mechanisms of ORR are shortened here.

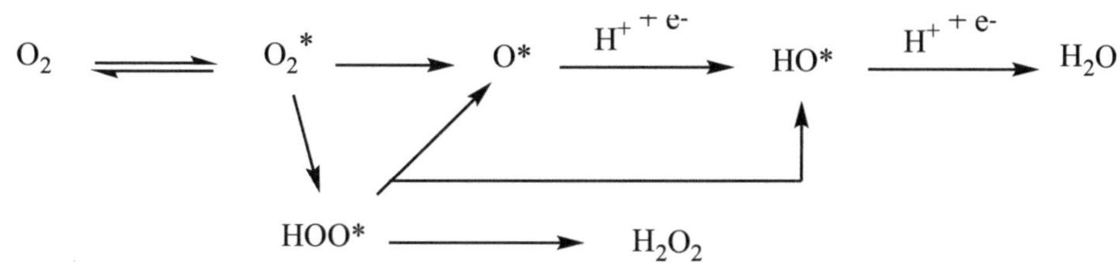

The heterogeneous electrocatalysis (ORR/OER) occurs at the catalyst–electrolyte–reactant triple point, where the binding energy of the reactive species (intermediates/reactant) has been calculated to determine the catalytic performance of the electrocatalyst. The French chemist Paul Sabatier made a principle with respect to the binding energy of the reactive species and catalytic performance of the electrocatalyst in a heterogeneous system called the Sabatier principle (also called the volcano plot). It states that the interaction between the surface of the catalyst and the reactant should be optimum; That is, it should be neither too strong nor too weak [17]. Because too weak interactions result in the failure of making bonds between them, causing no further reaction, and too strong interactions lead to kinetically slow dissociation of the resultant intermediates, the catalytic surface is not accessible by the reactant for further reaction. In the case of ORR, three oxygen-containing species are involved as an intermediate such as O*, OH* and OOH*. The binding energy of these reactive intermediates is the leading factor that decides the activity of the catalyst [18].

2.2 Reaction mechanism of OER

Water oxidation reaction or OER is the reversible reaction of ORR, which occurs at a higher positive overpotential than ORR; to attain the reversibility of each step involved, the ORR should be reversed. Moreover, it is the core electrochemical reaction in fuel cells, MABs and water splitting with their complementary reaction such as ORR and/or HER. It is a multi-step electron transfer reaction; each step requires some energy to overcome the energy barrier, drag the overall kinetics of OER and necessitate a large overpotential [19].

The proposed mechanism of OER on the metal oxide surface is as follows:

$$^* + H_2O \rightarrow {}^*OH + H + +e- \tag{9}$$

$$^*OH \rightarrow {}^*O + H + +e\text{-}or\,2\,{}^*OH \rightarrow {}^*O + * + H_2O \tag{10}$$

$$2\,{}^*O \rightarrow 2^* + O_2\,or\,{}^*O + H_2O \rightarrow {}^* + O_2 + 2H^+ + 2e^- \tag{11}$$

Although OER is relatively complex, it is considerably facile on the metal oxide surface rather than on the bare metal surface, since bare metals are more prone to oxidation at higher positive overpotentials. Moreover, the capability to exist in multiple oxidation states of metal oxides makes it a highly desirable candidate for OER, where the interaction between the metal and oxygen intermediates initiates the bond formation between them by change in the oxidation state of the metal. Therefore, metal oxides are the best choice of catalyst for OER. However, each metal oxide has a different OER mechanism on its surface; even oxides with identical element compositions due to the difference in the surface property of materials originated from the preparation method. The volcano plots relating the required overpotential to attain 1 mA current vs ΔG_O - ΔG_{OH} for OER have been constructed by the study of OER catalysis on a wide variety of metal oxide surfaces [20].

2.3 Bifunctional (ORR/OER) electrocatalysis

A single catalyst employment for the catalysis of two electrochemical reactions (both a reaction and its complementary reaction) is called bifunctional electrocatalysis. The catalyst used in the bifunctional electrocatalysis with the capability to catalyze two electrochemical reactions is termed as bifunctional electrocatalyst. The bifunctional electrocatalyst is primarily used to facilitate the electron transfer from the electrode to the reactant, which promotes oxygen electrode reactions (ORR/OER). A single catalyst employment to catalyze both OER/ORR, that is, a bifunctional catalyst is highly preferable particularly for energy storage and conversion devices, rather than employing two different catalysts separately. Either the catalyst itself or the catalyst decorated on the electrode surface is mostly used as a bifunctional electrocatalyst. The principal role of the electrocatalyst is to adsorb the reactant on its surface and develop the adsorbed reactant/intermediate, thereby promoting the kinetics of charge transfers from the electrode to the reactant. The equilibrium potential of the oxygen electrode reaction (ORR and OER) is 1.23 V versus reversible hydrogen electrode (RHE) [21]. However, the complicated mechanisms and sluggish kinetics of these reactions demand high overpotentials. In order to facilitate the oxygen electrode reaction, the bifunctional electrocatalyst must adhere to certain intrinsic structural features, that is, high inherent bifunctional catalytic activity; enlarged surface area; high electrical conductivity and favorable morphology with exposure of maximum active sites; a high electrochemical surface area; high chemical, electrochemical and mechanical stability; and good contact of electrochemical (electrode electrolyte) interfaces [22]. Although a material with aforementioned properties can perform as good bifunctional electrocatalyst, it has to be evaluated with certain electrochemical parameters for its comparison with other catalysts.

For a fair evaluation of the performance of electrocatalysts, all the electrochemical parameters must be taken into account such as the onset/overpotential, exchange current density, Tafel slope, turnover frequency (TOF), potential gap, number of electron transfers, amount of peroxide formation and electrochemical surface area [23]. Each and every parameter is very crucial and can provide much more insightful information about the catalyst material related to the mechanism of electrochemical reaction. For this reason, the detailed introduction of all the fundamental electrochemical parameters has been discussed elaborately in the following sections.

3. Efforts on the in-built possession of the catalyst materials to enhance the bifunctional activity

After the detailed discussion of the fundamental electrochemical parameters, it is now essential to highlight the significance of in-built possessions of the materials, which decide the properties of the catalyst materials. As stated earlier, a reputable candidate for electrochemical reaction must be recognized with certain properties such as intrinsic activity, density of active sites, conductivity, surface area, wettability (hydrophilicity) and stability. But all these properties are decided by their appropriate association with in-built possessions like phase, morphology, particle size, defects and strain. A slight change in these in-built possessions leads to the huge impact on the properties of catalyst materials. In addition, practically, there is no catalyst material with an adequate quantity of all these mentioned properties. To resolve this inadequateness, many efforts have been devoted on catalyst materials. A catalyst with optimum quantity of all the aforementioned properties can be attained only by a systematic fine-tuning of its in-built possessions such as phase, morphology, crystal facets, particle size, defects and strain. Therefore, to understand the linear thread connection between the in-built possessions, inherent properties and outperforming activity of the electrocatalyst, all these possessions have been discussed in more detail in this section.

3.1 Phase engineering

The smallest repeating domain containing identical (in both distance and space) arrangement of atoms is called a crystal lattice. The extent of a single or a group of crystal lattice can be defined as a phase, a key factor in deciding the properties of the materials. Each phase has its own property and the variation in phases causes dramatic changes in the properties especially in the intrinsic activity, density of active sites, conductivity, wettability and stability, thus having more impact on the bifunctional activity. Controlling materials with desirable phases is an art associated with synthetic methodology. Hence, it is recommended to choose a suitable methodology to prepare the materials with desirable phase selectivity, so that the extent of the activity can be attained [24]. Therefore, it has been concluded that the phase of the material is the primary in-built possession deciding almost all inherent properties, and hence, by choosing the appropriate methodology, the selectively particular phase can be attained.

3.2 Morphology engineering

An inherent property, surface area is often morphology dependent since different morphologies have different surface areas even for identical phases [25]. The electrochemical reaction takes place only at the interface, where surface atoms are in contact with electrolytes rather than interior atoms. Although all surface atoms are in contact with the electrolyte, only their certain portions are capable of catalyzing bifunctional reactions called active sites. The morphology with the exposure of the maximum number of active sites is the most opted for better bifunctional catalysis.

3.3 Defect engineering

Defect engineering is another way to tune the properties of the catalyst material for enhancing the catalytic activity; due to this reason, it is intentionally introduced into the lattice of the materials although it drags the stability of the catalyst materials. The introduction of defects on the lattice can tailor the intrinsic property, especially the electronic configuration, conductivity, chemical reactivity and stability by adjusting electron distribution on the lattice. Defects also lead to the formation of unsaturated atoms on the surface of the catalyst, which are capable of intensively catalyzing the reaction and hence are highly desirable to promote the oxygen electrode reactions [26]. Defects can exist in many forms, namely, structure distortion, oxygen vacancy, cation vacancy, anion vacancy and lattice defects. All the forms of defects lead to the enhancement of the catalytic activity somehow, mainly impacting the conductivity and density of active sites of the catalyst materials [27]. Caution must be taken while improving the activity of the catalyst by defect engineering since a larger extent of defect density leads to a fall in stability due to the increase in the number of dangling bonds in the material. And it is recommended to adopt an effective and feasible way to create defects, rather than following tedious processes and harsh conditions so as to retain the other properties of the catalysts.

3.4 Mixed metal engineering

As stated in the earlier section, an esteemed electrocatalyst must be effectual in its inherent properties that are intrinsic activity, density of active sites, conductivity, surface area, wettability and stability. But practically, it is rare that a single material has all these mandatory properties for effective catalysis especially for bifunctional (ORR/OER) catalysis; hence doping the foreign metal into the lattice of the host material is a fruitful way to implant additional properties which were absent or inadequate in the host material. Among the various properties, electrical conductivity and catalytic activity are the most essential properties of the non-noble catalyst that are mainly considered inadequate and need to be improved as far as possible. These two properties could be significantly improved by two tactics: one is the doping of an appropriate metal into the host material and another is the formation of a core-shell-type hybrid structure, where the core part must be highly conductive and the shell part should be capable of catalyzing the oxygen electrode reactions [28, 29]. The enriched activity of the resultant hybrid catalyst is mainly due to the synergistic effect of both the guest and host materials. Hence, the hybrid catalyst has the advantage to utilize certain properties from one part and remaining properties from other parts of the catalyst to exhibit the improved performance of the oxygen electrode reaction. Therefore, it has been understood that mixed metal engineering is a profitable way for selective embedding of certain properties into the electrocatalyst for boosted oxygen electrocatalysis.

3.5 Catalyst/conductive substrate engineering

The notable drawbacks of non-noble metal/oxide-based electrocatalysts are their poor electrical conductivity and their agglomeration during the catalysis process, which limit their bifunctional catalytic performance in oxygen electrocatalysis. The succeeding forward step to resolve these issues is the decoration of catalyst materials over the highly conductive substrate, which simultaneously limits the agglomeration and improves the electrical conductivity. Carbon-based materials are promising candidates that serve as a substrate by providing a facile electron transfer path and flat facets to adhere catalytically active non-noble metals/oxides. Moreover, non-noble metal/oxide-based electrocatalysts are efficient candidates for OER and not for ORR, which is another remarkable drawback; meanwhile, carbon-based materials are demonstrated to be effective catalysts for ORR. When combined, these two materials can effectively catalyze both the oxygen electrode reactions without any interruption due to synergistic effect, where the ORR part is taken care of by carbon-based materials and the OER part by catalytically active non-noble metal/oxide. So, again hybrid material (catalyst/carbon substrate) engineering is the solution, which resolves the ORR activity issue along with electrical conductivity and agglomeration issues. But carbon materials are prone to oxidize in highly positive OER overpotential regions, causing thermodynamic instability as another issue. Hence, it is determined that the hybrid material (catalyst/carbon substrate) engineering is an effective strategy to proliferate electrical conductivity and ORR activity. However, proper selection of carbon-based substrates (stable at higher positive overpotentials) and feasible way of incorporation lead to better catalytic performance and durability.

Therefore, in this section, all the in-built possessions of the catalyst materials and their impact on their properties have been elaborately discussed and the possible ways of fine-tuning them for the improvement of bifunctional catalysis are also suggested. This collective knowledge ignites the interest of the research community to formulate efficient bifunctional catalysts and take them to the next level.

4. Detailed chemistry and electrochemistry of MnO_2 and Co_3O_4

The oxides of manganese and cobalt still remain prominent alternatives for precious materials and have been extensively investigated in many growing areas such as lithium-ion batteries, supercapacitors and electrochemical sensors, as well as in electrocatalysis, due to their appreciable unique properties, affordability and environmental friendliness. So, before moving to recent explorations of these materials, the detailed properties and their influence in electrocatalysis have been discussed here more elaborately one by one.

4.1 Chemistry and electrochcmistry of MnO_2

Manganese oxide (MnO_2) is a two-dimensional (2D) layered semiconducting material. The outermost electronic configuration of Mn^{+2} is $3d^5\ 4s^0$, where the d orbital is in the half filled (unsaturated) state, which makes it more susceptible to loss or gain of electrons; this is the triggering property in the process of catalysis. So, it can behave as a good catalyst especially in the area of electrocatalysis. The major advantage with MnO_2 is the multivalent nature of manganese and its greater structural flexibility, due to which it can exist in more than 20 crystallographic forms like α-, β-, γ-, δ-, ε-, λ-types, amorphous MnO_2 (AMO) and others [30, 31]. All these crystallographic forms are formed by the basic unit MnO_6, where oxygen atoms are present at the top of eight surface body angles and manganese is located at the center. These crystal structures diverge from one another by way of linkage of their basic octahedral MnO_6 unit, which results in each crystal structure to have distinct properties. Along with many forms of crystal structures, MnO_2 is associated with many other advantages such as their capability to form many polymorphs, natural abundance, environmental compatibility, low cost and non-toxicity. There are many polymorphs of MnO_2: tunnel (1D), layered (2D) and spinel (3D) structures, prepared by dedication and great efforts. Nevertheless, most of the polymorphs of MnO_2 have an open tunnel structure, which can accommodate any of the small guest ions such as K^+, Ag^+, Na^+, Mg^{2+} etc. Moreover, in MnO_2, Mn centers are in mixed oxidation states (+3 and +4), where the charge neutrality is maintained by the assist of guest cations entrapped in the tunnels. The bifunctional activity of MnO_2 strongly depends on their crystal structure and morphology as well as their intrinsic properties. The bifunctional activity of MnO_2 with respect to the crystal structure follows the order α- > AMO > β- > δ-MnO_2 [32]. The superior OER activity of α-MnO_2 is attributed to the presence of mixed oxidation states of manganese (average oxidation state = 3.7), their capability to exist with enormous amounts of di-μ-oxo bridges as protonation sites and their suitable tunnel sizes ($0.46 \times 0.46\ nm^2$) affording high accessibility to electrolytes (reactant) as well as efficient charge transport. The excellent ORR activity (4 e^- transfer) of the same crystal structure is due to its expedient O_2 adsorption capability and enriched amount of higher oxidation state (+3, +4) of manganese on the surface, whereas other crystal structures follow 2 e^- transfer reduction pathways. Amorphous MnO_2 is the next better bifunctional catalyst because of the presence of excess oxygen defects and randomness. Besides, α-MnO_2 contains 2x2 tunnel structures along the c-axis made of double chain of the basic MnO_6 unit. The higher activity of the α-form is also supported by this tunnel and layered structures, having more number of edges and corner sharing of the MnO_6 unit. In addition, it is worth noting that the α-MnO_2 possesses higher activity per cost than Pt/C [33].

Although there is extensive information about MnO_2, the key factors influencing the OER and ORR activities are not well defined due to the lack of straightforward structure-related electrocatalytic activities. Many of the crystal structures of MnO_2 are still poorly understood. However, MnO_2 with various metal valences usually revealed different morphologies, which complicates the underlying relationship between metal valence and activity.

4.2 Chemistry and electrochemistry of Co_3O_4

The next superior electrocatalyst with competitive activity for bifunctional reaction is cobalt oxide (Co_3O_4) and is an extensive studied material especially for OER due to its excellent redox behavior, inexpensiveness, high abundance and great stability in alkaline medium. It is a spinel oxide, where oxygen atoms are arranged in a cubic close packed (CCP) matrix composed of two Co^{3+} and one Co^{2+} cations located in the octahedral (Oh) and tetrahedral (Td) lattice sites, respectively [34, 35]. The catalytic performance of spinel Co_3O_4 is deeply rooted in two factors such as exposed crystal facets and population of Td Co^{2+} and Oh Co^{3+} on it. The (110) facet of Co_3O_4 exhibits enhanced activity compared to other facets due to the existence of substantially higher density of the dangling bond and catalytically active Co^{3+} sites [36]. But the higher surface energy and existence of higher density of dangling bonds of this facet results in less possibility of the formation of this facet. The bifunctional activity of Co_3O_4 is oxidation state dependent; the higher occurrence of Co^{3+} in the octahedral environment is responsible for the superior performance of OER, whereas Co^{2+} in the tetrahedral environment is capable of releasing electrons under applied potential, which results in the formation of CoOOH, μ-OO peroxide intermediates over the surface of Co_3O_4, which is capable of acting as an active site in ORR. Although there are many efforts on Co_3O_4, the intrinsically poor electrical conductivity and relatively smaller surface area of the material still remain inadequate, which significantly drag the overall activity.

5. Recent exploration in Co_3O_4 and MnO_2

Transition metal oxides like manganese oxides α-MnO_2 [37, 38], β-MnO_2 [39], perovskite (i.e., $LaNiO_3$, $LaCoO_3$) [40] as well as spinel-based structures (Co_3O_4 and $NiCo_2O_4$) with multiple valences, serve as the most attractive non-precious metal catalysts for bifunctional OER/ORR electrocatalysis especially for the alkaline medium. This chapter particularly emphasizes the recent developments in metal oxides of manganese oxides and cobalt oxides, which are simple in composition yet very effective in promoting oxygen electrode reactions. The nature of active sites and their identification along with corresponding ORR/OER activities will be correlated with respect to their oxidation states, crystal structure, bulk structure and morphology with required experimental and computational analyses; all those concerns have been reviewed from the recent 10 years of literature. In addition to that, contrast results observed in those particular materials are discussed elaborately and possible strategies to further improve the activity of the catalyst have also been suggested at the end.

5.1 Recent exploration in MnO_2

Manganese oxides are one of the most widely investigated materials in bifunctional electrocatalysis and currently employed in commercial Zn air and alkaline batteries. The inspiration from the biological catalyst $CaMn_4O_x$ acts as an oxygen evolving centre in Photosystem II; Gorlin et al. developed a new catalyst α-Mn_2O_3 where Mn is in the +3 oxidation state, showing excellent bifunctional activity almost similar to the activities of precious catalysts like Pt, Ir and Ru [41]. To improve the activity of the catalyst, it is essential to understand the changes on the surface of the material during the process of OER and ORR. With this mind in their another study, a powerful in situ X-ray absorption spectroscopy technique was used to analyze the chemical nature of the surface during the process of OER and ORR [41]. Such an attempt reveals that the exposure of catalyst in ORR (0.7 V) and OER (1.8 V) potential causes the distortion of the $Mn_3^{II,III,III}$ O_4 phase and maximum (80%) oxidation of the catalytic surface to form a mixed $Mn^{III,IV}$ oxide. They also confirmed that the observed result is irrespective of the film thickness of the catalyst in both OER and ORR potential. At OER potential, $Mn^{III,IV}$ oxide is more dominant on the catalytic surface due to the electrochemical oxidation, indicating that it is the phase responsible for the observed OER activity rather than the $Mn_3^{II,III,III}$ O_4 phase [42]. Such an oxidation state changes upon the exposure to OER and ORR potential on α-Mn_2O_3, observed by the same group with the aid of ex situ X-ray photoelectron spectroscopy technique in their earlier studies [43, 44] and by Marcel Risch via in situ soft X-ray absorption spectroscopy study [45].

Moreover, the effect of surface manganese valence of MnO_2 on ORR activity is studied by Tang et al. [12]; from their observation, it has been understood that the oxidation state of the catalytic surface plays a crucial role in facilitating bi-functional catalysis. The main reason for the activity of manganese oxides is the presence of surface Mn^{+3} ion having Mn-O-Mn bonds, which are found to be more active for OER and are not found in the species having Mn^{+2}/Mn^{4+}. These labile Mn-O bonds allow the formation of surface Mn-OH_2 species and favor the cleavage of Mn-O_2 bonds, which increases the overall activity of the catalyst. Besides, in its electronic configuration, Mn(III) having one e_g^1 electron leads to the lattice distortion owing to the Jahn-Teller effect and contributing to the structural flexibility, promoting the catalytic activity of the material [46, 47].

The importance of surface facets on bifunctional activity has been discussed in another work [48], where different nanostructures of MnO such as nano-flowers, nanoparticles and nano-polypods are prepared with the exposure of different crystal facets. The detailed investigation of this study demonstrated that the maximum exposure of (100) facets of MnO nano-polypods largely promote the electrocatalytic activity in both OER and ORR compared to others facets. Therefore, it has been understood that (100) is the superior active facet of MnO_2 for both OER and ORR.

As mentioned earlier, MnO_2 can exist in large numbers of crystallographic forms, so it is essential to address the suitable crystal structure for encouraging electrolysis.

To resolve this issue, Meng et al. examined the influence of crystallographic structures of MnO_2 such as α-, β-, δ-MnO_2 and amorphous (AMO) MnO_2 on bifunctional activity. During the formation of various crystal structures of MnO_2, changes in morphologies such as nanoflakes, nanowires and nanoparticles were also obtained. The obtained results reveal that the bifunctional activity of MnO_2 is strongly dependent on the structures and follows the order $\alpha > AMO > \beta > \delta$-$MnO_2$ [32]. There is another study related to the crystal structure influence in water oxidation, which has been investigated under various conditions. In this study, the authors have prepared nine various MnO_x-based catalysts, that is, α_2-, β-, δ_2- and γ-MnO_2, Mn_3O_4, Mn_2O_3, L-MnO_2, γ-MnOOH, and R-MnO_2 and their water oxidation efficiency were analyzed and α-MnO_2 was found in alkali media as a superior one among all catalysts [49]. These studies confirm the influence of crystallographic orientation on catalytic activity, which is majorly controlled by the morphology of the catalyst. Moreover, Debart et al. explained the highest charge storage capacity, that is, 3000 mA g^{-1} of α-MnO_2, among other crystal structures such as β-MnO_2, γ-MnO_2, λ-MnO_2, Mn_2O_3 and Mn_3O_4 through the study of rechargeable lithium-air batteries [50]. They have also showed the surpassed catalytic activity of α- and β-MnO_2 nanowire compared to the bulk MnO_2, which is mainly due to the higher surface area and crystal structure. This further confirms the importance of the crystal structure and the nanoscale morphology of the electrocatalyst in bifunctional activity. Our group result also supports the above-stated information that α-MnO_2 nanowire turns out as the more preferable morphology and the crystal structure of MnO_2 for superior bifunctional activity over the other forms [37].

Another strategy to enhance ORR activity of MnO_2 is introduction of native oxygen defects without employment of any foreign additives, where the effort is devoted to identify the effect of oxygen nonstoichiometric on ORR activity in alkaline electrolytes. The study has introduced oxygen vacancy into the thermodynamically stable high purity rutile β-MnO_2 by a simple heat treatment method in argon atmosphere. And it is found that the oxygen defect bearing MnO_2 requires more positive overpotential, yields lower amount of peroxide in ORR and also facilitates the kinetics of OER [51].

In MnO_2, especially when we intended to tune metal centre valences (oxidation state), it lead to unavoidable changes in morphology along with changes in the oxidation state. Such circumstances usually add further complications to understanding of the actual relation between the metal valence and electrochemical activity. Although it is reported that MnOOH species is superiorly ORR active among manganese oxides [52], this study did not include the influence of morphology of the catalyst on activity. The influence of morphology on ORR activity is explained by two familiar examples, where nanoparticles and nanowires are the morphology. In MnO_2 nanoparticles, the activity increases in the order β- < λ- < γ- < α- ≈ δ-MnO_2 [53], whereas in MnO_2 nanowires, the activity follows the order λ- < β- < α-MnO_2 [54]. These results provide strong evidence for the influence of nanoscale morphology on ORR activity.

Although manganese oxide is a superior non-precious bifunctional catalyst, still it is unable to surpass the activity of precious metal catalysts due to its poor conductivity. The well-known approach to improve the conductivity of the material is decoration of metal oxides over carbon-based substrates. However, in the case of bifunctional electrocatalysis, the carbon substrate is not a preferable one due to the peroxide production and self-oxidation on OER condition. To resolve this issue, Ng et al. developed manganese oxide on a stainless-steel substrate (MnO_x-SS) through electrodeposition followed by calcination at 480°C. Calcination causes the phase conversion of MnO_x to Mn_2O_3 on stainless steel. Further, this MnO_x-SS material employed as an O_2 electrode in unitized regenerative fuel produces round-trip efficiencies of 42–45% at 20 mA cm^{-2} over 10 cycles that is excellent catalytic activity and durability in both OER and ORR even compared to the precious catalyst Pt/C-SS. The interaction between the stainless-steel substrate and catalyst significantly enhances the catalytic activity due to synergetic effect and prevents the agglomeration of catalysts by providing an appropriate platform [55].

The next approach to improve the activity is expanding the surface area and active sites by introducing porosity. The activity of the catalyst is directly proportional to the surface area due to the rise in the density of active sites. But there are many ways to introduce porosity, like the template-assisted method, surfactant-assisted micelle/inverse micelle sol-gel method, etc. However, each method is associated with its own merits and demerits. In the case of template-assisted process, although uniformity in the pore size and stability was achieved, the template had to be etched post synthesis, which added complexity to the method and was time consuming. However, although the surfactant-assisted method is a simple and cost-effective approach, there is no assurance for uniformity in the pore size.

In this view, Kuo et al. developed a highly effective mesoporous MnO_x material with a crystal structure of Mn_2O_3 via a simple inverse micelle template approach. This adopted methodology has the advantage of being a single-step synthesis, does not require any post-synthesis treatment, can be scalable and most importantly controls the formation of Mn_2O_3 with enriched Mn^{3+}. Then, the prepared mesoporous Mn_2O_3 material demonstrated superior catalyst property in both electrochemical water oxidation and photochemical processes with the highest TOF value [56].

Some other attempts have also been made to tune the Mn oxidation state, that is, the introduction of foreign elements into manganese oxide. It was found that the incorporation of foreign elements like gold nanoparticles [57, 58], Ca^{2+} [59] and cesium [60] into manganese oxide promotes the formation of Mn^{+3} ions, which significantly enhances the ORR activity.

Recently, Kang et al. have reported the influence of the interlayer distance of MnO_2 on OER activity by systematically varying the interlayer distance with the aid of intercalation of dissimilar-sized alkaline cations such as Li^+, Na^+, K^+, Rb^+ and Cs^+ between the layers and the observed interlayer distance values are 0.5, 0.6, 0.9, 1.05 and 1.5 nm, respectively. At the end, they concluded that the Cs^+ ion intercalated MnO_2 possesses a larger interlayer distance along with the accommodation of larger quantities of water molecules as a result of superiority toward OER activity, whereas Li^+ ion intercalated MnO_2 is inactive for OER due to their smaller interlayer distance and dehydrated structure [61]. In contrast to the aforementioned report, Kosasang et al. reported that the Li^+ intercalated MnO_2 is a superior candidate for both OER and ORR, where they intercalate the alkali cations such as Li, Na, K, Rb and Cs between the layers of MnO_2 and their bifunctional activity was examined. They have further ensured the observed result by the DFT calculations [62].

From the above revealed mechanistic insights, it is understood that developing an efficient method to facilitate both ORR and OER on a single active site is not an easy task due to the drastic difference in their overpotential as well as the rate-determining step (RDS). In fact, an efficient ORR active site naturally renders a poor OER activity and vice versa, making it challenging to maintain a proper balance for reversible OER/ORR catalysis on a single electrode surface. Due to this reason, typically employed bifunctional catalysts are a combination of precious metals and their oxides like Pt, Ru/RuO_2 and Ir/IrO_2, where ORR and OER reactions are taken care of by pure metal and metal oxides, respectively.

Therefore, a probable way to simultaneously catalyze both OER and ORR is to design bifunctional catalysts inherited with different active sites for ORR and OER separately. That can be achieved by the deliberate engineering of transition metal oxides, particularly their oxidation state, crystal structure, exposure of crystal plane and morphology are the efficient routes to develop an efficient inexpensive bifunctional electrocatalyst for oxygen electrode reactions.

5.2 Recent exploration in Co_3O_4 to enhance bifunctional catalysis

Another simple and most investigated metal oxide for bifunctional catalysis is Co_3O_4 due to its superior activity and durability. As mentioned earlier, Co_3O_4 is a spinel-type metal oxide, where Co^{2+} and Co^{3+} occupy the Td and Oh sites, respectively. The Co^{2+} tetrahedral sites are the active sites for ORR, and the Co^{3+} octahedral sites are the active sites for OER. The optimal amount of both ions would lead to overall bifunctional activity and will minimize the potential difference between the two reactions.

5.3 Effects of particle size and surface area

Esswein et al. elucidated the size dependence activity of Co_3O_4 crystallites on electrocatalytic OER in an alkaline medium [63]. They prepared cubic Co_3O_4 nanoparticle materials with the average size of 5.9, 21.1 and 46.9 nm. Then, the prepared materials were loaded onto a Ni foam support to evaluate the OER performance with a constant loading amount of 1 mg cm^{-2}. They attained 10 mA cm^{-2} current density at 328, 363 and 382 mV for small (5.9 nm), medium (21.1 nm) and large (46.9 nm) sized Co_3O_4 particles, respectively. The activities were correlated with the surface area of the isolated Co_3O_4 particles.

Menezes et al. established a method to produce nanochains of cobalt oxide (Co_3O_4) via low-temperature degradation of cobalt oxalate dehydrate [64]. In fact, they were able to display exceptional OER performance at low overpotentials in both basic and neutral media with the as-prepared Co_3O_4 nanochains. They also additionally prepared nanostructured Co_3O_4 materials by the solvothermal method and compared the activity of commercial Co_3O_4 of various morphologies. Remarkably, the ORR performance of carbon-supported Co_3O_4 nanochains displays remarkable activity compared to that of Pt. Surprisingly, they found that even though the nanochain Co_3O_4 prepared by the reverse micelle route possesses a lower BET surface area (12 $m^2\,g^{-1}$) than solvothermal Co_3O_4 (18 $m^2\,g^{-1}$) and commercial Co_3O_4 (49.4 $m^2\,g^{-1}$), it displays a low overpotential toward OER in both alkaline and neutral media.

5.4 Effect of mesoporosity and morphology

Sa et al. reported ordered mesoporous Co_3O_4 spinels with a gyroid mesostructure, obtained by the assistance of a KIT-6 mesoporous silica template through the nanocasting method, which emerged as one of the best methods to obtain ordered mesoporous structured materials. Particularly, silica supports such as KIT-6, which possess a double gyroid mesostructure, can deliver large active sites than that of analogous materials with a 2D hexagonal structure [65]. Furthermore, the ordered mesoporous Co_3O_4 spinels template from KIT-6, with a highly interconnected network structure, is expected to show enhanced stability under harsh catalytic or electrocatalytic reaction conditions.

5.5 Effect of the nature of hydrous oxide

Zhan et al. demonstrated the bifunctional activity of hydrothermally prepared Co $(OH)_2$ hexagonal nanoplates and cobalt oxides (CoO and Co_3O_4) in basic medium, where the hydroxide of cobalt ($Co(OH)_2$) displayed a superior activity than the oxides of cobalt (CoO and Co_3O_4) [66]. The bifunctional catalysis ability mostly shown by the OER/ORR potential difference (ΔE) could be achieved as lower as 0.87 V (RHE), comparable to that of metal-based catalysts, when the $Co(OH)_2$ nanoplates were anchored on N-doped reduced graphene oxide. In addition, both Co_3O_4 and CoO possess a lower n value of 2.6–2.7 for ORR, which testified that pristine cobalt oxide catalysts predominantly follow a two-electron pathway of ORR. $Co(OH)_2$ was therefore highly ORR active than other cobalt catalysts. Similarly, in the case of ORR, Co $(OH)_2$ was most OER active than Co_3O_4 and CoO even though it has a lower number of cobalt site and surface area. Therefore, it is understood that the hydroxide of cobalt is a better bifunctional catalyst than the oxides of cobalt.

5.6 Existing challenges and future directions

After the detailed review of each stairway of non-precious metal oxide (i.e., MnO_2, Co_3O_4), it is absolutely necessary to address the existing challenges and future working directions that could be highly supportive for the research community for finding more insightful information on the catalysts and their further development to enhance the bifunctional activity.

The exact mechanisms and underlying fundamental process of the oxygen electrode reactions are still unclear; they vary from material to material. There was a drastic difference at the electrode electrolyte interface of each electrocatalyst, and it possessed different binding energies with the reactant species. So it is highly recommended to adopt sophisticated in situ spectroscopic techniques to study the interface during the electrochemical reaction so that its exact mechanism could be identified. Along with this, the theoretic prediction (DFT calculation) of interface especially for the possible way of reactant species adsorption and desorption has to be developed. There is another study called post-experimental analysis that is the analysis of the catalyst after the employment of electrochemical reaction for a certain period. The deeper analysis and its comparison with its fresh nature provides more insightful information about the moiety, which is the exact reason for the catalytic activity. And finally, the collective analysis and interpretation of all these, that is, in situ spectroscopic study and theoretical prediction of the interface and post-experimental analysis, will lead to a loophole for the identification of the exact mechanism of the oxygen electrode reactions for the particular catalyst. The understanding of underlying mechanisms is the heart of knowledge, which is the key to trigger the researcher for the establishment of marvelous candidates for the oxygen electrode reactions.

Now the question is why the pinpoint mechanism of oxygen electrode reactions is not yet strongly declared even after knowing the route to identify it. The reason is that the existing issues in the aforementioned sequence of the route are (1) insufficient knowledge in theoretical prediction and (2) lack of the appropriate in situ spectroscopic technique and its integration into the electrochemical setup. Hence, it has been suggested that more efforts have to be devoted in the field of experimental characterization, and theoretical study leads to better fundamental understanding of electrochemical reactions. In addition, the integration of various characterization techniques into the electrochemical setup is also recommended to track the reaction.

If we look at the Co_3O_4 catalyst, it has a multi-oxidation state (Co^{2+} and Co^{3+}) compound, where there is still no clear-cut idea about the optimum level of Co^{2+}/Co^{3+} ratio for better performance of bifunctional activity. Moreover, there are some reports of Co_3O_4 highlighting that Co^{2+} and Co^{3+} are responsible for ORR and ORR, respectively. In recent years, some researchers have reported the exact opposite trend of responsibility in bifunctional reaction (ORR/ORR) with strong evidence, which creates more puzzles. In addition, Co_3O_4 is a good catalyst for OER but not for ORR; the intentional tuning on it for the improvement of ORR activity affects the OER performance. Therefore, it is recommended that a more in-depth study of this material has to be explored well rather than its advanced study.

In the case of MnO_2, it is a good candidate for ORR (can follow the four-electron transfer mechanism), but it is not so for OER. As stated earlier, the intentional tuning of one side of the reaction drastically affects the other side of the reaction. Hence, such a way of fine tuning the material for the catalytic enhancement of OER activity without affecting its ORR performance is highly desirable. Therefore, the existing challenges strongly direct the researcher to explore things in the area of understanding the mechanism of ORR/OER, tuning the composition of the existing atom of the catalyst to obtain optimum composition and development of novel synthetic approaches for selectively attaining certain properties of the catalyst.

6. Conclusions

A rechargeable metal-air battery with the aid of high theoretical energy output will be of prime competence for the carbon-based energy sources that we rely on hugely as of now. Metal-air batteries lack in commercial scale due to the limited kinetics behind OER and HER. The search for bi-functional catalysts that are earth abundant and kinetically facile is important to meet energy crisis. Previously, many breakthrough works have been done with superior activities, and among them, MnO_2 and Co_3O_4 have attracted researchers with their kinetic activities and structure-activity relationships. In this chapter, the important parameters judging the activities of both Co_3O_4 and MnO_2 such as phase and morphology engineering, defects and crystal facets engineering, strains and mixed metals oxide formations have been portrayed with proper investigations along with the fundamentals of water oxidation and reduction reactions. Moreover, the electrochemistry behind MnO_2 and Co_3O_4 has been investigated for the better understanding of catalyst reactions and mechanisms. This will be highly useful for developing and designing mixed oxides for the increased kinetics of rechargeable metal-air batteries with commercial production. This chapter urges the importance of MnO_2 and Co_3O_4 as better bi-functional catalysts with the introduction of fundamental electrochemistry, structure-activity relationship and finally the future directions in the field of metal-air batteries.

Acknowledgements

The authors thank the funding support from CSIR through the 12th FYP project MULTIFUN (CSC0101). S.V. and V.D. thank the UGC, New Delhi, for financial support through the SRF (21/06/2015(i)EU-V, Roll no 123184) and Rajiv Gandhi National Fellowship (RGNF, Grant No: F1-17.1/2016-17/RGNF-2015-17-SCTAM-21830), respectively. K.S. thanks, the financial support through the CSIR-HRDG, New Delhi, India, for SRF (Grant no: 31/20(159)2 k17).

Conflict of interest

The authors declare that there is no conflict of interest.

Author details

Selvaraj Venkateshwaran[1,2], Karuppiah Selvakumar[1], Velu Duraisamy[1,2] and Sakkarapalayam Murugesan Senthil Kumar[1,2]*

1 Electroorganic and Materials Electrochemistry (EME) Division, CSIR-Central Electrochemical Research Institute (CECRI), Karaikudi, Tamil Nadu, India

2 Academy of Scientific and Innovative Research (AcSIR), Ghaziabad, India

*Address all correspondence to: smsk_2k@yahoo.com

References

[1] Kim M, Wang C, Earnshaw J, Park T, Amirilian N, Ashok A, et al. Co, Fe and N co-doped 1D assembly of hollow carbon nanoboxes for high-performance supercapacitors. Journal of Materials Chemistry A. 2022;**2022**:24056-24063. DOI: 10.1039/d2ta06950d

[2] Sekar S, Preethi V, Saravanan S, Kim DY, Lee S. Excellent photocatalytic performances of Co3O4–AC nanocomposites for H2 production via wastewater splitting. Chemosphere. 2022;**286**:131823. DOI: 10.1016/j.chemosphere.2021.131823

[3] Fan X-Z, Du X, Pang Q-Q, Zhang S, Liu Z-Y, Yue X-Z. In situ construction of bifunctional N-doped carbon-anchored Co nanoparticles for OER and ORR. ACS Applied Materials & Interfaces. 2022;**14**: 8549-8556. DOI: 10.1021/acsami.1c21445

[4] Duraisamy V, Senthil Kumar SM. N and P dual heteroatom doped mesoporous hollow carbon as an efficient oxygen reduction reaction catalyst in alkaline electrolyte. International Journal of Hydrogen Energy. 2022;**47**:17992-18006. DOI: 10.1016/j.ijhydene.2022.03.284

[5] Duraisamy V, Krishnan R, Senthil Kumar SM. N-doped hollow mesoporous carbon nanospheres for oxygen reduction reaction in alkaline media. ACS Applied Nano Materials. 2020;**3**:8875-8887. DOI: 10.1021/acsanm.0c01639

[6] Ďurovič M, Hnát J, Bouzek K. Electrocatalysts for the hydrogen evolution reaction in alkaline and neutral media. A comparative review. Journal of Power Sources. 2021;**2021**:493. DOI: 10.1016/j.jpowsour.2021.229708

[7] Zhang T, Zhang B, Peng Q, Zhou J, Sun Z. Mo 2 B 2 MBene-supported single-atom catalysts as bifunctional HER/OER and OER/ORR electrocatalysts. Journal of Materials Chemistry A. 2021;**9**:433-441. DOI: 10.1039/D0TA08630D

[8] Cheng W, Lu XF, Luan D, Lou XW. NiMn-based bimetal–organic framework nanosheets supported on multi-channel carbon fibers for efficient oxygen electrocatalysis. Angewandte Chemie - International Edition. 2020;**59**: 18234-18239. DOI: 10.1002/anie.202008129

[9] Duraisamy V, Arumugam N, Almansour AI, Wang Y, Liu TX, Kumar SMS. In situ decoration of Co_3O_4 on N-doped hollow carbon sphere as an effective bifunctional oxygen electrocatalyst for oxygen evolution and oxygen reduction reactions. Colloids and Surfaces A: Physicochemical and Engineering Aspects. 2023;**656**:130347. DOI: 10.1016/j.colsurfa.2022.130347

[10] Selvakumar K, Duraisamy V, Venkateshwaran S, Arumugam N, Almansour AI, Wang Y, et al. Development of α-MnO_2 nanowire with Ni- and (Ni, Co)-cation doping as an efficient Bifunctional oxygen evolution and oxygen reduction reaction catalyst. ChemElectroChem. 2022;**2022**:9. DOI: 10.1002/celc.202101303

[11] He X, Yin F, Li Y, Wang H, Chen J, Wang Y, et al. NiMnO3/NiMn2O4 oxides synthesized via the aid of pollen: Ilmenite/spinel hybrid nanoparticles for highly efficient bifunctional oxygen electrocatalysis. ACS Applied Materials and Interfaces. 2016;**8**:26740-26757. DOI: 10.1021/acsami.6b08101

[12] Tang Q, Jiang L, Liu J, Wang S, Sun G. Effect of surface manganese valence of manganese oxides on the

activity of the oxygen reduction reaction in alkaline media. ACS Catalysis. 2014;**4**: 457-463. DOI: 10.1021/cs400938s

[13] Li S, Hao X, Abudula A, Guan G. Nanostructured Co-based bifunctional electrocatalysts for energy conversion and storage: Current status and perspectives. Journal of Materials Chemistry A. 2019;**7**:18674-18707. DOI: 10.1039/c9ta04949e

[14] Sarapuu A, Kibena-Põldsepp E, Borghei M, Tammeveski K. Electrocatalysis of oxygen reduction on heteroatom-doped nanocarbons and transition metal-nitrogen-carbon catalysts for alkaline membrane fuel cells. Journal of Materials Chemistry A. 2018;**6**: 776-804. DOI: 10.1039/c7ta08690c

[15] Xia W, Mahmood A, Liang Z, Zou R, Guo S. Earth-abundant nanomaterials for oxygen reduction. Angewandte Chemie - International Edition. 2016;**55**:2650-2676. DOI: 10.1002/anie.201504830

[16] Shao M, Chang Q, Dodelet JP, Chenitz R. Recent advances in electrocatalysts for oxygen reduction reaction. Chemical Reviews. 2016;**116**: 3594-3657. DOI: 10.1021/acs.chemrev.5b00462

[17] Ooka H, Huang J, Exner KS. The Sabatier principle in electrocatalysis: Basics, limitations, and extensions. Frontiers in Energy Research. 2021;**9**: 1-20. DOI: 10.3389/fenrg.2021.654460

[18] Raj CR, Samanta A, Noh SH, Mondal S, Okajima T, Ohsaka T. Emerging new generation electrocatalysts for the oxygen reduction reaction. Journal of Materials Chemistry A. 2016;**4**:11156-11178. DOI: 10.1039/c6ta03300h

[19] Anantharaj S, Ede SR, Sakthikumar K, Karthick K, Mishra S, Kundu S. Recent trends and perspectives in electrochemical water splitting with an emphasis on sulfide, selenide, and phosphide catalysts of Fe, Co, and Ni: A review. ACS Catalysis. 2016;**6**: 8069-8097. DOI: 10.1021/acscatal.6b02479

[20] McCrory CCL, Jung S, Ferrer IM, Chatman SM, Peters JC, Jaramillo TF. Benchmarking hydrogen evolving reaction and oxygen evolving reaction electrocatalysts for solar water splitting devices. Journal of the American Chemical Society. 2015;**137**:4347-4357. DOI: 10.1021/ja510442p

[21] Kulkarni A, Siahrostami S, Patel A, Nørskov JK. Understanding catalytic activity trends in the oxygen reduction reaction. Chemical Reviews. 2018;**118**: 2302-2312. DOI: 10.1021/acs.chemrev.7b00488

[22] Liu S, Wang Z, Zhou S, Yu F, Yu M, Chiang C, et al. Metal–organic-framework-derived hybrid carbon nanocages as a bifunctional electrocatalyst for oxygen reduction and evolution. Advanced Materials. 2017;**29**: 1700874. DOI: 10.1002/adma.201700874

[23] Vazhayil A, Vazhayal L, Thomas J, Ashok CS, Thomas N. A comprehensive review on the recent developments in transition metal-based electrocatalysts for oxygen evolution reaction. Applied Surface Science Advances. 2021;**6**: 100184. DOI: 10.1016/j.apsadv.2021.100184

[24] Worku AK, Ayele DW, Habtu NG, Yemata TA. Engineering Co_3O_4/MnO_2 nanocomposite materials for oxygen reduction electrocatalysis. Heliyon. 2021; **2021**:7. DOI: 10.1016/j.heliyon.2021.e08076

[25] Yin Z, Allado K, Sheardy AT, Ji Z, Arvapalli D, Liu M, et al. Mingled MnO_2

and Co_3O_4 binary nanostructures on well-aligned electrospun carbon nanofibers for nonenzymatic glucose oxidation and sensing. Crystal Growth and Design. 2021;**21**:1527-1539. DOI: 10.1021/acs.cgd.0c01299

[26] He J, Zou Y, Wang S. Defect engineering on electrocatalysts for gas-evolving reactions. Dalton Transactions. 2019;**48**:15-20. DOI: 10.1039/C8DT04026E

[27] Jia Y, Chen J, Yao X. Defect electrocatalytic mechanism: Concept, topological structure and perspective. Materials Chemistry Frontiers. 2018;**2**: 1250-1268. DOI: 10.1039/c8qm00070k

[28] Zhou D, Cai Z, Lei X, Tian W, Bi Y, Jia Y, et al. NiCoFe-layered double hydroxides/N-doped graphene oxide array colloid composite as an efficient bifunctional catalyst for oxygen electrocatalytic reactions. Advanced Energy Materials. 2018;**8**:1-7. DOI: 10.1002/aenm.201701905

[29] Balamurugan C, Song S, Jo H, Seo J. $GdFeO_3$ perovskite oxide decorated by group X heterometal oxides and bifunctional oxygen electrocatalysis. ACS Applied Materials and Interfaces. 2021;**13**:2788-2798. DOI: 10.1021/acsami.0c21169

[30] Sinha L, Pakhira S, Bhojane P, Mali S, Hong CK, Shirage PM. Hybridization of Co_3O_4 and α-MnO_2 nanostructures for high-performance nonenzymatic glucose sensing. ACS Sustainable Chemistry and Engineering. 2018;**6**:13248-13261. DOI: 10.1021/acssuschemeng.8b02835

[31] Ramesh S, Karuppasamy K, Kim HS, Kim HS, Kim JH. Hierarchical flowerlike 3D nanostructure of Co3O4@MnO2/N-doped graphene oxide (NGO) hybrid composite for a high-performance supercapacitor. Scientific Reports. 2018; **8**:1-11. DOI: 10.1038/s41598-018-34905-7

[32] Meng Y, Song W, Huang H, Ren Z, Chen SY, Suib SL. Structure-property relationship of bifunctional MnO_2 nanostructures: Highly efficient, ultra-stable electrochemical water oxidation and oxygen reduction reaction catalysts identified in alkaline media. Journal of the American Chemical Society. 2014; **136**:11452-11464. DOI: 10.1021/ja505186m

[33] Stoerzinger KA, Risch M, Han B, Shao-Horn Y. Recent insights into manganese oxides in catalyzing oxygen reduction kinetics. ACS Catalysis. 2015;**5**: 6021-6031. DOI: 10.1021/acscatal.5b01444

[34] Tong Y, Chen P, Zhou T, Xu K, Chu W, Wu C, et al. A bifunctional hybrid electrocatalyst for oxygen reduction and evolution: Cobalt oxide nanoparticles strongly coupled to B,N-decorated graphene. Angewandte Chemie—International Edition. 2017;**56**: 7121-7125. DOI: 10.1002/anie.201702430

[35] Hu Z, Hao L, Quan F, Guo R. Recent developments of Co_3O_4-based materials as catalysts for the oxygen evolution reaction. Catalysis Science & Technology. 2022;**12**:436-461. DOI: 10.1039/D1CY01688A

[36] Xu Y, Zhang F, Sheng T, Ye T, Yi D, Yang Y, et al. Clarifying the controversial catalytic active sites of Co_3O_4 for the oxygen evolution reaction. Journal of Materials Chemistry A. 2019;**7**: 23191-23198. DOI: 10.1039/c9ta08379k

[37] Selvakumar K, Senthil Kumar SM, Thangamuthu R, Kruthika G, Murugan P. Development of shape-engineered α-MnO_2 materials as bi-functional catalysts for oxygen evolution

reaction and oxygen reduction reaction in alkaline medium. International Journal of Hydrogen Energy. 2014;**39**: 21024-21036. DOI: 10.1016/j.ijhydene.2014.10.088

[38] Selvakumar K, Senthil Kumar SM, Thangamuthu R, Ganesan K, Murugan P, Rajput P, et al. Physiochemical investigation of shape-designed MnO_2 nanostructures and their influence on oxygen reduction reaction activity in alkaline solution. Journal of Physical Chemistry C. 2015;**119**: 6604-6618. DOI: 10.1021/jp5127915

[39] Selvakumar K, Kumar SMS, Thangamuthu R, Rajput P, Bhattacharyya D, Jha SN. 2D and 3D silica-template-derived MnO_2 electrocatalysts towards enhanced oxygen evolution and oxygen reduction activity. ChemElectroChem. 2018;**5**: 3980-3990. DOI: 10.1002/celc.201801143

[40] Suntivich J, May KJ, Gasteiger HA, Goodenough JB, Shao-Horn Y. A perovskite oxide optimized for oxygen evolution catalysis from molecular orbital principles. Science. 2011;**334**: 1383-1385. DOI: 10.1126/science.1212858

[41] Gorlin Y, Jaramillo TF. A bifunctional nonprecious metal catalyst for oxygen reduction and water oxidation. Journal of the American Chemical Society. 2010;**132**:13612-13614. DOI: 10.1021/ja104587v

[42] Gorlin Y, Lassalle-Kaiser B, Benck JD, Gul S, Webb SM, Yachandra VK, et al. In situ X-ray absorption spectroscopy investigation of a bifunctional manganese oxide catalyst with high activity for electrochemical water oxidation and oxygen reduction. Journal of the American Chemical Society. 2013;**135**:8525-8534. DOI: 10.1021/ja3104632

[43] Su HY, Gorlin Y, Man IC, Calle-Vallejo F, Norskov JK, Jaramillo TF, et al. Identifying active surface phases for metal oxide electrocatalysts: A study of manganese oxide bi-functional catalysts for oxygen reduction and water oxidation catalysis. Physical Chemistry Chemical Physics. 2012;**14**:14010-14022. DOI: 10.1039/c2cp40841d

[44] Gorlin Y, Jaramillo TF. Investigation of surface oxidation processes on manganese oxide electrocatalysts using electrochemical methods and ex situ X-ray photoelectron spectroscopy. Journal of The Electrochemical Society. 2012; **159**:H782-H786. DOI: 10.1149/2.017210jes

[45] Risch M, Stoerzinger KA, Han B, Regier TZ, Peak D, Sayed SY, et al. Redox processes of manganese oxide in catalyzing oxygen evolution and reduction: An in situ soft X-ray absorption spectroscopy study. Journal of Physical Chemistry C. 2017;**121**: 17682-17692. DOI: 10.1021/acs.jpcc.7b05592

[46] Robinson DM, Go YB, Mui M, Gardner G, Zhang Z, Mastrogiovanni D, et al. Photochemical water oxidation by crystalline polymorphs of manganese oxides: Structural requirements for catalysis. Journal of the American Chemical Society. 2013;**135**:3494-3501. DOI: 10.1021/ja310286h

[47] Zaharieva I, Chernev P, Risch M, Klingan K, Kohlhoff M, Fischer A, et al. Electrosynthesis, functional, and structural characterization of a water-oxidizing manganese oxide. Energy and Environmental Science. 2012;**5**: 7081-7089. DOI: 10.1039/c2ee21191b

[48] Kuo CH, Mosa IM, Thanneeru S, Sharma V, Zhang L, Biswas S, et al. Facet-dependent catalytic activity of MnO electrocatalysts for oxygen

reduction and oxygen evolution reactions. Chemical Communications. 2015;**51**:5951-5954. DOI: 10.1039/c5cc01152c

[49] Pokhrel R, Goetz MK, Shaner SE, Wu X, Stahl SS. The "best catalyst" for water oxidation depends on the oxidation method employed: A case study of manganese oxides. Journal of the American Chemical Society. 2015;**137**: 8384-8387. DOI: 10.1021/jacs.5b05093

[50] Débart A, Paterson AJ, Bao J, Bruce PG. α-MnO2 nanowires: A catalyst for the O2 electrode in rechargeable lithium batteries. Angewandte Chemie - International Edition. 2008;**47**: 4521-4524. DOI: 10.1002/anie.200705648

[51] Cheng F, Zhang T, Zhang Y, Du J, Han X, Chen J. Enhancing electrocatalytic oxygen reduction on MnO_2 with vacancies. Angewandte Chemie—International Edition. 2013;**52**: 2474-2477. DOI: 10.1002/anie.201208582

[52] Matsuki K, Kamada H. Oxygen reduction electrocatalysis on some manganese oxides. Electrochimica Acta. 1986;**31**:13-18. DOI: 10.1016/0013-4686(86)80054-8

[53] Cao YL, Yang HX, Ai XP, Xiao LF. The mechanism of oxygen reduction on MnO_2-catalyzed air cathode in alkaline solution. Journal of Electroanalytical Chemistry. 2003;**557**:127-134. DOI: 10.1016/S0022-0728(03)00355-3

[54] Cheng F, Su Y, Liang J, Tao Z, Chen J. MnO_2-based nanostructures as catalysts for electrochemical oxygen reduction in alkaline media. Chemistry of Materials. 2010;**22**:898-905. DOI: 10.1021/cm901698s

[55] Ng JWD, Tang M, Jaramillo TF. A carbon-free, precious-metal-free, high-performance O_2 electrode for regenerative fuel cells and metal-air batteries. Energy and Environmental Science. 2014;**7**:2017-2024. DOI: 10.1039/c3ee44059a

[56] Kuo CH, Mosa IM, Poyraz AS, Biswas S, El-Sawy AM, Song W, et al. Robust mesoporous manganese oxide catalysts for water oxidation. ACS Catalysis. 2015;**5**:1693-1699. DOI: 10.1021/cs501739e

[57] Gorlin Y, Chung CJ, Benck JD, Nordlund D, Seitz L, Weng TC, et al. Understanding interactions between manganese oxide and gold that lead to enhanced activity for electrocatalytic water oxidation. Journal of the American Chemical Society. 2014;**136**:4920-4926. DOI: 10.1021/ja407581w

[58] Kuo CH, Li W, Pahalagedara L, El-Sawy AM, Kriz D, Genz N, et al. Understanding the role of gold nanoparticles in enhancing the catalytic activity of manganese oxides in water oxidation reactions. Angewandte Chemie - International Edition. 2015;**54**: 2345-2350. DOI: 10.1002/anie.201407783

[59] Wiechen M, Zaharieva I, Dau H, Kurz P. Layered manganese oxides for water-oxidation: Alkaline earth cations influence catalytic activity in a photosystem II-like fashion. Chemical Science. 2012;**3**:2330. DOI: 10.1039/c2sc20226c

[60] Mosa IM, Biswas S, El-Sawy AM, Botu V, Guild C, Song W, et al. Tunable mesoporous manganese oxide for high performance oxygen reduction and evolution reactions. Journal of Materials Chemistry A. 2015;**4**:620-631. DOI: 10.1039/c5ta07878d

[61] Kang Q, Vernisse L, Remsing RC, Thenuwara AC, Shumlas SL, McKendry IG, et al. Effect of interlayer

spacing on the activity of layered manganese oxide bilayer catalysts for the oxygen evolution reaction. Journal of the American Chemical Society. 2017;**139**: 1863-1870. DOI: 10.1021/jacs.6b09184

[62] Kosasang S, Ma N, Wuamprakhon P, Phattharasupakun N, Maihom T, Limtrakul J, et al. Insight into the effect of intercalated alkaline cations of layered manganese oxides on the oxygen reduction reaction and oxygen evolution reaction. Chemical Communications. 2018;**54**:8575-8578. DOI: 10.1039/c8cc03775b

[63] Esswein AJ, Mcmurdo MJ, Ross PN, Bell AT, Tilley TD. Size-dependent activity of Co_3O_4 nanoparticle anodes for alkaline water electrolysis. Journal of Physical Chemistry C. 2009;**113**: 15068-15072. DOI: 10.1021/jp904022e

[64] Menezes PW, Indra A, González-Flores D, Sahraie NR, Zaharieva I, Schwarze M, et al. High-performance oxygen redox catalysis with multifunctional cobalt oxide nanochains: Morphology-dependent activity. ACS Catalysis. 2015;**5**:2017-2027. DOI: 10.1021/cs501724v

[65] Sa YJ, Kwon K, Cheon JY, Kleitz F, Joo SH. Ordered mesoporous Co_3O_4 spinels as stable, bifunctional, noble metal-free oxygen electrocatalysts. Journal of Materials Chemistry A. 2013;**1**: 9992-10001. DOI: 10.1039/c3ta11917c

[66] Zhan Y, Du G, Yang S, Xu C, Lu M, Liu Z, et al. Development of cobalt hydroxide as a bifunctional catalyst for oxygen electrocatalysis in alkaline solution. ACS Applied Materials and Interfaces. 2015;**7**:12930-12936. DOI: 10.1021/acsami.5b02670